NanoScience and Technology

NanoScience and Technology

The series NanoScience and Technology is focused on the fascinating nano-world, mesoscopic physics, analysis with atomic resolution, nano and quantum-effect devices, nanomechanics and atomic-scale processes. All the basic aspects and technology-oriented developments in this emerging discipline are covered by comprehensive and timely books. The series constitutes a survey of the relevant special topics, which are presented by leading experts in the field. These books will appeal to researchers, engineers, and advanced students.

Please view available titles in *NanoScience and Technology* on series homepage
http://www.springer.com/series/3705/

Gyu-Chul Yi
Editor

Semiconductor Nanostructures for Optoelectronic Devices

Processing, Characterization and Applications

With 221 Figures

Springer

Editor

Gyu-Chul Yi

Seoul National University, Department of Physics
Gwanak-ro 599, 151-747 Seoul
Korea, Republic of (South Korea)
gyuchul.yi@gmail.com

NanoScience and Technology ISSN 1434-4904
ISBN 978-3-642-22479-9 e-ISBN 978-3-642-22480-5
DOI 10.1007/978-3-642-22480-5
Springer Heidelberg Dordrecht London New York

Library of Congress Control Number: 2011941807

Printed on acid-free paper

Springer is part of Springer Science+Business Media (www.springer.com)

Preface

The goal in nanotechnology is to make high-performance nanodevices. For nanodevice fabrications, novel *bottom-up* approach, fabricating devices and systems by hierarchical assembly or controlled growth of nanoscale materials, has attracted tremendous interest. Because this bottom-up method allows single-crystalline nanostructure growth on a variety of substrates, the bottom-up method has been used to prepare high-quality nanomaterials even on amorphous glass, plastic, and graphene substrates. In the bottom-up approach, one-dimensional (1D) semiconductor nanostructures, including nanorods, nanowires, nanobelts, and nanotubes, are vital components for fabricating optoelectronic and photonic nanodevices. In particular, 1D semiconductor nanostructures such as nanowires, nanorods, and nanotubes open up significant opportunities for the fabrication of high-performance optoelectronic nanodevice. For the fabrication of high-efficiency optoelectronic devices including light-emitting diodes (LEDs) and solar cells, 1D heteroepitaxial nanostructures with well-defined crystalline interfaces must be essential building blocks since embedding quantum structures in individual nanostructures would enable novel physical properties such as quantum confinement to be exploited, such as the continuous tuning of spectral wavelength by varying the well thickness. Sophisticated optoelectronic nanodevices can be readily fabricated by composition and doping controls of semiconductor nanostructures. Furthermore, nanodevices based on vertically ordered 1D nanostructures permit extremely small size and ultrahigh density. Here, this book introduces the current status of semiconductor nanostructures for optoelectronic devices and outlines the processing and characterizations of semiconductor nanostructures and their optoelectronic device applications.

In Chaps. 1–6, current research activities related to the synthesis of 1D semiconductor nanostructures by various growth methods and their optoelectronic device applications are reviewed. Chapter 1 provides an overview of vapor–liquid–solid growth process, which has widely been employed for preparation of semiconductor nanowires. Using this technique, Si, Ge, GaAs, InP, GaP, ZnO, and GaN nanowires have been synthesized and several nanodevices including p–n junction semiconductor nanowire LEDs and solar cells have been fabricated. In Chaps. 2 and 3, catalyst-free metal-organic vapor phase epitaxy to prepare high purity semiconductor nanostructures is introduced. Here, the processes to control positions, conductivities, and compositions of nanostructures for fabricating coaxial nanostructure LEDs are also described. Chapter 4 describes synthesis methods and characteristics of AlN nanostructures for UV optoelectronic device applications. Chapter 5 reviews the research progress on the controlled synthesis of a wide variety of nanowire heterostructures such as branched heterostructures, which includes solution phase and template-based methods. Meanwhile, the semiconductor nanostructures can be hybridized with graphene, which has recently been attracting much attention as a novel nanomaterial system for flexible optoelectronic devices as details are described in Chap. 6.

In Chaps. 7 and 8, structural and optical characterizations of semiconductor nanomaterials and nanostructures are reviewed. Chapter 7 introduces research on structural properties of ZnO and GaN nanostructures using X-ray absorption fine structure. As described in Chap. 8, optical properties of semiconductor nanostructures were investigated using luminescence characterization techniques, which are nondestructive, nonintrusive, and sensitive to the presence of defects or impurities in nanomaterials.

The last three chapters describe nanodevice applications of 1D semiconductor nanostructures. In Chap. 9, lasing characteristics of single and assembled nanowires are reviewed. Chapter 10 introduces near-field optical evaluation and the use of nanorod quantum structures for nanophotonic devices such as a nanophotonic gate. Finally, Chap. 11 presents the overview of nanowire solar cell studies, and integration strategies for practical device applications.

This book entitled "Semiconductor Nanostructures for Optoelectronic Devices – Processing, Characterization and Applications" is being introduced to review the recent works in the field of 1D nanomaterials and their optoelectronic device applications. Each chapter is written by leading scientists in the relevant field. Thus, I hope that high-quality scientific and technical information is provided to students, scientists, and engineers who are, and will be, engaged in fabrications of semiconductor nanostructures and their optoelectronic device applications.

I extend my acknowledgment to Dr. Claus Ascheron of Springer-Verlag for his guidance and suggestions.

Seoul
Republic of Korea

Gyu-Chul Yi

Contents

Contributors

Ian H. Campbell Center for Integrated Nanotechnologies, Los Alamos National Laboratory, Los Alamos, NM 87545, USA, campbell@lanl.gov

Chuanwei Cheng Department of Physics, Tongji University, 1239 Siping Road, 200092, Shanghai City, China, cwcheng@tongji.edu.cn

Heon-Jin Choi Department of Materials Science and Engineering, Yonsei University, 50 Yonsei-ro, Seodaemun-gu, Seoul 120-749, Korea, hjc@yonsei.ac.kr

Shadi A. Dayeh Center for Integrated Nanotechnologies, Los Alamos National Laboratory, Los Alamos, NM 87545, USA, shadi@lanl.gov

Hong Jin Fan Division of Physics and Applied Physics, School of Physical and Mathematical Sciences, Nanyang Technological University, 21 Nanyang Link, 637371, Singapore, FANHJ@ntu.edu.sg

Takashi Fukui Graduate School of Information Science and Technology, Research Center for Integrated Quantum Electronics (RCIQE), Hokkaido University, Kita 14, Nishi 9, Kita-ku, Sapporo, Japan, fukui@rciqe.hokudai.ac.jp

Sang-Wook Han Department of Physics Education, Chonbuk National University, 567 Baekje-daero, Deokjin-gu, Jeonju-si, Jeollabuk-do 561-756, Korea, shan@jbnu.ac.kr

Xiao Hong Ji School of Materials Science and Engineering, South China University of Technology, Wushan RD., Tianhe District, Guangzhou, P.R.China, 510641 jxhong@scut.edu.cn

Daniel S. P. Lau Department of Applied Physics, The Hong Kong Polytechnic University, Hung Hom, Kowloon, Hong Kong, P.R. China, apsplau@polyu.edu.hk

Chul-Ho Lee Department of Materials Science and Engineering, POSTECH, San31, Hyoja-dong, Nam-gu, Pohang, Gyungbuk, Korea, chulholee@gmail.com

Dong Hyun Lee Department of Material Science and Engineering, Hanyang University, 222 Wangsimni-ro, Seongdong-gu, Seoul 133-791, Korea, azaleaz@hanyang.ac.kr

Jung Min Lee Department of Material Science and Engineering, Hanyang University, 222 Wangsimni-ro, Seongdong-gu, Seoul 133-791, Korea, dlwjdals2929@naver.com

Motoichi Ohtsu School of Engineering, University of Tokyo, 2-11-16 Yayoi Bunkyo-ku, 113-8656 Tokyo, Japan, ohtsu@ee.t.u-tokyo.ac.jp

Won Il Park Department of Material Science and Engineering, Hanyang University, 222 Wangsimni-ro, Seongdong-gu, Seoul 133-791, Korea, wipark@hanyang.ac.kr

Daniel E. Perea Center for Integrated Nanotechnologies, Los Alamos National Laboratory, Los Alamos, NM 87545, USA, d-perea@lanl.gov

S. Tom Picraux Center for Integrated Nanotechnologies, Los Alamos National Laboratory, Los Alamos, NM 87545, USA, picraux@lanl.gov

Katsuhiro Tomioka Graduate School of Information Science and Technology, Research Center for Integrated Quantum Electronics (RCIQE), Hokkaido University, Kita 14, Nishi 9, Kita-ku, Sapporo, Japan

Japan Science and Technology Agency (JST), PRESTO, Kawaguchi, Japan, tomioka@rciqe.hokudai.ac.jp

Takashi Yatsui School of Engineering, University of Tokyo, 2-11-16 Yayoi Bunkyo-ku, 113-8656 Tokyo, Japan, yatsui@ee.t.u-tokyo.ac.jp

Gyu-Chul Yi National Creative Research Initiative Center for Semiconductor Nanostructures, Department of Physics and Astronomy, Seoul National University, 599 Gwanak-ro, Gwanak-gu, Seoul 151–747, Korea, gyuchul.yi@gmail.com

Jinkyoung Yoo Center for Integrated Nanotechnologies, Los Alamos National Laboratory, Los Alamos, NM 87545, USA, trip297@gmail.com

Siu Fung Yu Department of Applied Physics, The Hong Kong Polytechnic University, Hung Hom, Kowloon, Hong Kong, P.R. China, Siu.Fung.Yu@inet.polyu.edu.hk

Chapter 1
Vapor–Liquid–Solid Growth of Semiconductor Nanowires

Heon-Jin Choi

Abstract Nanowires make possible to manipulate light in novel methods and thus are promising materials for advanced optoelectronics. To exploit the potential, the growth behavior has to be controlled since it dominates the physical and chemical states and, in turn, the optical properties of nanowires. In this chapter, the vapor–liquid–solid (VLS) mechanism for the growth and modulation of nanowires was discussed. The chapter first reviewed the fundamental aspects of the VLS mechanism. Then the state of the art of the growth and modulation of nanowires for optoelectronics were discussed from the point of view of the critical issues pertaining to this mechanism. Some examples of optoelectronic devices that had been fabricated based on the VLS mechanism were also reviewed in an effort to cover the cutting edge technology in this area. Lastly, a summary and several different perspectives on the VLS mechanism were presented.

1.1 Introduction

Nanowires are hair-like, one-dimensional (1D) nanomaterials with diameters in the sub-one hundred nanometer scale and lengths ranging from several hundreds of nm to as high as a few cm. Owing to their nanoscale dimensions in the radial direction, they have size confinement effects that give them novel physical properties as compared to bulk materials. Their one-dimensional geometry on the nanometer scale provides an extremely high surface area with a nanoscale radius of curvature and great mechanical flexibility with near theoretical strength. These properties are advantageous in many chemical and mechanical applications. The geometry also

H.-J. Choi (✉)
Department of Materials Science and Engineering, Yonsei University, 50 Yonsei-ro, Seodaemun-gu, Seoul 120-749, Korea
e-mail: hjc@yonsei.ac.kr

G.-C. Yi (ed.), *Semiconductor Nanostructures for Optoelectronic Devices*,
NanoScience and Technology, DOI 10.1007/978-3-642-22480-5_1,

provides anisotropic properties that should be interesting from the point of view of nanomaterials science and engineering. Their length, reaching as high as the cm scale, makes them easy to manipulate for device fabrication.

Nanowires are promising materials for advanced optoelectronics. In addition to the unique aspects of their physical, chemical, and mechanical properties, the size of these materials is comparable to visible light in wavelength from 400 to 650 nm. This implies that nanowires can be used to handle light on a nanometer scale and thus can be used as building blocks for advanced optoelectronics. Indeed, novel methods of the manipulation of light with nanowires, including nanoscale Fabry–Perrot mode stimulated emission, wave guiding of photons, random lasing action, highly efficient luminescence, and extremely sensitive photodetection, have recently been demonstrated. The concept of many advanced nanowire-based optoelectronic devices including light-emitting diodes (LEDs), lasers, optical sensors, photo diodes, and photovoltaic cells have also been demonstrated.

The physical and chemical states of nanowires dominate their optical properties. The length and diameter of nanowires as well as their alignment affect the emission and absorption properties. The composition, impurity, or doping level, defect concentration, crystal structure, growth direction, and nature of the facets are also critical to the emission and/or stimulated emission and absorption. It should be noted that these physical and chemical states are closely related to the growth of nanowires. Therefore, one must fully understand the growth behavior of nanowires and develop rational, reliable growth processes to exploit the potential of nanowires in optoelectronics.

Nanowires are a result of anisotropic, 1D crystal growth on a nanometer scale. Therefore, the key issue related to the growth of nanowires is how to induce 1D crystal growth in a controlled manner. Regarding this, many approaches have been studied, including the use of the metal-catalyst-assisted vapor–liquid–solid (VLS) mechanism, the vapor–solid (VS) mechanism, and the template-assisted (TA) mechanism. Among these, the VLS mechanism is the most widely used owing to its simplicity and versatility when applied in many semiconductor systems.

This chapter reviews the growth of semiconductor nanowires by the VLS mechanism in the area of optoelectronics. As mentioned earlier, the growth process is critical to the physical and chemical state of nanowires and thus their optical properties. Therefore, a review of the growth process may be helpful so as to facilitate the preparation of superior nanowires for optoelectronics. This chapter focuses on the VLS mechanism. This may, however, limit our viewpoint regarding the growth of nanowires, as other mechanisms are also available. However, the VLS mechanism is a mainstay at present. Therefore, it may be sufficient to review the state of the art of this area. This chapter seeks to explain the understanding of what the VLS mechanism is as well as the manner in which better nanowires can be grown for optoelectronics. Accordingly, the chapter first reviews the fundamental aspects of the VLS mechanism. The growth of nanowires and a number of critical issues pertaining to VLS mechanism follow. Some examples of optoelectronic devices that have been fabricated based on the VLS mechanism are also reviewed in an effort to cover the cutting edge technology in this area. Lastly, a summary and several different perspectives on the VLS mechanism are presented.

1.2 VLS Mechanism for One-Dimensional Crystal Growth

The VLS mechanism is a 1D crystal growth mechanism that is assisted by a metal catalyst. It results in the creation of whiskers, rods, and wires. 1D crystal growth was initially developed nearly 50 years ago in the Si industry and the mechanism was suggested for wider use by Wagner in 1964 [1]. Figure 1.1 shows a schematic of the VLS mechanism. In this mechanism, the metal catalyst forms liquid alloy droplets at a high temperature by adsorbing vapor components. For some reason, e.g., temperature or vapor pressure fluctuation, the alloy is further supersaturated; i.e., it becomes a solution in which the actual concentration of the components is higher than the equilibrium concentration. It then drives the precipitation of the component at the liquid–solid interface to achieve minimum free energy of the alloy system. Accordingly, the 1D crystal growth begins, and it continues as long as the vapor components are supplied. Because vapor (carries solid components), liquid (catalyst alloy), and solid (precipitated one-dimensional structures) phases are involved, it is known as the VLS mechanism. At a glance, one can know that the size and position of the catalyst are related to the diameter and position of the 1D structures, as the liquid phase is confined to the area of the precipitated solid phase.

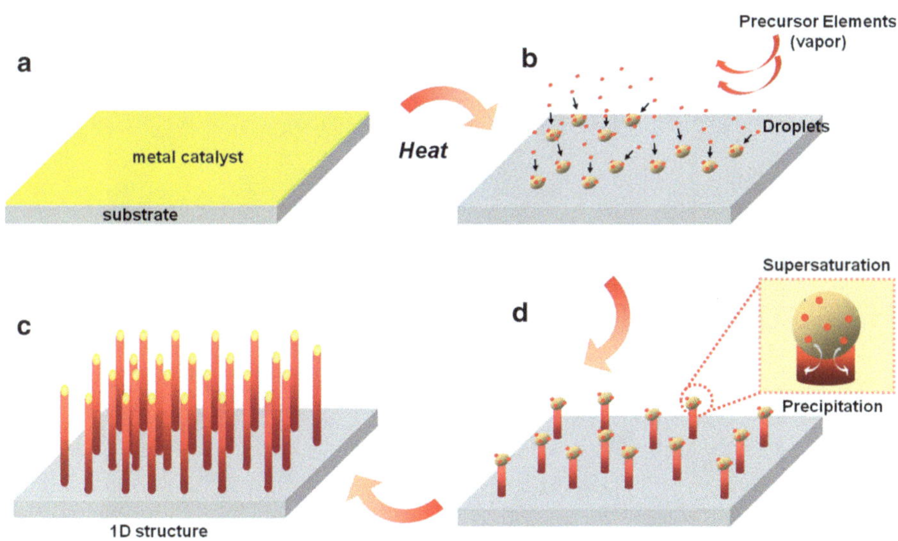

Fig. 1.1 Growth of 1D structures by VLS mechanism

The mechanism works at a high temperature at which the metal catalyst forms a liquid alloy. Therefore, chemical processes that occur at high temperatures, such as chemical vapor deposition (CVD), molecular beam epitaxy (MBE), laser ablation (LA) and carbothermal reduction (CR), are generally used in conjunction with the mechanism. Occasionally, metal catalysts sometimes work in a solid state in a vapor or liquid phase environment in a process termed the VSS (vapor–solid–solid) or LSS (liquid–solid–solid) mechanism.

Since the 1970s, the mechanism has been used to grow various types of whiskers on the micrometer or mm scale. A typical example is SiC whiskers, which are excellent reinforcements for high-strength, high-toughness ceramic or metal composites [2]. In this application, larger whiskers have a better reinforcing effect; thus, the catalyst size is made to be as large as possible, up to more than 10 μm, to grow large-diameter SiC whiskers, as shown in Fig. 1.2.

The mechanism was then noted for the growth of 1D structures on a nanometer scale, i.e., nanowires, in the 1990s, and the feasibility of this was demonstrated by several groups, including the Lieber group at Harvard University, the Yang group at the University of California Berkeley, and the Samuelson group at Lund University. As this mechanism was slated to become a core method for the growth of semiconductor nanowires, unambiguous experimental evidence of this mechanism was required for further study. Regarding this, Yu directly observed the growth of Ge nanowires by using an in situ high-temperature transmission electron microscope [3]. The findings of Yu's study showed that there are three well-defined stages in the VLS mechanism: alloying (note that the catalyst in Fig. 1.3a–c becomes larger as the Ge component dissolves and becomes alloyed with Au), precipitation of Ge (the bright area in c and d), and axial growth (extended structures in e and f).

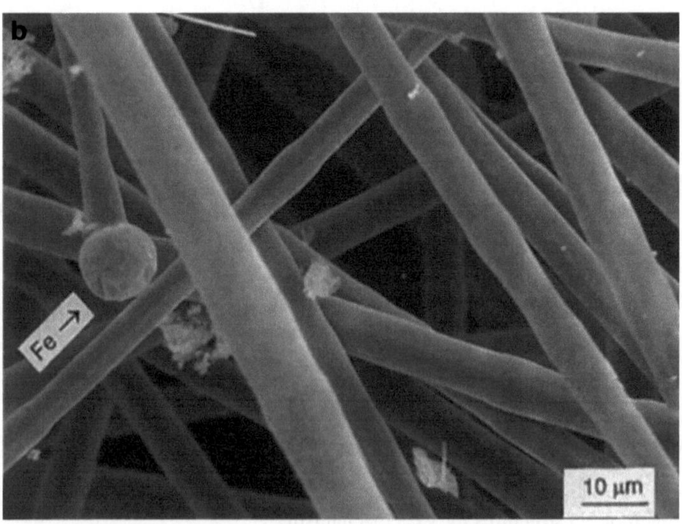

Fig. 1.2 SiC whiskers grown by VLS mechanism using Fe as catalyst (after [2])

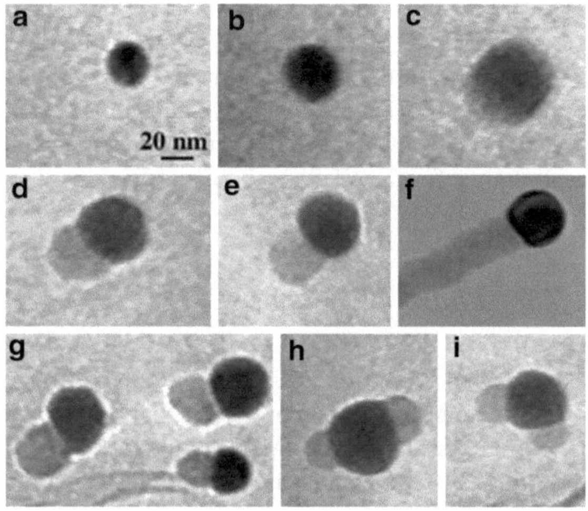

Fig. 1.3 Direct observation of growth of 1D Ge structures by VLS mechanism using Au as catalyst (after [3])

This observation clearly supports the proposed VLS mechanism, which is shown in Fig. 1.1. Other observations regarding the growth of Si nanowires on the substrate further confirmed the working of the VLS mechanism with the assistance of a liquid catalyst [4].

The brief history of the VLS mechanism implies that it can be generally used for the growth of many 1D structures, from the nm to even the mm scale. It also shows the rising of new technology from old technology as a good example of the progress of science and technology from previous studies. In fact, previous studies have established some fundamental aspects, as discussed below, which are essential for growing 1D structures using a catalyst.

1.2.1 Requirements for Metal Catalyst

Metal catalysts are essential in the VLS mechanism, but not all metals can work. These meet the following requirements: (1) It must form a liquid solution with a component of the solid phase. (2) The solubility limit of the catalyst component in the liquid phase must be much higher than that in the solid phase (i.e., $K = C_s/C_l < 1$, where C_s is the solubility limit in the solid phase and C_l is the solubility limit in the liquid phase). Under this condition, the catalyst easily leads to the formulation of the liquid alloy with little contamination in the solid phase. (3) The vapor pressure (V_p) of the catalyst component over the liquid alloy should be small. Otherwise, the catalyst will evaporate and eventually disappear in the course of growth. (4) It must be inert to chemical reactions. Otherwise, a reaction could deprive it of its catalytic function. (5) It must not make an intermediate solid. Otherwise, the intermediate solid will also deprive it of its catalytic function [1].

Previous studies have revealed that some metals meet the requirements. Generally, noble and transition metals work well with the VLS mechanism. For example,

Au works well for the growth of a 1D structure of group IV materials (e.g., Si and Ge), oxides (e.g., ZnO), and III–V semiconductors. Transition metal such as Ni and Fe also work for the growth of group IV materials (e.g., Si, Ge, and SiC) and III–V semiconductors. Naturally, many studies have focused on these metals to grow 1D structures. However, it should be noted that many other metals can also be developed as a catalyst for the VLS mechanism [5].

1.2.2 Phase Diagram

Because the adsorption, dissolving, mixing, diffusion, and precipitation in the liquid phase are thermodynamic processes that work toward equilibrium, a phase diagram is useful to predict how a catalyst will work. Figure 1.4 shows the phase diagram of the Au–Si system, which can be referred to regarding the growth of 1D Si structures with Au [6]. The diagram indicates that the minimum growth temperature for Si should be higher than the eutectic point of the system (364°C). The diagram also indicates that the composition of the Au–Si alloy above the eutectic point will follow the liquidus line (solid line) that denotes equilibrium between the solid and liquid phase. Therefore, the composition of the liquid alloy can be found at the liquidus line (point A) at a given temperature (1,100°C in the diagram). However, some temperature or vapor pressure fluctuation over the liquid alloy dissolves more Si

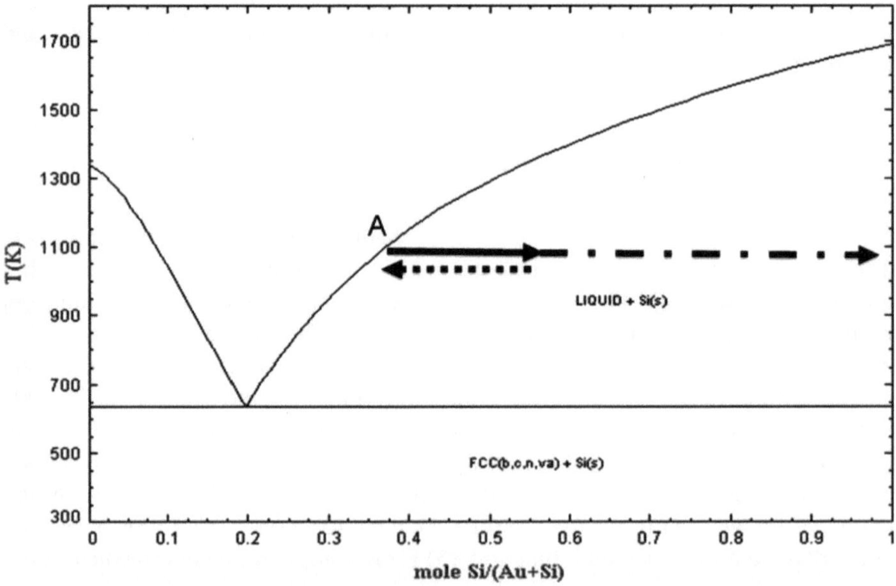

Fig. 1.4 Phase diagram of Au–Si system with indication of the composition of liquid alloy catalyst in the course of growth of 1D structures by VLS mechanism

than the equilibrium composition and renders it into a supersaturated state. As a result, the composition of the alloy goes beyond the equilibrium composition, moving to the right side of the liquidus line (solid arrow). This supersaturation state is thermodynamically a nonequilibrium and unstable state and thus drives the precipitation of the solid phase from the supersaturated liquid alloy until an equilibrium state is reached. The composition will then move back to the left and reach the liquidus line (dashed arrow). The composition of the precipitated solid phase corresponds to that of the phase boundary and thus is pure Si according to the diagram (dashed dot arrow). Meanwhile, the composition of the alloy goes beyond the equilibrium composition again as Si is dissolved from the vapor and drives the additional growth of 1D structures with the precipitation at the interfaces.

It should be noted that the currently available phase diagrams are constructed from bulk systems. Because the thermodynamic properties of a nanosystem are wholly different from those of a bulk system, the phase diagram of the type of nanometal catalyst that we are interested in should differ from that of the bulk system. Indeed, Eli and Peter Sutter investigated the equilibrium composition of nanoscale Au–Ge alloy droplets at the tips of Ge nanowires and found that the equilibrium composition of these droplets deviates significantly from that of the bulk alloy (Fig. 1.5) [7]. Adhikari et al. also investigated Au-catalyzed Ge nanowire and constructed a binary Au–Ge phase diagram that shows the catalyst size-dependent liquidus temperature [8]. These phenomena may be due to the critical role of the surface energy in the nanosystem [7–10]. Regardless of the cause of the deviation, it should be considered that the different thermodynamic equilibrium of the nanosystem causes discrepancies between the actual growth behavior of nanowires through a nanometal catalyst and predictions on the basis of a diagram of the bulk system.

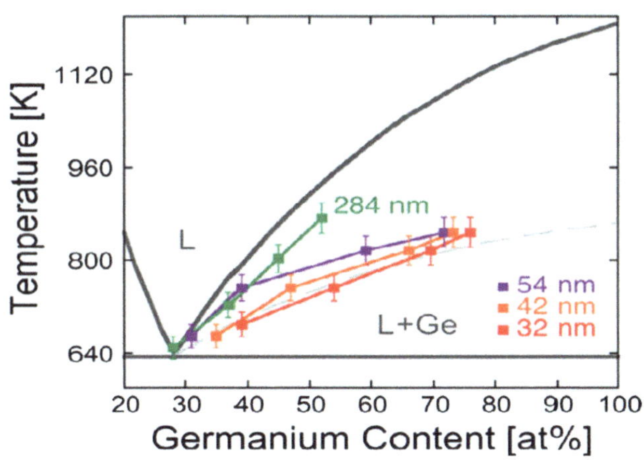

Fig. 1.5 Au–Ge binary alloy phase diagram. The *solid gray curves* represent the Au and Ge liquidus and solidus lines, respectively. *Squares* represent measurements of the temperature-dependent Ge content of Au–Ge alloy drops at the tip of Ge nanowires (after [7])

1.2.3 Kinetics and Rate-Determining Step

Three phases (gas, liquid, and solid) and two interfaces (gas/solid and liquid/solid) are involved in the VLS mechanism. In these complex system, the kinetics of the VLS mechanism consists of four steps: (1) mass transport in the gas phase; (2) chemical reaction at the vapor–liquid interface; (3) diffusion in the liquid phase; and (4) incorporation of atoms in a crystal lattice (Fig. 1.6) [11–14]. Identification of the rate-determining step among these is important to control the overall kinetics of the VLS mechanism. However, this is complicated, as three phases, two interfaces, and chemical reactions are involved [11]. Nevertheless, it may be possible to draw some insight based on the experimental results. As an example, the rate-determining step for the growth of 1D Si structures with an Au catalyst could be postulated as follows: Among the steps, step (3) can be excluded, because atoms diffuse in liquid metals very quickly [12] and thick nanowires or whiskers do not grow more slowly than those that are thinner, while the shape of the liquid droplet is maintained as nearly hemispherical and thus retains a longer diffusion length [11]. Step (1) can also be excluded because the diffusion coefficient in the gaseous phase usually follows the following power law: $D = D_0(T/T_0)^n(P/P_0)$, $n = 1.5 \sim 2$ [11, 12]. Therefore, the growth rate should follow the power law. However, this is not the case in many cases [11–14]. The primary evidence for regarding step (2) as the rate-limiting step is that the growth rate is proportional to the partial pressure of the reactant gas. However, this does not fully support the argument given that the growth process consists of two activated steps in series [11]. The dependence of the growth rate on the reactant vapor concentration is not in itself evidence that any of the steps is the rate-determining step. Rather, it simply reflects the dependence of the growth rate on supersaturation. Therefore, the rate-determining step would be step (4), the incorporation of atoms in a crystal lattice. It should be noted that the rate-determining step can be changed by the materials involved in the kinetics and by the processing conditions. Therefore, it should be carefully postulated by

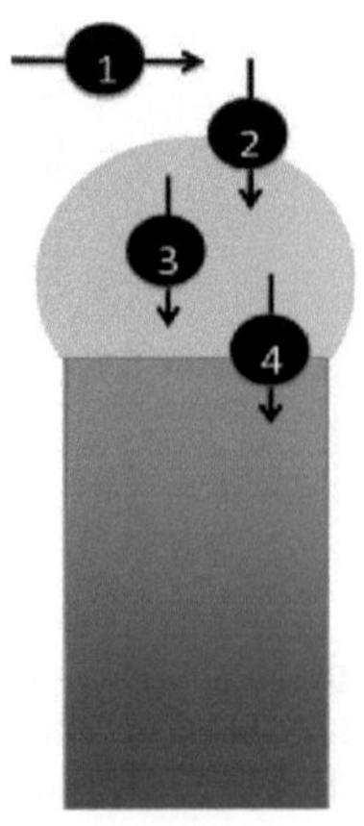

Fig. 1.6 Kinetic steps in VLS mechanism: (1) mass transport in the gas phase; (2) chemical reaction on the vapor–liquid interface; (3) diffusion in the liquid phase; and (4) in corporation of atoms in a crystal lattice

as many experimental results and data as possible. In fact, the postulation here is possible because a considerable amount of experimental data regarding the growth of Si whiskers, rods, and recently nanowires is available in the literature.

1.2.4 Size of the Metal Catalyst

As described earlier, nanowires can be grown using a nanometer-sized metal catalyst because the diameter of a 1D structure is confined by the size of the catalyst. Indeed, the bulk of previous studies demonstrated the growth of nanowires using nanofilms (that convert to nanoliquid droplets at a high temperature due to surface tension) or nanoparticles. However, it is difficult to decrease the size of the catalyst and in turn the diameter of nanowires in an unlimited manner. Thermodynamically, the minimum radius of a liquid metal droplet is given as [15]

$$R_{\mathrm{m}} = \frac{2V_{\mathrm{l}}}{RT\ln(s)}\sigma_{\mathrm{lv}}, \tag{1.1}$$

where V_{l} is the molar volume of the droplet, σ_{lv} is the liquid–vapor surface energy, and s is the degree of supersaturation of the vapor. According to this equation, using a smaller catalyst requires a higher degree of supersaturation. However, the chemical potential of the component in the metal–alloy catalyst becomes high as the size of the catalyst decreases due to the Gibbs–Thompson effect:

$$\Delta\mu = \frac{2\gamma}{r}. \tag{1.2}$$

Here, $\Delta\mu$ is the chemical potential difference of the component species in the liquid droplet, γ is the surface energy, and r is the radius of curvature of the droplet. Therefore, dissolving a vapor component into a liquid alloy becomes increasingly difficult as the size decreases, making it difficult to reach supersaturation states that sufficiently induce the growth of nanowires. Indeed, it is known that the growth of 1D structures with diameters of several tenths of nm is feasible; however, ensuring a smaller diameter (e.g., sub-10 nm) is difficult due to the thermodynamic limitations associated with the use of a nanocatalyst.

An additional difficulty that arises when downsizing a catalyst comes from the manipulation of metal nanoparticles or droplets. It is well known that nanoparticles have strong van der Waals attractive forces and thus agglomerate into larger particles. Furthermore, Ostwald ripening occurs between nanoparticles at high temperature. Ostwald ripening is a spontaneous process that occurs because larger particles are more energetically favorable. Accordingly, nanoparticles tend to transform into large particles to attain a lower energy state if the temperature is high enough to induce diffusion of the metal component. Because the van der Waals attractive forces and Ostwald ripening lead to the formation of larger droplets, larger diameter 1D structures are often grown from a nanometal catalyst. Thus, metal nanoparticles have to be carefully separated from each other in the course of the preparation, positioning on the substrate, and heating for the growth of 1D structures.

1.3 Growth of Nanowires by the VLS Mechanism and Current Issues for Optoelectronics

1.3.1 Growth of Semiconductor Nanowires by the VLS Mechanism

Si nanowires are a typical case of the growth of nanowires by the VLS mechanism with the assistance of a metal catalyst. Typically, Si nanowires are grown by Au because this metal meets the requirements for a VLS catalyst. The eutectic point of the Au–Si system is low, the phase relationship is simple and the system is stable at high temperatures. Figure 1.7 shows Si nanowires grown by the VLS mechanism using Au as a catalyst and $SiCl_4$ as the Si precursor [16]. The SEM images shown in Fig. 1.7a reveal Si nanowires with diameters of $\sim$100 nm and lengths of several μm that were grown on a Si substrate coated with Au film with a thickness of 2 nm. It also shows that the nanowires grew vertically on the substrate. It is well known that an epitaxial relationship between the substrate and nanowires can be attributed to the vertical growth of nanowires [17]; thus, the vertical growth here indicates that Au yields epitaxial interfaces between the nanowires and Si substrate. The transmission electron microscopy (TEM) images and selected area electron diffraction (SAED) patterns in Fig. 1.7b clearly show that the Si nanowires are single-crystalline nanowires without any structural defects. Alloy globules of Au–Si formed at the tips of the nanowires. An energy-dispersive spectroscopy (EDS) analysis across the catalyst–nanowire interface at the tip of an individual nanowire shows that Au operates as a VLS catalyst in this case. Note that an Au-rich globule formed when the Au catalyst was used according to the phase diagram of the Au–Si binary system (Fig. 1.4). Characterization of the catalyst globule/nanowire interfaces by high-resolution transmission electron microscopy (HRTEM) reveals

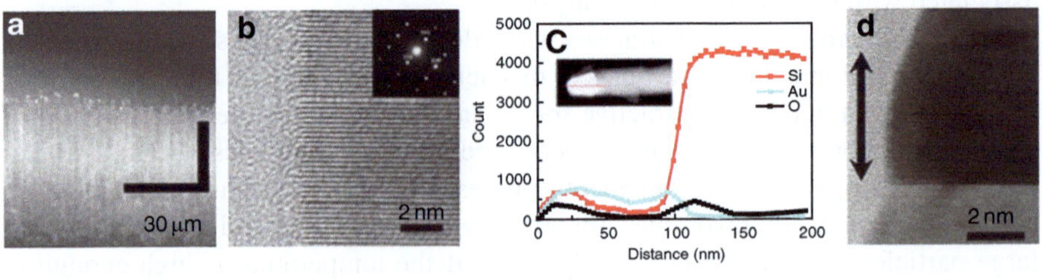

Fig. 1.7 Si nanowires grown by VLS mechanism using Au as catalyst. The nanowires were grown in the [110] direction for (110) Si substrates. (**a**) SEM image of Si nanowires, (**b**) TEM image and diffraction pattern, (**c**) EDS analysis of catalyst, (**d**) HRTEM image of the interface (after [16])

Fig. 1.8 (**a**) Ge, (**b**) GaN, and (**c**) ZnO nanowires grown by VLS mechanism using metal catalyst of Au, Ni, and Au, respectively. All these nanowires are single crystal

sharp interfaces between Si and catalyst (Fig. 1.7d). This indicates that there were no intermediate compounds. Taken as a whole, these outcomes indicate that Au is ideal catalyst that satisfies the requirements of a VLS catalyst.

Other semiconductor nanowires can also be grown by the VLS mechanism. Figure 1.8 shows Ge, GaN, and ZnO nanowires grown by the VLS mechanism using Au or Ni. The diameter and length of these nanowires typically range from 50 to 100 nm and tens of micrometers, respectively, depending on the size of the catalyst and the growth time. An HRTEM image of these nanowires indicates that they are single-crystalline in nature without defects or secondary phases. In fact, most nanowires grown by the VLS mechanism are of a single-crystal nature. These nanowires typically grow in the direction corresponding to the closest packing plane; however, this depends on factors such as the diameter of nanowires, the substrate used, the catalyst, and the processing conditions.

1.3.2 Issues Associated with the VLS Mechanism for Optoelectronics

As shown above, semiconductor nanowires can be grown by the VLS mechanism. However, some issues have to be addressed before the potential of nanowires can be exploited in the area of optoelectronics. By considering the nature of the VLS mechanism in conjunction with the findings of previous studies, these issues can be summarized as metal catalyst and structural modulation including control of the diameter, vertical growth, creation of coaxial and/or longitudinal heterostructure nanowire (COHN or LOHN), and compositional modulation including the alloying and doping of nanowires.

1.3.2.1 Metal Catalyst

Many metals are successfully used as VLS catalysts for the growth of nanowires. While they have important advantages for the controlled growth of nanowires, there are some issues from the point of view of semiconductor engineering. For example, Au as a catalyst is used for the growth of many nanowires, including Si, Ge, ZnO, GaN, and GaAs. However, it inevitably becomes contaminated into the nanowires as a result of contact between the liquid Au alloy and the semiconductor at a high temperature [18]. This contamination can increase the impurity level in the band gap and thus degrade the optical properties of nanowires. Furthermore, Au is not compatible with current CMOS processes and thus is limited in terms of how it can be introduced into the CMOS process. The very high diffusivity of Au induces migration of this metal on the surface of Si nanowires, making it difficult to control the growth precisely [19]. Accordingly, efforts to find a suitable catalyst that will not degrade the optical properties of nanowires and that can be used in the CMOS process have been made.

For Si nanowires, Al is considered as a promising candidate because the Al–Si binary phase diagram is similar to that of Au–Si. Indeed, limited studies have shown the successful growth of Si nanowires using Al [20, 21]. For example, Ke et al. grew Si nanowires by thermal CVD using Al as the catalyst [20]. The Al, however, oxidized quickly, showing that a well-designed, airtight growth process is necessary. In their work, Ke et al. used high H_2 and SiH_4 partial pressures to suppress the Al oxidation.

The other candidate for Si nanowires is Pt. Pt is a noble metal and thus has similar physical properties to Au. Indeed, Pt also successfully works as a catalyst for the growth of Si nanowires [16]. Figure 1.9 shows scanning electron microscopy (SEM) images and TEM images of Si nanowires grown using Pt as a catalyst. The size of the Si nanowires is comparable to that of nanowires grown by Au. In this case, they

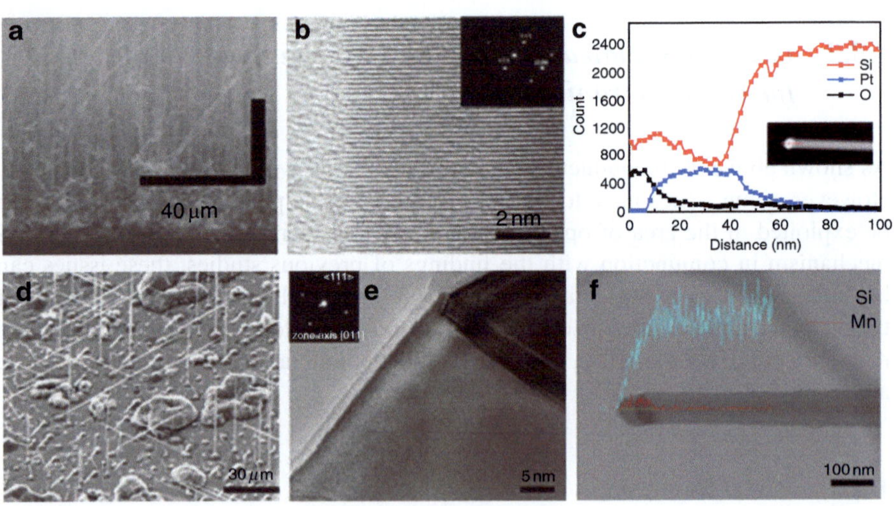

Fig. 1.9 Si nanowires grown by VLS mechanism with Pt (**a, b, c**) (after [16]) and Mn catalyst (**d, e, f**). (**a**), (**d**) SEM images of Si nanowires, (**b**), (**e**) TEM images of Si nanowires, (**c**), (**f**) EDS analysts of catalyst

grew vertically on the substrate. Hence, Pt yields an epitaxial interface between the nanowires and the Si substrate. It appears that the growth rate is generally faster than it is with Au under the same conditions. The Si nanowires are single-crystalline structures without any structural defects. Alloy globules of Pt–Si form at the tips of the nanowires. An EDS analysis across the catalyst–nanowire interface at the tip of an individual nanowire clearly shows that the Pt operated as a VLS catalyst, similar to Au.

Transition metals can also be considered as the VLS catalyst for the Si nanowires. For example, Mn can be used to grow Si nanowires. As shown in Fig. 1.9, the Si nanowires grown with Mn as a catalyst are tens of μm in length and tens of nm in diameter. Compared to other metal catalysts, such as Au or Pt [16], Mn catalysts lead to rather slow growth under the same growth conditions. The nanowires also have metal globules at their tips and, as shown in Fig. 1.9, consist of a Si–Mn alloy containing about 64% Si and 36% Mn. These globules clearly indicate that the nanowires were grown by the VLS mechanism with the assistance of Mn. In addition, there is no Mn in the nanowire body. These results indicate that many other metals can also be explored to grow Si nanowires. In fact, Ag, Bi, Cd, Co, Cu, Dy, Fe, Ga, Gd, Mg, Ni, Os, Pb, Pd, Te, Ti, and Zn can be explored as VLS catalysts for Si nanowires, as summarized by Shimit et al. [5].

GaN and ZnO nanowires are typically grown using a transition metal such as Ni and Au, respectively. As mentioned earlier, these metal catalysts can also increase the impurity level in the band gap and thus degrade the optical properties of nanowires. However, the M (metal catalyst)–Ga–N or the M–Zn–O system is complex compared to M–Si system. Therefore, it is likely that the number of catalysts available for these nanowires is limited as compared to Si nanowires. While many catalysts should have been explored for GaN nanowires, successful growth has not been reported thus far. In the case of ZnO nanowires, Sn has been reported as a VLS catalyst [22].

With regard to the formation of the impurity level in the band gap [23], the self-catalytic VLS mechanism is interesting. This mechanism works with a liquid catalyst that is formed in an in situ mode. For example, it was found that InAs nanowires can be grown without a metal catalyst [24]. Closer investigation revealed that a liquid In catalyst which forms from the substrate leads the growth of nanowires. Vertically aligned InP nanowires were also grown without a metal catalyst. It was revealed that temperature and precusor ratio control induces indium droplets to form on the surface and act as nucleation sites for nanowire growth [25]. This self-catalytic VLS mechanism is not commonly used, but it has the potential to grow nanowires without contamination of the catalytic components in some semiconductor systems.

1.3.2.2 Structural Modulation

To realize novel optical properties and newly fabricated devices, structural modulation of nanowires, such as control of the diameter, alignment, and growth-position,

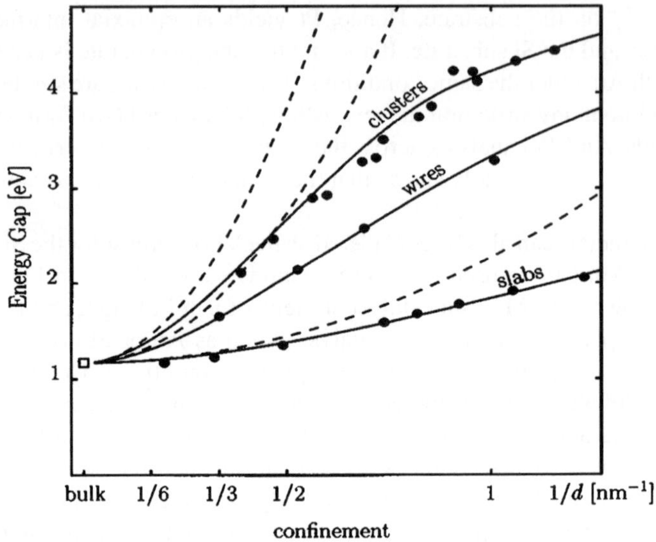

Fig. 1.10 Band-gap modulation of Si structures as a function of size (after [26])

is needed. Creating heterostructures in the radial or longitudinal direction is also crucial.

Diameter control: The optical properties of nanowires are dependent on their diameter. For example, the band gap, which determines the wavelength of luminescence, of semiconductor nanowires is renormalized by the diameter. Theoretical predictions of this type of band gap modulation are shown in Fig. 1.10, where the band gap of Si nanowires becomes wider as the diameter approaches the sub-10 nm scale [26]. Such small diameters also change the semiconductor characteristic of Si from an indirect to a direct band gap [27]. These findings imply that the optical properties of nanowires with both indirect and direct band gaps can be tuned by decreasing their diameter down to their Bohr exciton radius.

However, as described in Sect. 1.2, downsizing of the catalyst raises the chemical potential of the liquid alloy droplet and thus makes it difficult to grow nanowires with such a small diameter. This is thus a challenging issue from the point of view of thermodynamics. It should also be noted that these types of very thin nanowires have been grown using the VLS mechanism, though the optical properties in these cases have not been reported [28]. Therefore, these types of very thin nanowires can be grown for the development of indirect-band-gap nanowire-based optoelectronics.

With diameter control of direct-band-gap semiconductors such as GaN or ZnO, nanowires can also be created with novel optical properties such as multicolor luminescence by the quantum confinement effect. Indeed, such a novelty has already been demonstrated in GaN, InP and ZnO nanowires in which shifts in the emission spectra occur due to radial quantum confinement [29–31]. However, the growth of these nanowires reliably on the sub-10 nm scale using the VLS mechanism remains a challenging issue.

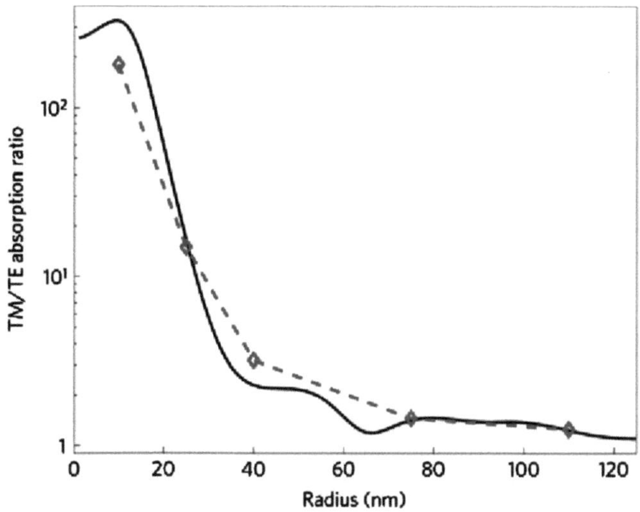

Fig. 1.11 Polarization dependence of the absorption efficiency, Q, and its dependence on Ge nanowire diameter. Theoretical calculations (*solid line*) and experimental measurements (*symbols*) on the ratio of the absorption efficiencies for transverse-magnetic- and transverse-electric-polarized 633 nm light (after [32])

Similar to the luminescence, the absorption of nanowires is dependent on the diameter. In a study of Ge nanowires [32], it was revealed that the absorption of nanowires is strongly dependent on the diameter, as shown in Fig. 1.11; due to the heavy transverse-magnetic/transverse-electric degeneracy in larger wires, the polarization dependence sharply drops as the diameter increases.

Vertical growth: A vertical array of nanowires provides novel optical properties that are advantageous for many applications, including nanolasers, LEDs, photovoltaic cells, and field emitters. For example, theoretical analysis indicates that vertical nanowire arrays have much lower reflectance compared to thin films. In a high-frequency regime, nanowire arrays have higher absorption than their thin film counterparts. In low-frequency regime, nanowire arrays absorb less but can be designed to approach the level of the film by changing the filling ratio [33]. Due to the low reflectance, it is possible to develop highly efficient photovoltaic devices. Other novel functions, such as direct conduction paths for photogenerated carriers, the creation of natural waveguiding cavities, and field emissions from arrayed atomicscale sharp tips can also be expected from the vertical nanowire array. The use of this type of array also enables the preparation of three-dimensional optoelectronic architectures.

The vertical growth of Si nanowires by the VLS mechanism is relatively easy, as many previous studies have demonstrated, as Si wafers are readily available as a substrate. Because the most rational approach to grow nanowire vertically is to establish an epitaxial relationship between the nanowire and the substrate,

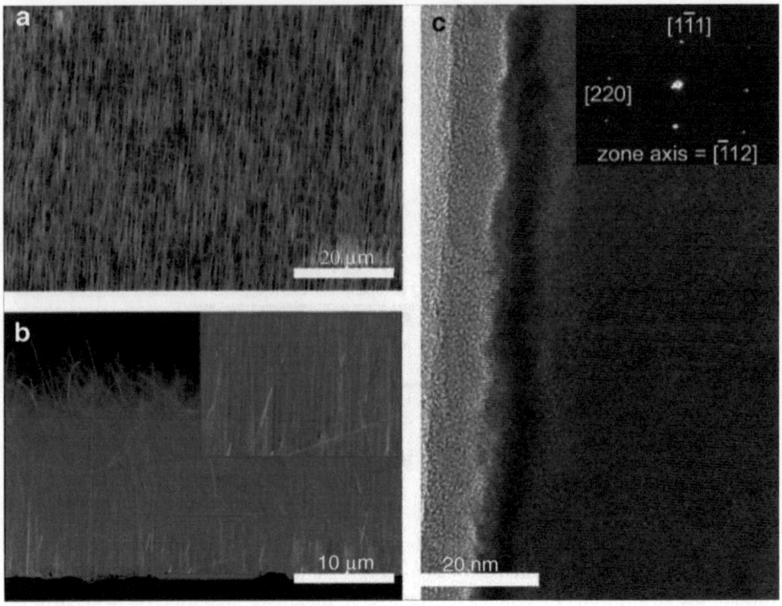

Fig. 1.12 SEM images of Mn:Ge nanowires vertically grown on Ge (111) substrate in a (**a**) 45°-tilted and (**b**) cross-section view. The inset of (**b**), the region indicated by a square in (**b**), shows the vertical growth more clearly. (**c**) HRTEM image of the nanowire grown on Ge (111). SAED pattern in upper right of (**c**) confirms that the growth direction of the nanowires is [111] (after [34])

the availability of the same material as the substrate is a great advantage for a homoepitaxial relationship and vertical growth (refer Figs. 1.7 and 1.9).

Ge nanowires can also be grown vertically with a Ge substrate. For example, Kim et al. successfully grew Ge nanowires vertically on Ge(111) substrates by the VLS mechanism using Au as a catalyst [34]. As shown in Fig. 1.12a, most of the Ge nanowires grown on Ge[111] are arrayed vertically. Figure 1.12b and its inset also show SEM images of vertically grown Ge nanowires in a cross-section view, indicating the epitaxial growth of the nanowires in the [111] direction. In a structural analysis using HRTEM and SAED, the SAED pattern confirmed that the single-crystalline nanowires grew in the [111] direction (Fig. 1.12c).

As shown from the growth of Si and Ge nanowires, a homoepitaxial relationship is a rational approach for the vertical growth of nanowires. Therefore, the best way to achieve vertical growth of III–V or oxide nanowires is through the respective use of an III–V or oxide substrate. Figure 1.13 shows GaN nanowires grown vertically on GaN substrates (more precisely, GaN film deposited onto sapphire substrates). This figure indicates that the epitaxial relationship can be a success for the vertical growth of III–V nanowires by the VLS mechanism.

It should be noted that such substrates are not available in many cases. Furthermore, these types of substrates (e.g, GaN or ZnO) will raise the fabrication cost and thus complicate the mass production of devices. In this regard, the vertical growth

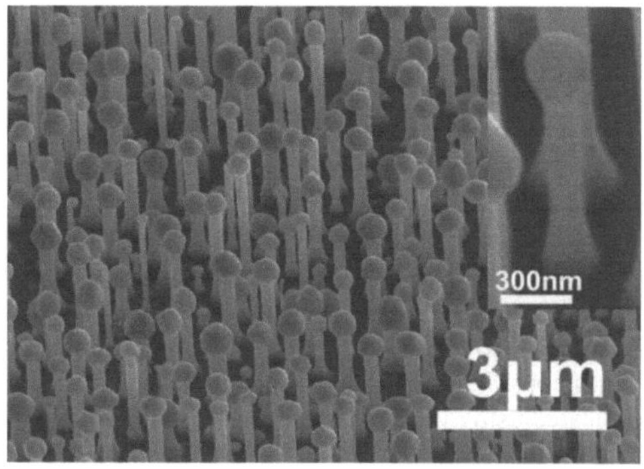

Fig. 1.13 Vertical growth of GaN nanowires on GaN film deposited sapphire substrate by establishing homoepitaxial relationship between nanowires and substrate. The growth direction of nanowires is same as the orientation of substrate, <0001>

of nanowires through a heteroepitaxial relationship between the nanowires and substrates has been studied. For example, Kuykendall et al. grew a crystallographic alignment in high-density GaN nanowire arrays using $LiAlO_2$ or MgO [35]. It was revealed that the selection of single-crystal substrates is critical for achieving deterministic control of the growth direction; i.e., a close match of both the symmetry and lattice constant between the substrate and GaN is essential for successful heteroepitaxy, and this is anticipated to influence the nanowire growth direction strongly. For example, the oxygen sublattice in the (100) plane of $\gamma-LiAlO_2$ has twofold symmetry, which matches the twofold symmetry of the (100) plane of wurtzite GaN well. The lattice constants $a = 5.17$ Å and $c = 6.28$ Å of $\gamma-LiAlO_2$ represent a close match of the lattice constants $c = 5.19$ Å and two times $a = 3.19$ Å of GaN, respectively. In contrast, the (111) plane of MgO has threefold symmetry and an interatomic separation distance of 98 Å for atoms in the (111) plane. This is a good match for the threefold symmetry of the (001) plane of GaN and the lattice constant $a = 3.19$ Å. As a result, these two substrates result in the selective growth of GaN nanowires in the $[1\bar{1}0]$ and [001] directions, respectively. The growth of GaN nanowires vertically on sapphire or Si substrates has also been investigated [36, 37].

Another area of study has been the vertical growth of other III–V group nanowires, such as GaAs and GaP. The approach for these nanowires is also based on the heteroepitaxial relationship. For example, GaP nanowires were grown vertically on Si substrates by utilizing a small lattice mismatch of less than 0.4% relative to Si [38]. It should be noted that epitaxial vertical growth was also achieved in the growth of GaAs nanowires on Si substrates with an interface with a larger lattice mismatch ($\sim$4.1%) [39]. A detailed analysis indicated that stacking faults in the nanowires influence the epitaxial growth of nanowires with a large mismatch. This implies that epitaxial growth can be achieved even under the condition of a large lattice mismatch between the nanowires and the substrate when a stress mediation mechanism is utilized.

Vertical growth of ZnO nanowires was achieved by utilizing a heteroepitaxial relationship. For example, ZnO nanowires can be grown vertically on the (110) plane of a sapphire substrate through matching with the (0001) plane of ZnO [40].

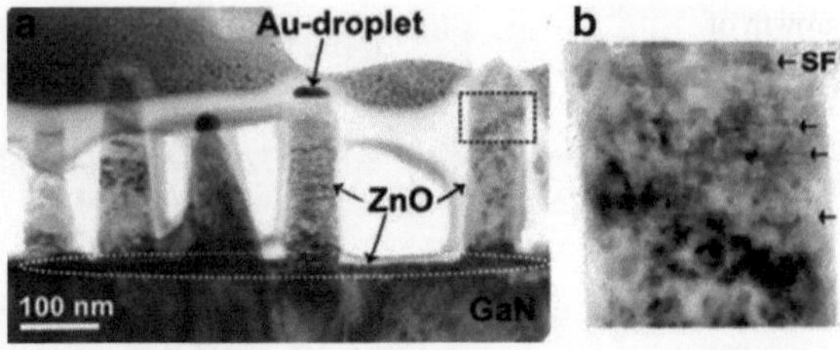

Fig. 1.14 (**a**) TEM cross-sectional image of the ZnO nanowires on the GaN substrate. A nearly continuous interfacial layer of ZnO separating vertical ZnO nanorods and GaN substrate is indicated using *white dashed line*. (**b**) Magnified view of the rectangular area in (**a**) (after [42])

Though the (110) plane of sapphire is symmetric by twofold and the ZnO c-plane is symmetric by sixfold, the a-axis of ZnO and the c axis of sapphire are mismatched at less than 0.08%. Such a coincidental matchup along the sapphire [0001] direction leads to the vertical heteroepitaxial growth of ZnO nanowires. Vertically aligned ZnO nanowires can also be grown on other substrates, for example, GaN, $Al_{0.5}Ga_{0.5}N$, and AlN substrates [41]. In these substrates, near-perfect vertical alignment of ZnO nanowires was observed, as both ZnO and the substrates have the same wurtzite structure and because the deposited ZnO nanowires are confined in their six equivalent $< 01\bar{1}0 >$ directions and grow along the [0001] direction, following the substrate crystal orientation precisely.

Although a few studies have successfully demonstrated the vertical growth of III–V semiconductor and ZnO nanowires, it is also true that the vertical growth of nanowires was found to be unsuccessful in many studies. In those cases, it is unclear whether an epitaxial relationship between the nanowires and substrates was established because an interfacial layer typically formed between the nanowires and the substrate. These interfacial layers appear to have been deposited by the VS mechanism in advance of the growth of the nanowires by the VLS mechanism. Indeed, closer investigation of the interfaces between the ZnO nanowires and the GaN substrate showed this type of interface [42] (Fig. 1.14). This issue, the formation and effect of the interfaces, should be addressed in detail to achieve the heteroepitaxial as well as homoepitaxial vertical growth of nanowires.

One of the means of overcoming this difficulty is the activation of a metal catalyst for the VLS mechanism at a low temperature. Figure 1.15 shows the growth of GaN nanowires by the VLS mechanism. The left image shows nanowires grown using Ni as a catalyst. These grew randomly due to the formation of an interfacial layer. However, the right image shows the vertical growth of nanowires when using a cocatalyst, e.g., Au–Ni. This cocatalyst creates a liquid alloy at a low temperature and works as a VLS catalyst in advance of the deposition of the layer by the VS mechanism.

Coaxial heterostructure nanowire (COHN): The optical properties can be manipulated by creating COHN. For example, COHN can confine the photons as well as the electrons and thus yield improved performance of semiconductor lasers and LEDs with a variety of applications that require coherent light sources with low

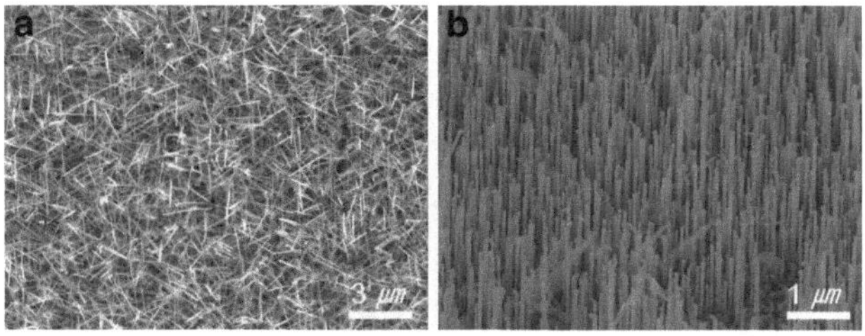

Fig. 1.15 Vertical growth of GaN nanowires on the GaN deposited sapphire substrate by VLS mechanism using cocatalyst on the substrate

power consumption and the capability of high-speed digital modulation. It is also ideal for photovoltaic cells due to the large junction area that extends along the entire length of the nanowire with carrier separation in the radial direction, which promises high efficiency. Furthermore, an electrical injection in COHN can be carried out efficiently with more flexibility in device designs that should be far superior to other structures for electrical injection optoelectronics [43].

Based on the VLS mechanism, COHN can be fabricated by two approaches. The first of these involves a combination of the VLS and VS mechanism in which nanowires using the VLS mechanism are initially grown in the longitudinal direction, after which a shell is deposited in the radial direction by the VS mechanism. A typical example of this approach is GaN/InGaN COHN multiquantum-well structures fabricated by the subsequent deposition of an In/AlGaN layer on the surface of GaN nanowires grown by the VLS mechanism (Fig. 1.16) [44]. During this process, the composition of the shell is controlled by the temperature; a multiquantum-well structure with as many as 26 wells can be fabricated with COHN.

The other approach to fabricate COHN is through the use of a self-organization mode in a one-step VLS mechanism. Figure 1.17 shows GaN/AlGaN COHN grown by spontaneous phase separation within the Ga–Al–N alloy nanowire system [45]. Bright-field transmission electron microscopy (BFTEM) images of the nanowire show a "dark" GaN core and a "bright" AlGaN sheath along the axis of nanowires with smooth surfaces and sharp interfaces. The cores have diameters in the range of 5–40 nm, whereas the sheath thickness is in the range of 50–200 nm. The driving forces for this self-organizing process are the strain that develops in the nanowires. The curved surfaces of the materials are subjected to local pressure ΔP (and, in turn, stresses) according to the Laplace equation $\Delta P = \gamma[(1/r_1) + (1/r_2)]$, where r_1 and r_2 are two principal radii of curvature in a given point of a surface and γ is the surface tension. For nanowires, r_1 and r_2 correspond to the curvature in the radial and longitudinal directions, respectively; the former equals the radius of the nanowires, whereas the latter can be assumed to be infinite. Therefore, ΔP generated on the curved surface of the nanowires can be estimated as $\Delta P = \gamma / r_{\text{nanowire}}$, where r_{nanowire} is the nanowire radius. Owing to the positive curvature of the surfaces, compressive stresses in the inward radial direction arise and become significant

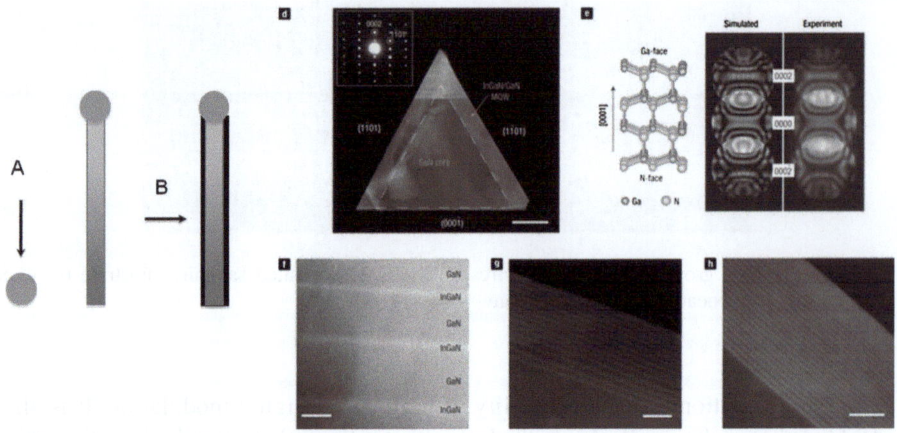

Fig. 1.16 (*Left*) Strategy for the fabrication of coaxial heterostructure nanowires by longitudinal growth by VLS mechanism followed by shell deposition by VS mechanism. A and B components are supplied for longitudinal growth and shell deposition, respectively (*Right*) Structural analysis of In/AlGaN coaxial heterostructure nanowires (after [44])

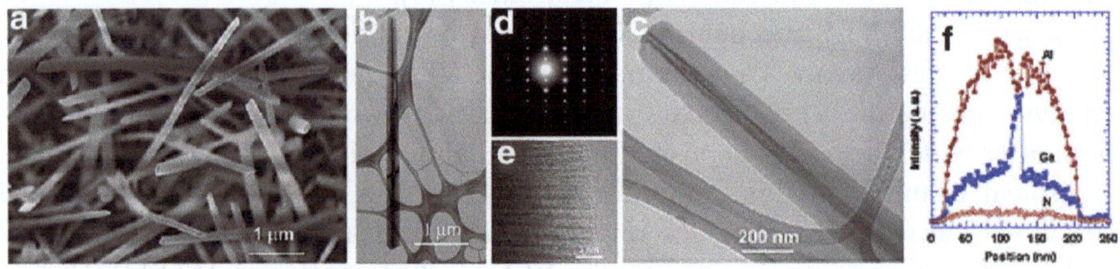

Fig. 1.17 GaN/AlGaN COHN by self-organization mode. (**a**) SEM image of nanowires, (**b**)–(**e**) TEM images with diffraction pattern, (**f**) compositional analysis of nanowires (after [45])

as the size decreases. Similar phenomena can be found in the overgrowth of semiconductor alloy thin films, where spontaneous phase separation is accompanied by surface roughening (i.e., the formation of curved surfaces). Such a strain-induced self-ordering process can also lead to irregular spatial compositional modulation in the GaInAs nanowire system [46].

Si COHN can also be prepared in the self-organization mode. Figure 1.18 shows a TEM image of Si–Er COHN and a schematic illustration of the self-organization process for creating single-crystalline Si COHN arrays on (111) Si substrates [47]. In this mode, Si and Er are supplied to Au catalysts and form an Au–Si–Er alloy. One-dimensional epitaxial growth of nanowires then occurs with the precipitation of supersaturated Si and Er. Meanwhile, Er is extracted to the outside, where it reacts with oxygen and forms a Si–Er–O amorphous layer. An $Er_2Si_2O_7$ layer is then generated on the sheath by phase separation, leaving the amorphous SiO_2 layer inside. Further optical characterization showed strong luminescence with a wavelength of 1.54 μm. These outcomes indicate that COHN can be used to

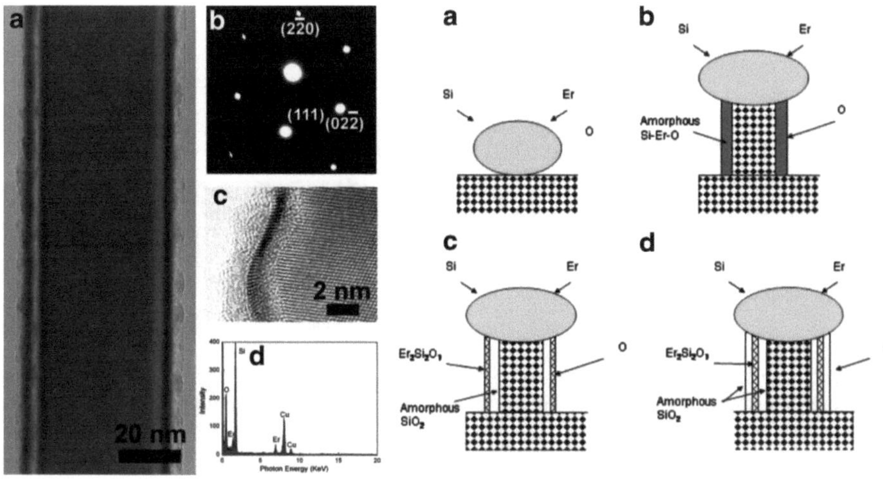

Fig. 1.18 (*Left*) TEM images of optically active COHN. (*Right*) The self-organization mechanism of COHN (after [47])

realize novel functionalities of semiconductor nanowires, for example, the optical activation of indirect-band-gap Si nanowires.

Longitudinal heterostructure nanowires (LOHN): The optical properties of nanowires can also be modulated through the fabrication of LOHN. For example, LOHNs can serve as quantum structures in the longitudinal direction that show enhanced optical properties due to the quantum confinement effect. The photoluminescence and electroluminescence can also be improved by creating a LOHN p–n junction [48].

LOHN can be prepared by the VLS mechanism with a catalyst or by using the VS mechanism without a catalyst, though only the former method is reviewed here. Figure 1.19 shows the strategy for the growth of LOHN involving the feeding of the components for the nanowires through a catalyst sequentially [48]. Figure 1.19 also shows TEM and EDS images of GaP/GaAs LOHN grown using this strategy. A compositional analysis showed that heterostructures were formed in the course of VLS growth. LOHNs of Si/SiGe and InAs/InP system could also be fabricated by this approach [49,50].

One of the issues related to the growth of LOHN by the VLS mechanism involves the interfaces between the heterostructures. Generally, optoelectronic devices require sharp interfaces in terms of their compositions or structures. However, the interfaces in LOHN using the VLS mechanism are typically not sharp enough due to the precipitation of the solid phase from the liquid phase, where the diffusion rate of the components is rather slow as compared to the vapor phase. Recent study, however, shows some promising results to overcome this limitation. Figure 1.20 (left) shows InAs/InP nanowire heterostructures that were grown with chemical beam epitaxy (CBE) on InAs (111) substrates using Au nanoparticles as a catalyst [51]. This figure shows very thin heterostructures in the longitudinal

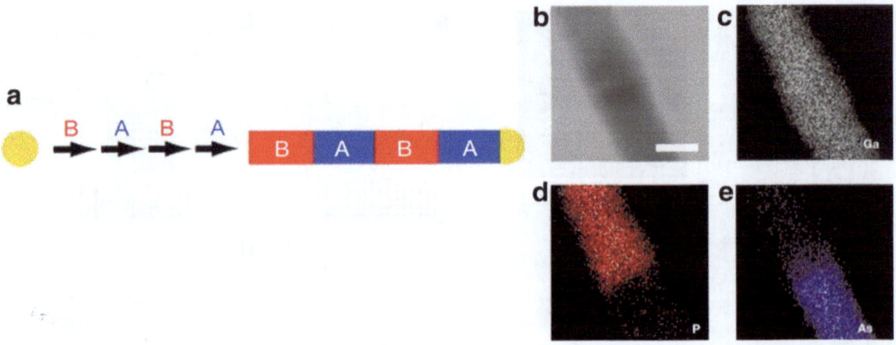

Fig. 1.19 (*Left*) Basic approach for the growth of longitudinal heterostructure nanowires using VLS mechanism. (*Right*) Compositional analysis of GaP/GaAs longitudinal heterostructure nanowires (after [48])

Fig. 1.20 (*Left*) TEM image of InAs/InP longitudinal heterostructure nanowires. (*Right*) TEM image of InP longitudinal heterostructure nanowires (after [51,52])

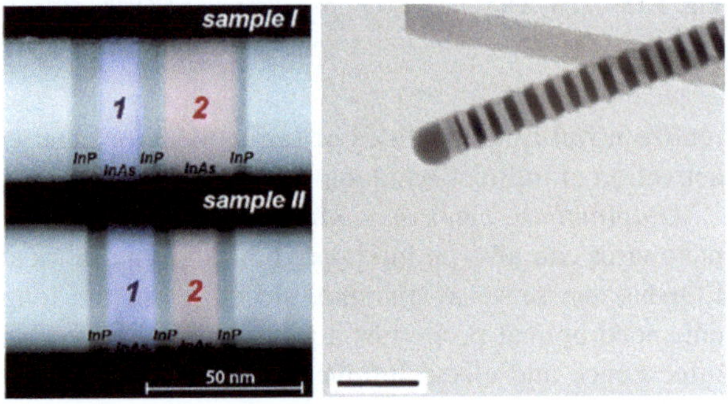

direction that display a quantum confinement effect. Additionally, sharp interfaces were achieved by low growth rates together with the rapid switching of an indium source (TMIn). These abrupt changes between sections with different compositions, e.g., InAs, InP, or InAsP, along the nanowire with atomically sharp interfaces between them are typically challenging to achieve through the liquid phase. It is thus suggested that LOHN with sharp interfaces can be fabricated under controlled conditions.

Additionally, different approaches for the growth of LOHN have been studied [52]. Figure 1.20 (right) shows the InP superlattice of LOHN. To create heterostructures, InP nanowires were grown first from colloidal gold particles by the VLS mechanism. In the course of doing this, the nanowires grow in a zinc-blended crystal structure with a supply of diethyl zinc in a vapor form. When sufficient Zn is supplied, twin planes that exhibit constant spacing for a given Zn concentration and wire diameter appear, with twinning superlattice structures as a result. These LOHNs form because the doped Zn decreases the activation barrier for two-dimensional nucleation growth of zinc-blended InP and therefore promotes the crystallization of the InP nanowires in the zinc-blend instead of the commonly found wurtzite crystal structure. It is not clear whether this approach can be applied to other semiconductor nanowires; however, it deserves further study as a new fabrication means of LOHN with sharp interfaces.

1.3.2.3 Compositional Modulation

Compositional modulation including alloying and doping is essential for the manipulation of the optical properties of semiconductors nanowires. For example, the photoluminescence or electroluminescence from III–V semiconductors can be tuned by modulating the composition in In–Al–Ga–N or In–Al–Ga–As in a quaternary-based system. The absorption of SiGe nanowires can be tuned by the composition in a binary-based system. This type of compositional modulation is thus important to exploit the potential of these materials in the area of optoelectronics.

Alloying of nanowires: IV and III–V semiconductors create a complete solid solution; thus, a binary, ternary, or quarterly alloy (e.g., $Si_{1-x}Ge_x$, $In_xGa_{1-x}As$, or $Al_{1-x}In_yGa_{1-x-y}N$) can be prepared. In line with this, compositional modulation of IV and III–V semiconductor nanowires has been studied.

The alloying of Si nanowires with Ge is readily achieved in the course of growth by the VLS mechanism. For example, $Si_{1-x}Ge_x$ alloy nanowires ($x = 0 \sim 0.3$) can be grown on Si (111) substrates using Au as a catalyst and $SiCl_4$ and Ge powders as precursors (Fig. 1.21) [53]. The resulting $Si_{1-x}Ge_x$ nanowires are well aligned on the substrates, and the diameter of these nanowires typically ranges from 50 to 100 nm. The composition of the $Si_{1-x}Ge_x$ nanowires can be varied by changing the growth temperature or the substrate distance (or both) from the Ge powder, which acts as a Ge precursor. Figure 1.21d shows a typical HRTEM image of the nanowires. The single-crystalline nature with a thin layer of native oxide can be seen in the HRTEM image. The SAED pattern recorded along the [001] zone axis, as shown in Fig. 1.21e, indicates that the nanowires grew in the [110] direction. Figure 1.21f shows the relative composition of Si and Ge in the $Si_{1-x}Ge_x$ nanowires through an EDS analysis, as shown for the $Si_{0.95}Ge_{0.05}$ and $Si_{0.7}Ge_{0.3}$ nanowires. This shows the compositional homogeneity of each nanowire with an EDS line scan. The profiles in both the radial and axial directions of the wire show that the composition of the native oxide is primarily SiO_x. No evidence of phase inhomogeneity was found (Fig. 1.21g, h); that is, no obvious Ge segregation within the nanowire was observed, as often found in thin film CVD. These outcomes imply that the alloying of Si and Ge was appropriate, leading to random substitutional alloy nanowires.

Alloying in III–V semiconductor nanowires is especially important in optoelectronic applications because it creates emissions with various wavelengths, from ultraviolet to the infrared region. Indeed, complete composition tunability of nanowires in the In–Ga–N system has been achieved [54]. In this study, single-crystalline $In_xGa_{1-x}N$ nanowires across the entire compositional range from $x = 0$ to 1 were grown by low-temperature halide CVD. It also showed tunable emission

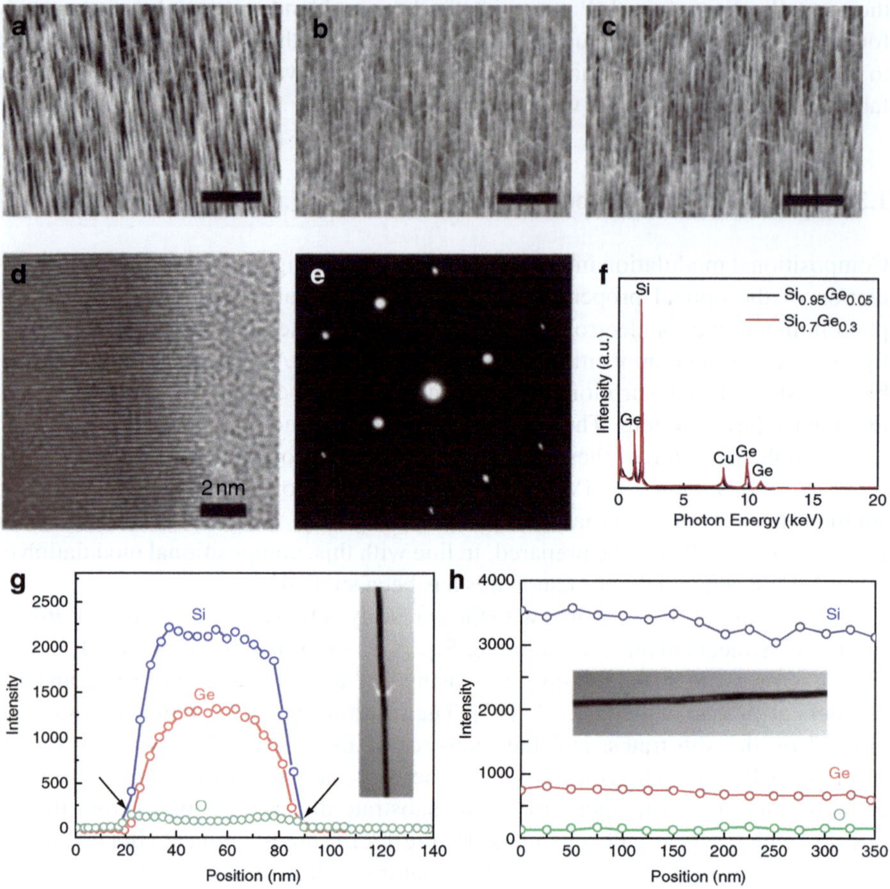

Fig. 1.21 $Si_{1-x}Ge_x$ nanowires grown by VLS mechanism. SEM images of (**a**) $Si_{0.95}Ge_{0.05}$, (**b**) $Si_{0.85}Ge_{0.15}$, (**c**) $Si_{0.7}Ge_{0.3}$ nanowires aligned on the Si (111) substrates, (**d**) Typical HRTEM image of controlled growth $Si_{1-x}Ge_x$ nanowires, showing the single-crystalline and defect-free nature, (**e**) SAED pattern, taken along the [001] zone axis that confirms the diamond structure of the wire with [110] growth direction, (**e**) typical EDS spectra of $Si_{0.95}Ge_{0.05}$ and $Si_{0.7}Ge_{0.3}$ nanowires. EDS line profiles in both radial (**g**) and axial (**h**) directions, showing that composition of native oxide is primarily SiO_x and any evidence of phase inhomogeneity is not found (after [53])

from the near-ultraviolet to the near-infrared region. However, these nanowires were grown without a catalyst. It should be noted that other studies pertaining to the alloying of III–N nanowires using the VLS mechanism have not been reported. This may due to the difficulty of alloying of ternary- and quaternary-based systems through a liquid–metal phase. Unlike alloying directly from the vapor phase, the kinetics of alloying through the liquid phase is typically limited by the thermodynamics and kinetics of the system. However, our recent study demonstrates that compositional modulation of InGaN nanowires can be achieved

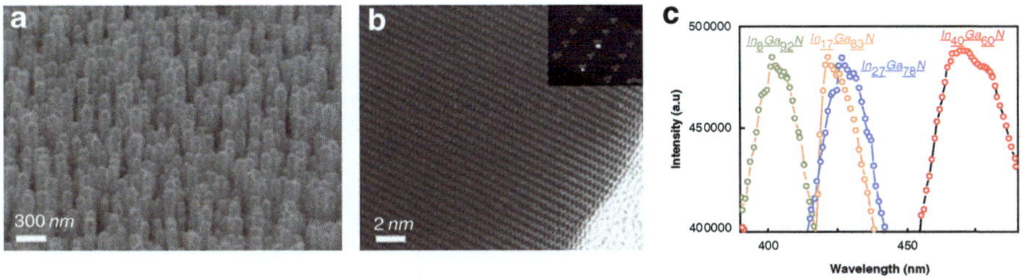

Fig. 1.22 InGaN nanowires grown by VLS mechanism and their PL. (**a**) SEM image of nanowries, (**b**) TEM image with diffraction pattern, (**c**) PL of InGaN nanowires with their composition

with the VLS mechanism. Figure 1.22 shows InGaN nanowires grown on a substrate using a cocatalyst of Au–Ni. The nanowires in this case grew on the substrate vertically. Figure 1.22b shows a typical HRTEM image of the nanowires. The single-crystalline nature can be seen in the HRTEM image. The SAED pattern recorded along the [001] zone axis, as shown in Fig. 1.22b, indicates that the nanowires grew in the [0001] direction. The PL measurements shown in Fig. 1.22c indicate that composition tunability and thus band-gap modulation is feasible through the VLS mechanism.

Alloyed InGaAs nanowires have been demonstrated by the VLS mechanism (Fig. 1.23). Kim et al. grew InGaAs nanowires using the VLS mechanism with Au as a catalyst [55]. Though the composition of these nanowires is not uniform along the longitudinal direction, their finding clearly indicates that compositional modulation can be achieved during the growth process.

Doping of nanowires: To realize electrical injection into a nanowire for optoelectronics, doping of electronic impurities is essential. The functionalization of nanowires for advanced optics, e.g., realizing magnetism in semiconductor nanowires for spin LEDs, also requires functional (e.g., magnetic) impurity doping.

Doping was previously demonstrated in Si nanowires [56]. The process is simple after supplying a dopant through vapor with a metal organic CVD precursor or the laser ablation of a solid target during the course of VLS growth. It results in n- and p-type Si nanowires. In a similar approach, Mg-doped, p-GaN nanowires can be prepared using magnesium nitride (Mg_3N_2) as a Mg source [37]. In this study, the doping concentration was controlled by changing the separation distance between the doping source and the substrate (Fig. 1.24, left). As shown in Fig. 1.24 (right), this was effective and changed the resistance of the nanowires.

In addition to electronic impurities, the doping of functional impurities has been studied. One typical example is magnetic impurity doping into semiconductor nanowires to realize magnetism in the semiconductor and spin-related optoelectronics. Regarding this, Mn was doped into Ge nanowires by transporting germanium chloride ($GeCl_4$) and manganese dichloride ($MnCl_2$) onto an Au-coated silicon substrate [34]. Despite the Mn doping, an analysis revealed that the nanowires maintain their single-crystalline nature without defects or secondary phases. Compositional

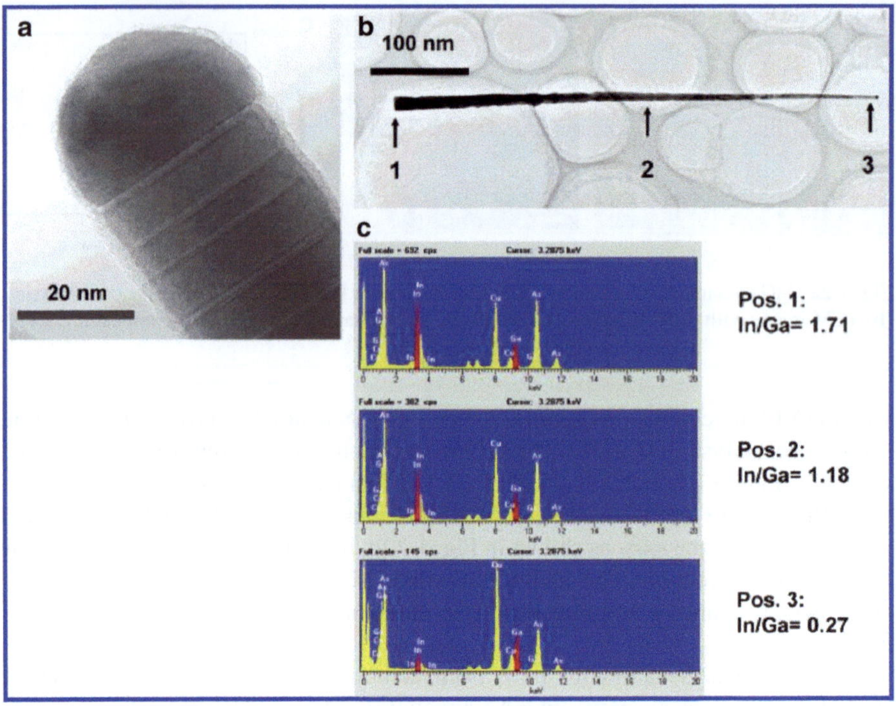

Fig. 1.23 InGaAs nanowires grown by VLS mechanism (after [55])

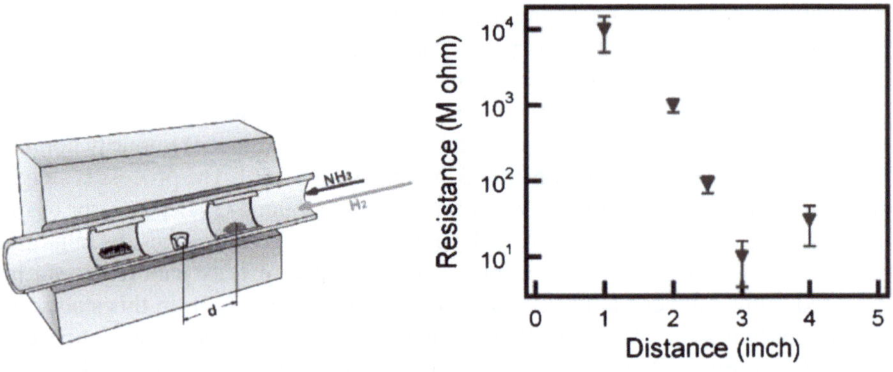

Fig. 1.24 (*Left*) Schematic for the apparatus for doping of GaN nanowires. (*Right*) The resistance of nanowires as a function of separation between doping source and substrate (after [37])

analysis indicated that the average Mn concentration measured from ten nanowires was ca. 1.5%. This result indicates that magnetic impurities can be doped into IV semiconductor nanowires by the VLS mechanism. Magnetic impurity doping into GaN nanowires was also achieved. For example, Seong et al. doped Mn or Cu in GaN nanowires up to several at% without sacrificing the single-crystalline nature by using the VLS mechanism [57, 58]. In these works, single-crystalline magnetic semiconductor $Ga_{1-x}Cu(\text{or}\,Mn)_xN$ nanowires were synthesized using a Ni catalyst deposited onto c-plane sapphire substrates in a chemical vapor transport system. Solid metallic Ga and Cu(or Mn)Cl powder, and NH_3 were used as the Ga and dopant and nitrogen source. Figure 1.25a shows an SEM image of GaCuN nanowires grown on the substrate. The diameter and length of these nanowires are from 10 to 100 nm and tens of micrometers, respectively, and they have a triangular structure (inset in Fig. 1.25a). Figure 1.25d shows an HRTEM image. The single-crystalline nature without defects or secondary phases can be seen in all of the HRTEM images. The SAED pattern of the wire, as shown in Fig. 1.25b, indicates that the nanowires grew in the [1–100] direction, perpendicular to the (1000) crystal plane. Figure 1.25c shows the representative Cu concentration as determined by an EDS analysis. The average Cu concentration measured from ten nanowires was ca. 1 to 4% depending on the processing conditions. Further characterization indicates that room-temperature ferromagnetism was achieved from Cu- or Mn-doped GaN semiconductor nanowires. These outcomes demonstrate that n- and p-doping as well as magnetic doping with large ions can be achieved. It should be noted that a high level of doping as compared to thin films can be achieved in the nanowires. This may be due to uniaxial stress in the nanowires, which may be attributed to higher solubility to dopants.

Although limited studies successfully showed the doping of nanowires, doping nonetheless remains a challenging issue owing to the lack of any background on the incorporation of impurities into nanowires. In this regard, Perea et al. carried out atomic-scale direct measurements of dopant concentrations in arbitrary regions of individual nanowires using atomic probe tomography [59]. They found that differences in precursor decomposition rates between the liquid catalyst and the solid nanowire surface give rise to a heavily doped shell surrounding an under doped core (Fig. 1.26). This finding shows that doping in nanowires is different from bulk and thin films and thus requires additional characterization to accumulate the quantitative data necessary to understand the doping mechanism in many different types of nanowires.

1.4 Devices Based on the VLS Mechanism

Based on the VLS mechanism, nanowire optoelectronics can be developed, given that many conceptual devices have already been demonstrated. These conceptual devices were investigated based on a single nanowire or an array of nanowires. Single-nanowire devices are fabricated by the process as follows. First, the

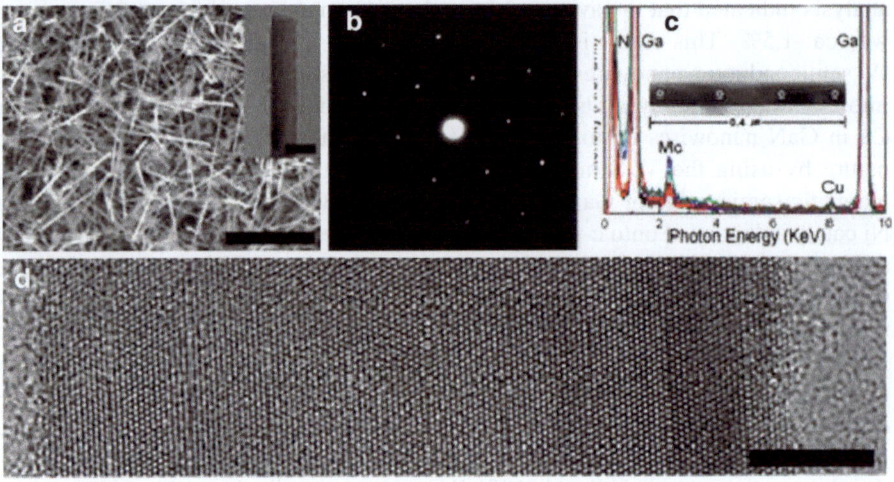

Fig. 1.25 Synthesis and structural characterization of $Ga_{1-x}Cu_xN$ nanowires. (**a**) Typical SEM image of $Ga_{1-x}Cu_xN$ nanowires grown on the sapphire substrate. *Inset* is TEM image of the nanowire showing a triangular structure. The *scale bars* in (**a**) and *inset* are 5μm and 50 nm, respectively. (**b**) SAED pattern of the nanowire, recorded on the [0001] zone axis. (**c**) EDS spectra collected from different positions within the $Ga_{1-x}Cu_xN$ nanowires as marked with O. Inset is a TEM image of a nanowire. The spectra show essentially the same compositions without any evidence of phase inhomogeneity. (**d**) HRTEM image of a nanowire with diameter of 50 nm. The *scale bar* is 2 nm (after [58])

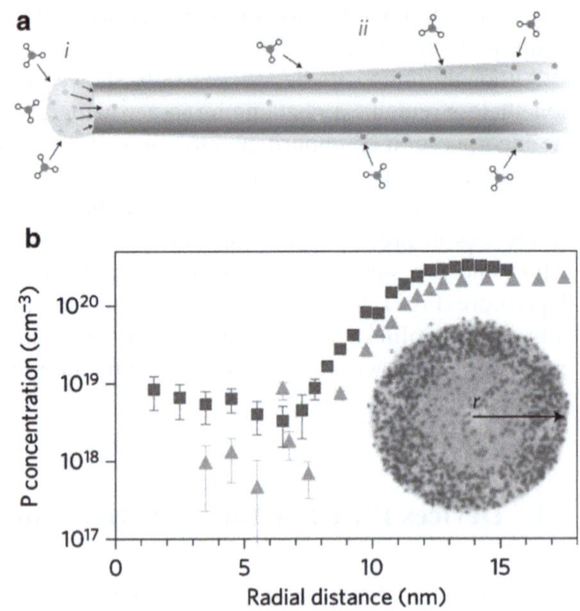

Fig. 1.26 Dopant incorporation pathways and distribution. (**a**) Schematic representation of dopant incorporation pathways via the catalyst (i) and surface decomposition (ii). (**b**) Radial plot of phosphorus concentration for germanium nanowires. The *inset* shows the path along which the concentration was measured (after [59])

nanowires are grown on the substrate. The nanowires are then dispersed in a liquid medium and assembled ex situ on the substrates, where the electrodes are prepatterned. Fabrication of the top electrodes, serving as a passivation layer, is then carried out to the degree necessary. A typical illustration can be found in the preparation of InP nanowire optoelectronic devices [60]. In this process, InP nanowires were grown using Au as a catalyst, and Te and Zn were added as precursors at 1%. After the growth, the nanowires were collected and dispersed in ethanol and then deposited onto oxidized silicon substrates, with conductive silicon used as a back gate. Electrical contact for the NWs was done using electron beam lithography. Ni/In/Au contact electrodes were then thermally evaporated. By this process, a single-nanowire device can be fabricated. Moreover, to fabricate cross-nanowires as a junction structure, layer-by-layer deposition was used. In this process, first a dilute solution of one type of nanowire is deposited on the substrate and the positions of individual nanowires are recorded. In the second step, a dilute solution of another type (for example, p-type) of nanowires is deposited, and the positions of the crossed n- and p-type nanowires are recorded. Metal electrodes are then defined. The inset in Fig. 1.27b shows the crossed nanowire p–n junction, while Fig. 1.27 shows an optical image of the electroluminescence characteristics of the junction.

By further modulating the nanowires during the course of their growth, more versatile devices can be fabricated. One example is the multicolor emission of a nanowire LED [61]. Figure 1.28a shows the concept of single-nanowire structures that consist of an inner n-type GaN core and sequentially deposited i-InGaN, i-GaN, p-AlGaN, and p-type GaN shells for modulation of the luminescence. In this structure, the n-type GaN core and the p-type GaN outer shell serve as electron and hole injection layers, respectively. The $In_xGa_{1-x}N$ layer provides a tunable band gap as well for the efficient radiative recombination of injected carriers, while the wider band gap and lower index of refraction of the AlGaN cladding layer can enhance the confinement of both carriers and photons in the InGaN active layer. Based on this concept, $n-GaN/In_xGa_{1-x}N/GaN/p-AlGaN/p-GaN$ nanowire radial heterostructures were grown by longitudinal growth using a catalyst followed by controlled shell deposition onto the nanowire core. During the deposition process, the composition of the InGaN layer was systematically tuned and could therefore be used to define the band gap of the InGaN and the corresponding emission energy. Through these modulations, the nanowire devices yielded electroluminescence (EL) with red shifts in their emission peak after increasing the In composition with high quantum efficiency. This approach was also used for the growth of a single COHN for photovoltaic cells [62].

Nanowire array optoelectronic devices were fabricated in an in situ mode using the as-grown nanowires on the substrates. Figure 1.29 shows the typical process of the fabrication of a nanowire array for optoelectronics, such as LEDs [57]. The process starts with the growth of the nanowires on the substrates. In this step, the nanowires can be modulated structurally (e.g., diameter control) or compositionally (e.g., doping of electronic or magnetic impurities) to improve the performance of the device. In the example in Fig. 1.29, quasivertically aligned GaN nanowires

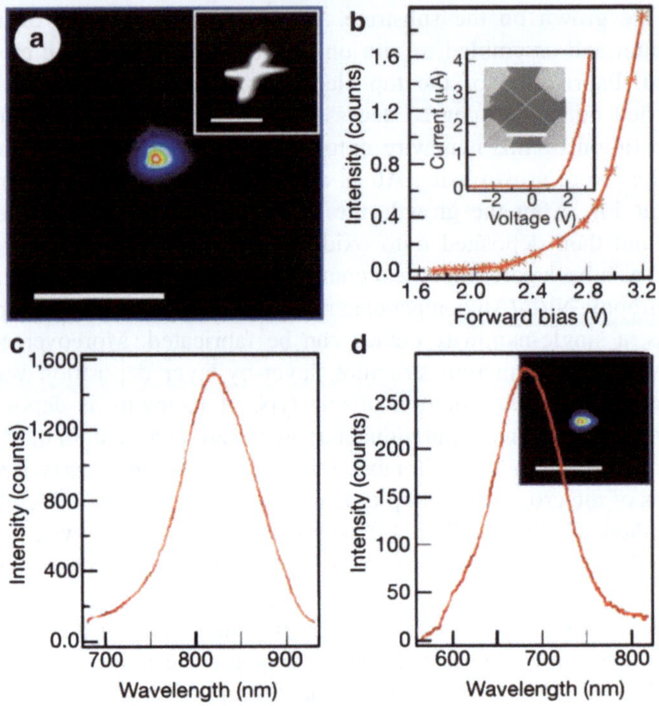

Fig. 1.27 Optoelectrical characterization of nanowire p–n junctions. (**a**) Electroluminescence (EL) image of the light emitted from a nanowire p–n junction. *Inset*: photoluminescence (PL) image of the junction, (**b**) EL intensity versus voltage. *Inset*: $I \pm V$ characteristics, (**c**) EL spectrum of the junction shown in (**a**), (**d**) EL spectrum recorded from a second forward-biased crossed nanowire p–n junction. *Inset*: EL image showing that the EL originates from the junction region (after [60])

were grown on n-SiC (0001) substrates using Ni catalysts (Fig. 1.29a). Here, an n-SiC substrate was used to construct a p–n junction between the nanowires and the substrates and also for the alignment of the GaN nanowires. During the growth process, the nanowires were compositionally modulated by doping with Mn to achieve magnetism as well as p-type characteristics. Therefore, the p–n junctions formed at the interface between the nanowires and the substrates while the diameter and length were controlled by the thickness of the Ni films and the growth time. The fabrication of electrodes on top of the nanowires and on the bottom of the substrates for electrical injections was done subsequent to the growth process. In this case, ohmic contacts were achieved by evaporating Ni/Au and Ni bilayers on the nanowires and substrates, respectively, followed by rapid thermal annealing. Transport measurements showed well-defined current rectification characteristic of p–n diodes (Fig. 1.29c). The I–V data recorded from the nanowires and substrate were symmetric and thus can attribute the rectification to the p–n junction between the nanowires and substrate and not to the junction between the nanowires and

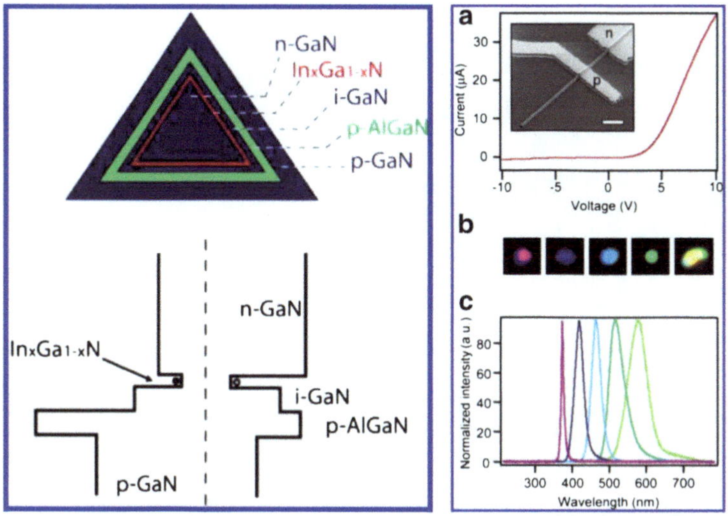

Fig. 1.28 (*Left*) Cross-sectional view of a nanowire heterostructure and the corresponding energy band diagram. (*Right*) (**a**) Current versus voltage data recorded on a nanowire device. *Inset*: field emission scanning electron microscopy image of a representative nanowire device. (**b**) Optical microscopy images collected from around p-contact of nanowire LEDs in forward bias, showing purple, blue, greenish-blue, green, and yellow emission, respectively. (**c**) Normalized EL spectra recorded from five representative forward-biased multicolor nanowire LEDs (after [61])

the metal contacts. Electroluminescence spectra measurements of these junctions showed a dominant emission peak centered at 430 nm, which is consistent with the PL of the nanowires (Fig. 1.29e).

A good example that shows a combination of the CMOS process with the structural and compositional modulation of nanowires for realizing nanowire-based LEDs is contained in the report by Sevenson et al. [63] (Fig. 1.30). In this work, i-GaAs nanowires were grown on a p-Si/p-GaP substrate vertically by the VLS mechanism using Au as catalyst. The GaAs nanowires were grown vertically on the GaP and Si substrates due to the epitaxial relationship between the nanowires and the substrates (Fig. 1.31). For the positioning of the nanowires, standard lithography techniques were used. After the growth of the nanowires, InGaP shells with n-type doping were deposited on the surface of the nanowires and p–i–n junction structures were then formed in the nanowires. The fabrication of the electrodes was done via a standard CMOS process. The established LED functionality of these devices demonstrated that modulation of nanowires by the VLS mechanism can be carried out in the CMOS process for advanced optoelectronics.

Other optoelectronic devices can be developed based on the VLS mechanism. For example, the growth of III–V nanowires on Si has also been demonstrated for the fabrication of GaN nanowire-based solar cells. In this study, vertically aligned Mg-doped GaN nanorods were epitaxially grown on an n-type Si substrate. The result showed a good high-photocurrent density, high-energy conversion efficiency,

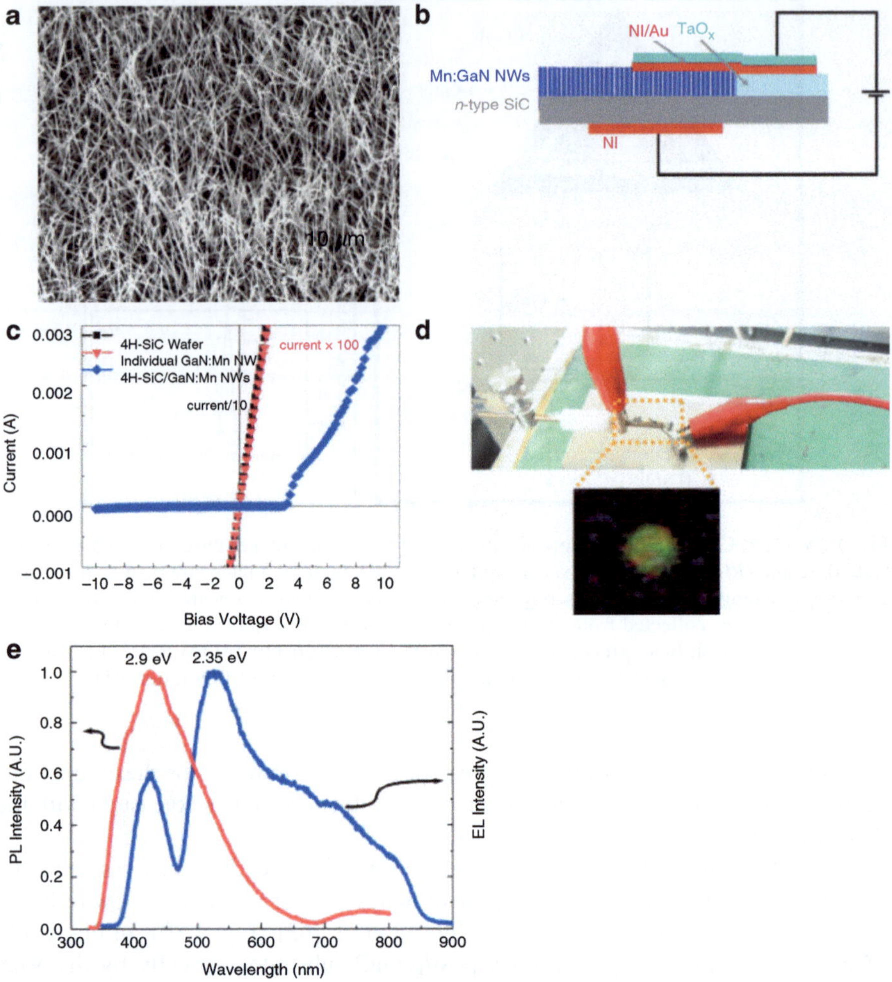

Fig. 1.29 GaN:Mn nanowire LED. (**a**) SEM image of the quasivertical nanowire arrays on substrate, (**b**) schematic illustration of the LED structures, (**c**) $I-V$ behavior of n-SiC substrate/GaN:Mn nanowire junction, (**d**) image of light-emitting interface. *Top* and *bottom* images show the device configuration of the nanowire-based LED structure and an optical image of the emitting device, respectively, (**e**) EL spectrum from nanowire LED, and PL spectrum from GaN:Mn nanowires measured at room temperature using He–Cd laser as excitation source (after [57])

and reduced light loss due to reflection [64]. Though the interfaces were not clearly characterized, this outcome also demonstrates that vertically aligned nanowire p–n junctions can be fabricated by the VLS mechanism. A Si nanowire-based photodetector has also been investigated similarly through the growth of Si nanowires on a substrate, and a photodetector was fabricated in a manner similar to that used with LED or photovoltaic cells [65].

1.5 Summary and Perspectives

The VLS mechanism has been successfully used for the growth of nanowires for optoelectronics. The uniqueness that has made this mechanism a mainstay for the growth of nanowires is its simplicity. The VLS mechanism can be realized by simply adding a metal catalyst to the crystal growth process. Therefore, it can be easily adapted in many conventional semiconductor fabrication processes. Because this mechanism has been rediscovered for the growth of nanowires, most of the studies related to it have investigated metal catalysts to the point that this mechanism is now considered a general method that can be used for the growth of various nanowires. The feasibility of the structural and compositional modulation of nanowires further fuels this mechanism as workhorse in this field.

However, it is true that several issues have to be addressed before the potential of the VLS mechanism can be exploited in the future. One of these issues is the establishment of the nanothermodynamics and kinetics for the nanoliquid–solid system. The thermodynamic and kinetic data from bulk materials have been used to explain the growth of nanowires thus far. However, the growth that occurs in nanosystems in which the thermodynamics and kinetics are different is itself quite different to that in bulk systems. Therefore, frameworks such as size effect on the phase relationship, role of surface energy on the stability of 1 D nanostructures, and diffusion kinetics in liquid on a nanometer scale have to be established. Such fundamentals are essential to grow nanowires rationally. It can also pave the way for the preparation of nanowires in an unprecedented size range of sub-10 nm. The compatibility of metal catalysts to the CMOS process should be addressed for device fabrication. Though nanowires can be grown easily with a metal catalyst in a semiconductor fabrication process, some of them are not compatible with the CMOS process. The possible unintended contamination of a catalyst component in nanowires should also be addressed. A few studies have been done on this issue; however, more quantitative data have to be published to overcome this issue

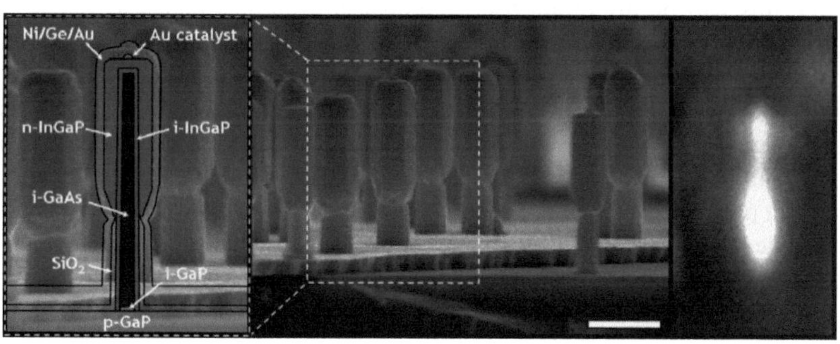

Fig. 1.30 Side-view scanning electron microscopy (SEM) image showing nanowire LEDs. *Left inset*: sketch drawing of the device structure. *Right inset*: side-view CCD camera image showing electroluminescence (EL) from a single-nanowire LED structure (after [63])

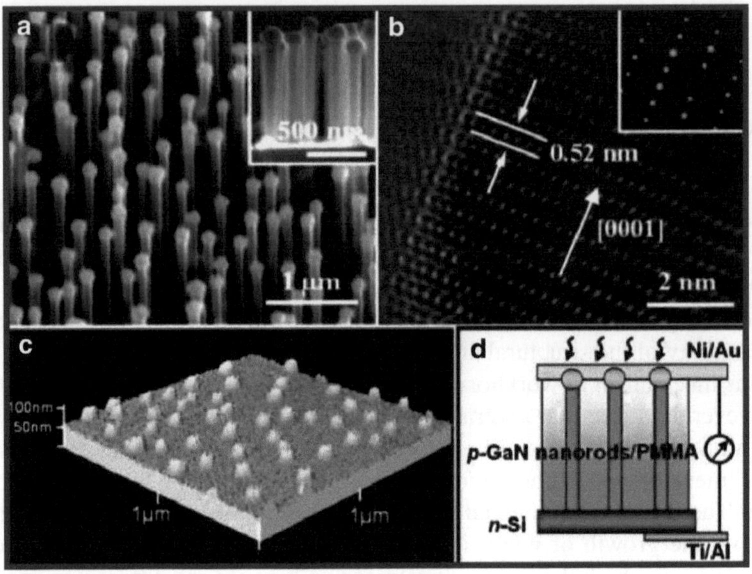

Fig. 1.31 (**a**) Tilt top view SEM image of Mg-doped GaN nanorod arrays. *Inset* shows cross-sectional SEM image. (**b**) HRTEM image of GaN nanorod and its corresponding SAED pattern (*inset*). (**c**) AFM image of GaN nanorod tips exposed above the photoresist layer. (**d**) A schematic of the p-GaN nanorod/n-Si heterojunction photovoltaic cell (after [64])

and in turn to achieve the better optical properties of nanowires. Regarding these, the correlation of the final optical properties with the structure and composition of nanowires may reveal the more sensitive processing parameters for the VLS mechanism. These parameters can be studied quantitatively as well as qualitatively to establish the growth process of nanowires in a predictable, reliable manner for optoelectronic applications. The scaling up of the growth process is another issue that should be addressed. This issue has not been investigated thoroughly as yet; however, this will be important for the industrialization of nanowires near future. To achieve this, the many steps required for the growth of nanowires should be standardized to achieve reliability of the growth process.

Since the 1990s, the VLS mechanism has played a critical role in the creation of the science of nanowires, which has fueled the advance in many optoelectronic fields. The many studies carried over the past two decades have built the foundation of the great potential of this mechanism. To step forward in this area, the mechanism has to be developed from the point of view of industrialization of the nanowires. It will fuel the scientific research in the field of nanowires as well as the creation of advanced optoelectonic market.

Acknowledgements This work was supported by a grant from the National Research Laboratory program and Pioneer Program through the Korea Science and Engineering Foundation funded by the Ministry of Education, Science & Technology.

References

1. S. Wagner, W.C. Ellis, Appl. Phys. Lett. **4**, 89 (1964); in *Whisker Technology*. ed. by A.P. Levitt (John Wiley and Sons, Inc., New York, 1970)
2. H.-J. Choi, J.-G. Lee, J. Mat. Sci. **30**, 1982 (1995)
3. Y. Wu, P. Yang, J. Am. Chem. Soc. **123**, 3165 (2001)
4. F.M. Ross, J. Tersoff, M.C. Reuter, Phys. Rev. Lett. **95**, 146104 (2005)
5. V. Schmidt, J.V. Wittemann, S. Senz, U. Gosele, Adv. Mat. 21, 2681 (2009)
6. T.B. Massalski, P.R. Submanian, H. Okamoto, *Binary Alloy Phase Diagrams*, 2nd ed., vol. 1 (ASM International, Materials Park, OH, 1998)
7. E.A. Sutter, P.W. Sutter, ACS Nano **4**, 4943 (2010)
8. H. Adhikari, A.F. Marshall, I.A. Goldthorpe, C.E.D. Chidsey, P.C. McIntyre, ACS Nano **1**, 415 (2007)
9. E.J. Schwalbach, P.W. Voorhees, Nano Lett. **8**, 3739 (2008)
10. S. Kodambaka, J. Tersoff, M.C. Reuter, F.M. Ross, Science **316**, 729 (2007)
11. E.I. Givargozov, J. Cryst. Growth **31**, 20 (1975)
12. T.I. Kamins, R.S. Williams, D.P. Basile, T. Hesjedal, J.S. Harris, J. Appl. Phys. **89**, 1008 (2001)
13. K. Lew, J.M. Redwing, J. Cryst. Growth **254**, 14 (2003)
14. J. Kikkawa, Y. Ohno, S. Takeda, Appl. Phys. Lett. **86**, 123109–1 (2005)
15. M.H. Huang, Y.Wu, H. Feick, N. Tran, E. Weber, P. Yang, Adv. Mat. **13**, 113 (2001)
16. H. Jeong, T.E. Park, H.K. Seong, M. Kim, U. Kim, H.J. Choi, Chem. Phys. Lett.**467**, 331 (2009)
17. T. Stelzner, G. Andra, E. Wendler, W. Wesch, R. Scholz, U. Gosele, S. Christiansen, Nanotechnology **17**, 2895 (2006)
18. J.E. Allen, E.R. Hemesath, D.E. Perea, J.L. Lenscch-Falk, Z.Y. Li, F. Yin, M.H. Gass, P. Wang, A.L. Bleloch, R.E. Palmer, L.J. Lauhon, Nat. Nanotechnol., **3**, 168. (2008)
19. J.B. Hannon, S. Kodambaka, F.M. Rossand, R.M. Tromp, Nature **440**, 69 (2006)
20. Y. Ke, X. Weng, J.M. Redwing, C.M. Eichfeld, T.R. Swisher, S.E. Mohney, Y.M. Habib, Nano Lett. **9**, 4494 (2009)
21. Y. Wang, V. Schmidt, S. Senz, U. Gosele, Nat. Nanotechnol. **1**, 186 (2006)
22. P.X. Gao, Y. Ding, Z.L. Wang, Nano Lett. **3**, 1315 (2003)
23. J. Yoo, Y.-J. Hong, S. An, G.-C. Yi, B. Chon, T. Joo, J.-W. Kim, J.-S. Lee, Appl. Phys. Lett. **89**, 043124–1 (2006)
24. B. Mandl, J. Stangl, E. Hilner, A.A. Zakharov, K. Hillerich, A.W. Dey, L. Samuelson, G. Bauer, K. Deppert, A. Mikkelsen, Nano Lett. **10**, 4443 (2010)
25. L. Gao, R.L. Woo, B. Liang, M. Pozuelo, S. Prikhodko, M. Jackson, N. Goel, M.K. Hudait, D.L. Huffaker, M.S. Goorsky, S. Kodambaka, R.F. Hicks, Nano Lett. **9**, 2223 (2009)
26. B. Delleyand, E.F. Steigmeier, Appl. Phys. Lett. **67**, 2370 (1995)
27. C. Harris, E.P. O'Reilly, Physica E **32**, 341 (2006)
28. Y. Wu, Y. Cui, L. Huynh, C.J. Barrelet, D.C. Bell, C.M. Lieber, Nano Lett. **4**, 433 (2004)
29. J. Goldberger, R. He, Y. Zhang, S. Lee, H. Yan, H.-J. Choi, P. Yang, Nature **422**, 599 (2003)
30. M.S. Gudiksen, J. Wang, C.M. Lieber, J. Phys. Chem. B **106**, 4036 (2002)
31. X. Wang, Y. Ding, C.J. Summers, Z.L. Wang, J. Phys. Chem. B **108**, 8773 (2004)
32. L. Cao, J.S. White, J.-S. Park, J.A. Schuller, B.M. Clemens, M.L. Brongersma, Nat. Mater. **8**, 643 (2009)
33. L. Hu, G. Chen, Nano Lett. **7**, 3249 (2007)
34. U. Kim, T.-E. Park, I. Kim, H.-K. Seong, M.-H. Kim, J. Chang, J.-G. Park, H.-J. Choi, J. Appl. Phys. **106**, 123903–1 (2009)
35. T. Kuykendall, P.J. Pauzauskie, Y. Zhang, J. Goldberger, D. Sirbuly, J. Denlinger, P. Yang, Nat. Mater. **3**, 524 (2004)
36. Y.B. Tang, Z.H. Chen, H.S. Song, C.S. Lee, H.T. Cong, H.M. Cheng, W.J. Zhang, I. Bello, S.T. Lee, Nano Lett. **8**, 4191 (2008)

37. Z. Zhong, F. Qian, D. Wang, C.M. Lieber,Nano Lett. **3**, 343 (2003)
38. T. Mårtensson, C.P.T. Svensson, B.A. Wacaser, M.W. Larsson, W. Seifert, K. Deppert, A. Gustafsson, L.R. Wallenberg, L. Samuelson, Nano Lett. **4**, 1987 (2004)
39. X.-Y. Bao, C. Soci, D. Susac, J. Bratvold, D.P.R. Aplin, W. Wei, C.-Y. Chen, S.A. Dayeh, K.L. Kavanagh, D. Wang, Nano Lett. **8**, 3755 (2008)
40. M.H. Huang, S. Mao, H. Feick, H. Yan, Y. Wu, H. Kind, E. Weber, R. Russo, P. Yang, Science **292**, 1897 (2001)
41. X. Wang, J. Song, P. Li, J.H. Ryou, R.D. Dupuis, C.J. Summers, Z.L. Wang, J. Am. Chem. Soc. **127**, 7920 (2005)
42. I. Levin, A. Davydov, B. Nikoobakht, N. Sanford, P. Mogilevsky, Appl. Phys. Lett. **87**, 103110–1 (2005)
43. D. Li, C.Z. Ning, Nano Lett. **8**, 4234 (2008)
44. F. Qian, Y. Li, S. Gradecak, H. Park, Y. Dong, Y. Ding, Z. Wang, C.M. Lieber, Nat. Mater. **7**, 791 (2008)
45. H.-J. Choi, J.C. Johnson, R. He, S.-K. Lee, F. Kim, P. Pauzauskie, J. Goldberger, R.J. Saykally, P. Yang, J. Phys. Chem. B, **107**, 8721 (2003)
46. P.D. Markowitz, M.P. Zach, P.C. Gibbons, R.M. Penner, W.E. Buhro, J. Am. Chem. Soc. **123**, 4502 (2001)
47. H.-J. Choi, J.H. Shin, K. Suh, H.-K. Seong, H.-C. Han, J.-C. Lee, Nano Lett. **5**, 2432(2005)
48. M.S. Gudiksen, L.J. Lauhon, J. Wang, D.C. Smith, C.M. Lieber, Nature **415**, 617 (2002)
49. M.T. Bjork, B.J. Ohlsson, T. Sass, A.I. Persson, C. Thelander, M.H. Magnusson, K. Deppert, L.R. Wallenberg, L. Samuelson, Nano Lett. **2**, 87 (2002)
50. Y. Wu, R. Fan, Peidong Yang, Nano Lett., 2, 83 (2002)
51. A. Fuhrer, L.E. Froberg, J.N. Pedersen, M.W. Larsson, A. Wacker, M.-E. Pistol, L. Samuelson, Nano Lett.**7**, 243 (2007)
52. R.E. Algra, M.A. Verheijen, M.T. Borgstrom, L. Feiner, G. Immink, W.J.P. VanEnckevort, E. Vlie, E.P.A.M. Bakkers, Nature, 456, 369–372 (2008)
53. H.-K. Seong, E.-K. Jeon, M.-H. Kim, H. Oh, J. Lee, J.-J. Kim, H.-J. Choi, Nano Lett. **8**, 3656 (2008)
54. T. Kuykendall, P. Uurich, S. Aloni, P. Yang, Nat. Mater. **6**, 951 (2007)
55. Y. Kim, H.J. Joyce, Q. Gao, H.H. Tan, C. Jagadish, M. Paladugu, J. Zou, A.A. Suvorova, Nano Lett. **6**, 599 (2006)
56. Y. Cui, X. Duan, J. Hu, C.M. Lieber, J. Phys. Chem. B **104**, 5213 (2000)
57. H.-J. Choi, H.-K. Seong, J. Chang, K.-I. Lee, Y.-J. Park, J.-J. Kim, S.-K. Lee, R. He, T. Kuykendall, Peidong Yang, Adv. Mater. **17**, 1351 (2005)
58. H.-K. Seong, J.-Y. Kim, J.-J. Kim, S.-C. Lee, S.-R. Kim, U. Kim, T.-E. Park, H.-J. Choi, Nano Lett. 7, 3366 (2007)
59. D.E. Perea, E.R. Hemesath, E.J. Schwalbach, J.L. Lensch-Falk, P.W. Voorhees, L.J. Lauhon, Nat. Nanotechnol. **4**, 315 (2009)
60. X. Duan, Y. Huang, Y. Cui, J. Wang, C.M. Lieber, Nature **409**, 66 (2001)
61. F. Qian, S. Gradecak, Y. Li, C.-Y. Wen, C.M. Lieber, Nano Lett. **5**, 2287 (2005)
62. Y. Dong, B. Tian, T.J. Kempa, C.M. Lieber, Nano Lett. **9**, 2183 (2009)
63. C.P.T. Svensson, T. Martensson, J. Tragardh, C. Larsson, M. Rask, D. Hessman, L. Samuelson, J. Ohlsson, Nanotechnology **19**, 305201 (2008)
64. Y.B. Tang, Z.H. Chen, H.S. Song, C.S. Lee, H.T. Cong, H.M. Cheng, W.J. Zhang, I. Bello, S.T. Lee, Nano Lett. **8**, 4191 (2008)
65. J. Bae, H. Kim, X.-M. Zhang, C.H Dang, Y. Zhang, Y.J. Choi, A. Nurmikko, Z.L Wang, Nanotechnology **21**, 095502 (2010)

Chapter 2
Catalyst-Free Metal-Organic Vapor-Phase Epitaxy of ZnO and GaN Nanostructures for Visible Light-Emitting Devices

Chul-Ho Lee and Gyu-Chul Yi

Abstract In this chapter, we present a review of current research activities related to ZnO and GaN nanostructures and their heterostructures for visible light-emitting devices. For the preparation of high-quality nanostructures, catalyst-free metal-organic vapor-phase epitaxy has been used because the catalyst-free method offers accurate doping and composition control required for optoelectronic device fabrication. Here, we discuss the catalyst-free growth mechanism, reliable and reproducible position control of ZnO and GaN nanostructures, and their visible light emitter applications.

2.1 Introduction

One-dimensional (1D) semiconductor nanostructures, including nanorods, nanowires, nanobelts, and nanotubes, are vital components for fabricating optoelectronic and photonic nanodevices [1–3]. In particular, nanorod heterostructures exhibiting the quantum-confinement effect have been produced by accurate control of both composition and layer thickness [4–6]. The ability to embed quantum structures in a single nanorod would enable exploring novel physical properties such as quantum confinement and tuning spectral wavelength continuously by varying the well thickness. Further advantages of nanorod quantum structures include the miniaturization

C.-H. Lee
Department of Materials Science and Engineering, POSTECH, San31, Hyoja-dong, Nam-gu, Pohang, Gyungbuk, Korea
e-mail: chulholee@gmail.com

G.-C. Yi (✉)
National Creative Research Initiative Center for Semiconductor Nanostructures, Department of Physics and Astronomy, Seoul National University, 599 Gwanak-ro, Gwanak-gu, Seoul 151–747, Korea
e-mail: gyuchul.yi@gmail.com

G.-C. Yi (ed.), *Semiconductor Nanostructures for Optoelectronic Devices*, NanoScience and Technology, DOI 10.1007/978-3-642-22480-5_2,

of optoelectronic devices and the presentation of a platform to investigate novel principles of functional operations inherent to nanoscale optoelectronics and photonics. The light–matter interaction at the nanoscale is important for fabricating optoelectronic and photonic nanodevices and understanding their operation; thus, quantum effects in nanorod quantum structures are key features in nanoscale optoelectronics and photonics [3, 7, 8].

The nanorod quantum structures can be used for fabricating sophisticated nanoscale photonic and optoelectronic devices. For axial structures, both carriers and excitons can be confined in the quantum-well (QW) layer along the axial direction of a nanorod, which is useful for optoelectronic and photonic devices such as nanophotonic switches [3] and light-emitting diodes (LEDs) [9–13]. Radial (or coaxial) nanorod quantum structures with well-defined interfaces are used as the principal components in nanoscale light-emitting devices [7] and high-speed electronic devices [14, 15]. Radial nanorod quantum structures also play important roles as both optical interconnects and functional units in fabricating optoelectronic and photonic nanodevices [16–18].

Over the past several years, much effort has been devoted to developing various semiconductor nanorods and their heterostructures for nanodevice applications. Metal catalyst-assisted vapor–liquid–solid (VLS) [19–21] and catalyst-free methods [22–26] have been used widely. Among numerous nanorod growth methods, catalyst-free metal-organic vapor-phase epitaxy (MOVPE) has several advantages over catalyst-assisted methods. First, the catalyst-free method enables to grow high-purity, single-crystalline semiconductor nanorods [27, 28]. Second, the MOVPE with computer-controlled reactant gas-delivery system makes it possible to control layer thicknesses in a sub-nanometer scale; this is highly advantageous for growing composition-modulated nanorod heterostructures with sharply defined interfaces [4, 6, 29]. In nanorod heterostructures of wide band gap materials, such as ZnO and GaN, the QW layer thickness must be smaller than a few nanometers for effective quantum confinement of charge carriers, which can be achieved with the MOVPE system [4]. Third, vertically aligned ZnO nanorods can be grown on various substrates, including Al_2O_3 [22], Si [27], GaN [29], graphene [30], glass, polymer, and metal [31]. MOVPE growth temperature of ZnO nanorods was as low as 400°C. Furthermore, position-controlled, selective growth of ZnO and GaN nanorod arrays on Si substrates was recently demonstrated [23, 32, 33], making an important breakthrough for the functional integration of optoelectronic and photonic nanodevices on Si.

This chapter reviews recent research activities related to catalyst-free MOVPE of ZnO and GaN nanostructures and their heterostructures, including quantum structures for visible light-emitting device applications. The main text of this chapter is organized into three sections. The next section briefly introduces catalyst-free growth of ZnO nanorods, and the mechanism of catalyst-free growth of ZnO nanorods is also described. In the second section, position-controlled growth of ZnO and GaN nanostructures and the heterostructures required for the integration of devices are reviewed. The last section focuses on the fabrication of LEDs based on coaxial nanorod quantum structures. Finally, conclusions and perspectives on ZnO and GaN nanostructures for LED applications are presented.

2.2 Catalyst-Free MOVPE of ZnO Nanorods

ZnO has drawn considerable attention for visible to ultraviolet (UV) light-emitting device applications, primarily because of its direct band gap energy as large as 3.4 eV, large exciton binding energy of 60 meV, band gap tunability, high thermal and chemical stability, and the feasibility for wet chemical etching [34, 35]. For such applications, ZnO nanostructures offer marked advantages in terms of material quality over their thin-film counterparts. Generally, it is extremely difficult to prepare a high-quality heteroepitaxial thin film over the entire surface of a large substrate if there are large mismatches in the lattice constants and thermal expansion coefficients between the thin film and the substrate [29, 36, 37]. Thus, only limited types of single-crystal substrates have been used for high-quality ZnO thin-film growth [38]. However, because nanoepitaxy can overcome material incompatibility, single-crystalline ZnO nanostructures, such as nanorods, nanowires, and nanotubes, have been fabricated successfully on many substrates using solution- or vapor-phase approaches [22, 27, 39–41].

A wide variety of growth methods have been developed to prepare ZnO nanostructures, which are classified into three representative preparation methods: metal-catalyst-assisted VLS process [42], catalyst-free vapor-phase epitaxy [22,27], and wet chemical synthesis [39]. Among them, the VLS process was first developed by Wagner and Ellis to fabricate microscale Si whiskers in the 1960s [43]. As details in the VLS growth of nanowires are described in Chap. 1, a typical VLS process starts with the dissolution of vapor-phase reactants into nanosized liquid droplets of a metal catalyst, followed by precipitation of single-crystalline nanostructures from the saturated metal catalyst [44]. Typically, Au, Cu, and Sn have been used as metal catalysts, and GaN(0001) or $Al_2O_3(11\bar{2}0)$ as a substrate for the growth of ZnO nanorods [42, 45, 46]. Meanwhile, catalyst-free MOVPE was employed to grow high-quality ZnO nanorods and their heterostructures because the method does not require a metal catalyst for nanorod formation, which is presented in the following section. Finally, wet chemical synthesis requires very low growth temperatures below 100°C compared with vapor-transport methods; thus, an amorphous glass or even plastic can be used as a substrate, providing the potential for low-cost and large-scale processing [39, 47].

The basic principle of catalyst-free MOVPE for ZnO nanorod growth is almost identical to that for ZnO thin-film growth: transport of precursor metal-organic molecules by a carrier gas onto a heated substrate and subsequent surface chemical reactions with oxygen gas on a substrate. As depicted in Fig. 2.1a, the MOVPE process consists of several steps. First, reactants in the form of precursors are carried into a reactor through a gas-handling manifold. Second, the reactant flows generate a boundary layer that governs the transport of mass, momentum, and energy. Third, the reactants diffuse and adsorb on the heated substrate, inducing nucleation. Fourth, the diffused reactants decompose thermally and react to form ZnO nanorods on the substrate. Simultaneously, by-products of reaction desorb. For typical ZnO nanorod growth, highly oriented ZnO nuclei are formed along the c-axis and then grow as

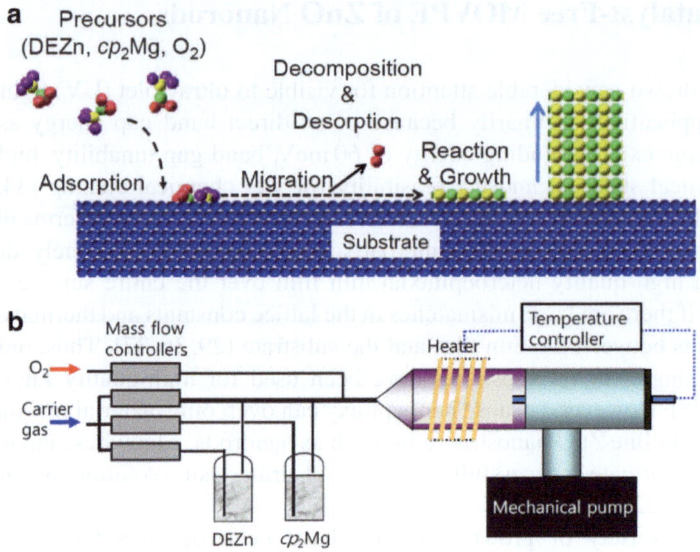

Fig. 2.1 Schematic illustrations of (**a**) growth process and (**b**) MOVPE system for growth of ZnO nanorods

nanorods with a growth direction normal to the surface, because of their higher growth rate along the *c*-axis direction of ZnO.

A schematic of the MOVPE process for ZnO nanorod growth is shown in Fig. 2.1b. The reactants for ZnO growth were diethyl-zinc ($Zn(C_2H_5)_2$; DEZn) and oxygen. To prevent premature reactions, each reactant was supplied into the reactor through separate gas lines. DEZn was carried into a quartz reactor by Ar. On a heated substrate in the reactor, DEZn reacted with oxygen to produce ZnO nanorods. Thus, the general chemical reaction of MOVPE of ZnO is as follows:

$$Zn(C_2H_5)_2 + 7O_2 \rightarrow ZnO + 5H_2O + 4CO_2.$$

Oxygen and DEZn flow rates were in the ranges of 20–100 sccm and 0.5–5 sccm, respectively, at a DEZn bubbler temperature of –15°C, respectively. Typical growth temperatures were in the range of 400–800°C. During growth, the reactor was kept at a low pressure of 5 Torr. Prior to ZnO nanorod growth, a very thin ZnO buffer layer was grown at a low temperature, which improved the vertical alignment of the ZnO nanorods. Details of the growth parameters of ZnO buffer layers on Si substrates are the same as those on sapphire substrates, except for DEZn flow prior to ZnO growth [22, 27]. That is, at the initial ZnO growth stage, only DEZn with a carrier gas flowed for a short time prior to ZnO growth to prevent oxidation of Si substrate surfaces.

Catalyst-free MOVPE enables the growth of ZnO nanorods on many different substrates. As shown in Fig. 2.2, ZnO nanorods with good vertical alignment

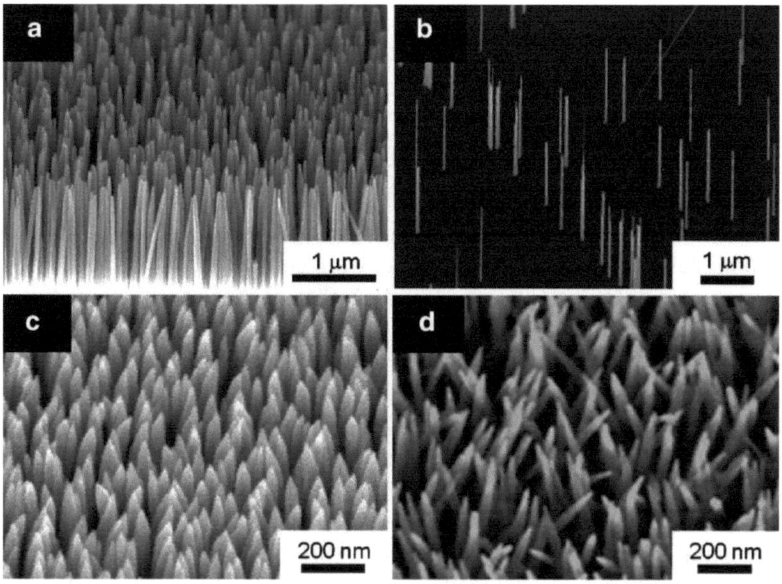

Fig. 2.2 SEM images of vertically aligned ZnO nanorod arrays grown on various substrates including (**a**) $c - Al_2O_3$, (**b**) graphene layers, (**c**) amorphous SiO_2, and (**d**) platinum metal

were grown on various substrates, including single-crystalline $Al_2O_3(0001)$ [22], graphene [30], amorphous SiO_2, and even polycrystalline platinum/Si substrates [27, 48]. All ZnO nanorods were oriented with their c-axes along their hexagonal prismatic axes and had similar diameters of 50 ± 10 nm and number densities of $\sim 2 \times 10^{10}/cm^2$, regardless of the substrate. Such vertical nanorod arrays with uniform diameters and heights are suitable for reliable, large-area fabrication of light-emitting devices [49].

The nanorod growth mechanism was investigated by theoretical calculations of a series of surface and interface formation energies of ZnO crystals using the Vienna ab initio simulation program [50]. The first-principles calculations were conducted with ultrasoft pseudopotentials [51] using plane waves up to a cutoff energy of 29.1 Ry (396 eV). For some calculations, the cutoff energy was increased to 33.0 Ry to ensure convergence of the results. The exchange-correlation potential was described within the generalized gradient approximation parameterized by Perdew and Wang [52], and Brillouin-zone integrals were determined through summations over sufficiently dense meshes of special points, at least 30 **k**-points per 1×1 surface unit cell. All surfaces were represented by a periodically repeating symmetric slab consisting of several atomic layers and were separated by a vacuum region with a thickness ranging from 12.9 to 15.3 Å. Slabs with 10–13 atomic layers (containing up to 60 atoms) were used for the (0001) and $\{11\bar{2}0\}$ surfaces, and 18-30 atomic layers (containing also up to 60 atoms) were used for the $\{10\bar{1}0\}$ and $\{10\bar{1}1\}$ surfaces. For polar (0001) surfaces, a dipole correction was used to prevent artificial

Table 2.1 Surface formation energies of wurtzite ZnO

ZnO crystal planes	$\{10\bar{1}0\}$	$\{11\bar{2}0\}$	$\{10\bar{1}1\}$	(0001)
Surface formation energy $[J/m^2]$	0.91	1.64	1.58	1.74

electrostatic interactions between the repeating units. To simulate the underlying bulk structure, the slab lattice constant was set equal to the theoretical equilibrium bulk value in a direction parallel to the surface, and the atomic positions in two or three atomic layers in the center of the slab were kept fixed at their bulk values.

The first-principles calculations revealed that each surface of wurtzite ZnO, consisting of several planes of $\{10\bar{1}0\}$, $\{11\bar{2}0\}$, $\{10\bar{1}1\}$, and (0001), had a different surface energy [53]. As summarized in Table 2.1, the surface formation energy of the (0001) plane was the highest of the planes of wurtzite ZnO, with that for $\{10\bar{1}0\}$ much smaller than $\{11\bar{2}0\}$ and (0001). This indicates that there is a substantial energy gain for the formation of nanorods rather than thin films due to the reduced surface area of ZnO(0001). That is, ZnO has a tendency to minimize the area of the (0001) plane for minimized total surface energy.

Upon using a substrate with an isotropic surface energy, no constraint toward 1D ZnO nanorod growth is provided. Accordingly, ZnO nuclei can occur randomly across the entire substrate surface during initial growth and subsequently transform into nanorods with a reduction in the surface formation energy. However, use of a substrate with highly anisotropic surface energies may allow us to control growth mode and morphology of ZnO. To prove this argument, we prepared facet-controlled GaN micropatterns with highly anisotropic surface energies and grew ZnO on the GaN micropatterns [48].

Figures 2.3a, b show scanning electron microscopy (SEM) images of ZnO nanostructures grown on GaN micropyramids. Only one ZnO nanotube was grown on each tip of GaN micropyramids. More precise information on the crystal structure and the relevant growth mode of ZnO is shown in a cross-sectional transmission electron microscopy (TEM) image of a single ZnO nanotube grown atop a GaN micropyramid. As more clearly shown in Fig. 2.3c, a thin, coexisting ZnO film has formed on the inclined $\{10\bar{1}1\}$ sidewalls of the GaN micropyramid, with an individual ZnO nanotube located at the GaN(0001) tip. After 1 h growth, the length of the ZnO nanotube was ~950 nm, and the thickness of the ZnO thin film was ~50 nm. Furthermore, Fig. 2.3d shows a high-resolution TEM lattice image for the outlined area of Fig. 2.3c, with arrows I_A and I_B indicating the interfaces between ZnO and GaN. The *c*-plane lattice slabs of GaN and ZnO are parallel, showing that both the ZnO nanorod and the thin film have grown heteroepitaxially on the GaN micropatterns without the formation of any significant structural defects.

The role of the substrate crystal plane in the formation of either a ZnO nanorod or a thin film was investigated theoretically via a series of the first-principles calculations [50]. For the heteroepitaxial growth of ZnO on GaN substrate, the interface formation energy also needs to be considered along with the surface energy because the epitaxial relationship between ZnO and GaN strongly affects both the ZnO growth mode and morphology. Table 2.2 summarizes the calculations

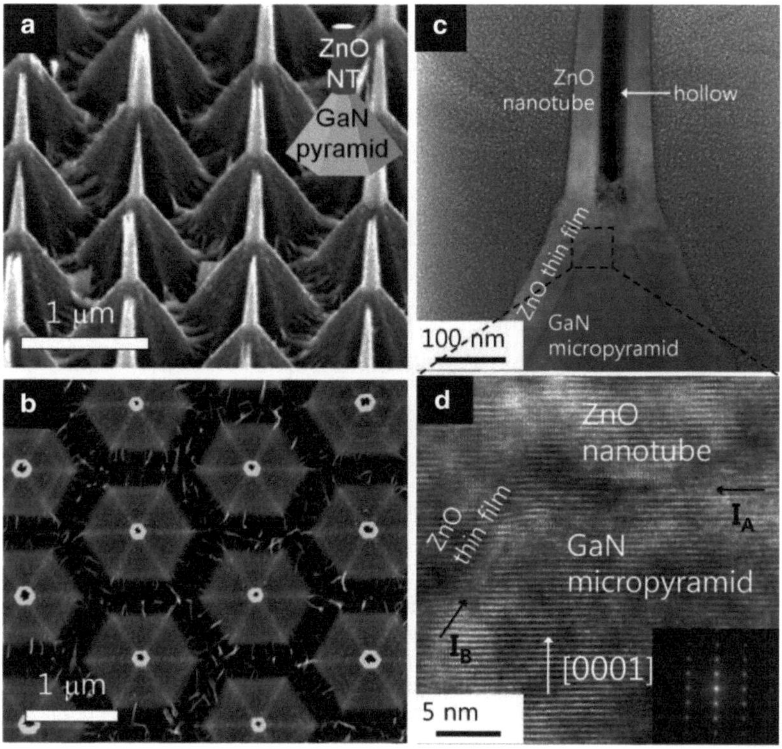

Fig. 2.3 SEM images of ZnO nanotubes atop the GaN micropyramid patterns in (**a**) bird's-eye and (**b**) top views. (**c**) Cross-sectional annular dark-field scanning TEM image of a single ZnO nanotube grown atop a GaN micropyramid. ZnO thin-film formation has occurred on the sidewalls of GaN$\{10\bar{1}1\}$. (**d**) HR-TEM image of the selected area in (**c**) showing the epitaxial growth of the ZnO nanotube and thin film on the GaN micropyramid. The arrow I_A (I_B) shows the interface between the ZnO nanotube (thin film) and the topmost plane (sidewall) of the GaN micropyramid. Adapted with permission from [48]. Copyright 2009 Royal Society of Chemistry

of the surface formation energies of fundamental crystal planes of ZnO and GaN and the interface formation energies of a ZnO epitaxy on GaN. At the GaN(0001) surface, the growth of ZnO nanorods with $\{10\bar{1}0\}$ sidewalls is preferred to reduce the surface formation energy of ZnO(0001). For ZnO growth on the GaN$\{11\bar{2}0\}$ plane, however, the GaN$\{11\bar{2}0\}$ surface energy of $1.53\,\mathrm{J/m^2}$ is reduced to yield an interface formation energy of $1.24\,\mathrm{J/m^2}$ following initial deposition of the ZnO$(11\bar{2}0)$ film. An even smaller surface formation energy for the ZnO$(11\bar{2}0)$ plane of $1.02\,\mathrm{J/m^2}$ results in a ZnO film rather than nanorods on the GaN$\{11\bar{2}0\}$ plane. These calculations are consistent with the experimental results shown in Fig. 2.3. The theoretical calculations also indicate that GaN micropyramids with smooth sidewalls should be used for selective nanorod growth. If the surfaces of side walls have a combination of small (0001) ledges and $\{10\bar{1}0\}$ vertical walls, vertically aligned nanorods can be grown even on the sidewalls of GaN micropyramids.

Table 2.2 Theoretically calculated surface formation energies of GaN and ZnO crystals

Crystal planes	$(00\bar{1}1)$	$\{11\bar{2}0\}$	$\{10\bar{1}1\}$	$\{10\bar{1}0\}$
Surface formation energy of ZnO [J/m^2]	1.91	1.02	1.57	1.01
Surface formation energy of GaN [J/m^2]	2.64	1.53	1.76	1.40
Interface formation energy of ZnO on GaN [J/m^2]	2.69	1.24	1.37	1.13

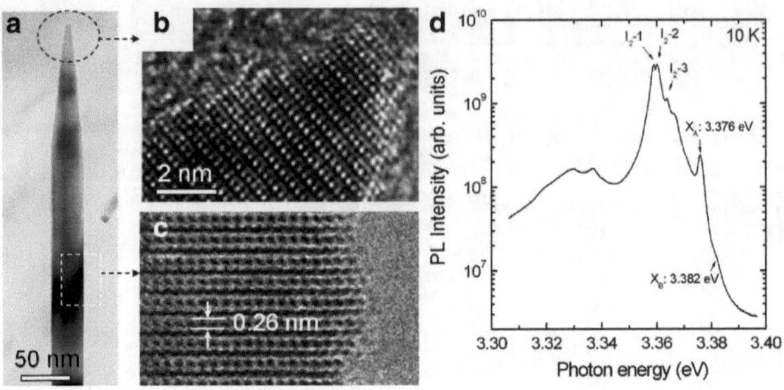

Fig. 2.4 Microstructural and optical characteristics of ZnO nanorods grown by catalyst-free MOVPE. (**a**) Low-magnification TEM image shows sharp tip morphology. (**b**) High-resolution TEM image of the nanorod tip shows clear lattice image without any significant defects or metal nanoparticles commonly observed at the end of nanowires grown by VLS method. (**c**) High-resolution scanning TEM image of nanorod stem. (**d**) High-resolution PL spectra of ZnO nanorods measured at 10 K

ZnO nanorods, grown by catalyst-free MOVPE, exhibited excellent structural characteristics, as investigated by TEM in Fig. 2.4 [27, 28]. The TEM analysis revealed that the nanorod tips had a sharp morphology and a uniform diameter along the nanorod stems. Figure 2.4b shows a TEM image of a ZnO nanorod tip, revealing that metallic nanoparticles, which are commonly observed at the end of nanowires grown by the catalyst-assisted VLS method, were not observed at the end of ZnO nanorods. As shown in Fig. 2.4c, high-resolution scanning TEM images of the nanorod stem showed that lattice spacing between the adjacent planes was 0.26 nm, corresponding to the d-spacing of ZnO(0002) planes. More importantly, extended crystal defects, such as dislocations, were rarely observed in ZnO nanorods.

Photoluminescence (PL) spectroscopy was employed to investigate the optical characteristics of ZnO nanorods. As shown in Fig. 2.4d, the low-temperature PL spectrum exhibited a free-exciton peak and well-resolved neutral-donor-bound excitonic emission with a very narrow full-width-at-half-maximum (FWHM) value of 1–3 meV. The observation of a sharp free-exciton peak for ZnO nanorods at 10 K strongly suggests that the nanorods were of high optical quality and crystallinity despite their large surface-to-volume ratio. The ability of catalyst-free MOVPE to grow high-quality ZnO nanorods on various substrates makes it possible to fabricate high-performance nanostructure-based LEDs on transparent and/or conducting substrates that are practical for low-cost, large-scale fabrication.

Complementary doping in a controlled manner is necessary to use ZnO nanorods as functional components in LEDs. H, Ga, and Al are typically used as n-type dopants for ZnO nanostructures. For example, tunable n-type conductivity of ZnO nanowires using chemical vapor deposition with Ga_2O_3 powder as a Ga source was reported [54]. However, the homogeneous doping of Ga in ZnO nanowires causes the degradation of both optical and structural characteristics. Such unintentional effects can be undesirable for many electronic and optoelectronic device applications.

As an alternative approach, Yoo et al. demonstrated a modulation-doping technique to control electrical conductivity and enhance the electrical characteristics of the devices [55]. As schematically illustrated in Fig. 2.5a, for the modulation-doping process, gas flow was interrupted during growth, instead of flowing reactants and dopant gases continuously. High-resolution TEM revealed that modulation-doped ZnO nanorods maintained their single crystallinity and [0001] orientation with no structural deformation (Fig. 2.5b). Good electrical characteristics of modulation-doped ZnO nanorod field-effect transistors (FETs) were observed: the overall conductance increased systematically with increasing Ga dopant concentrations (Fig. 2.5c) and the transfer characteristic (I_{ds}–V_g curve) in Fig. 2.5d also exhibited

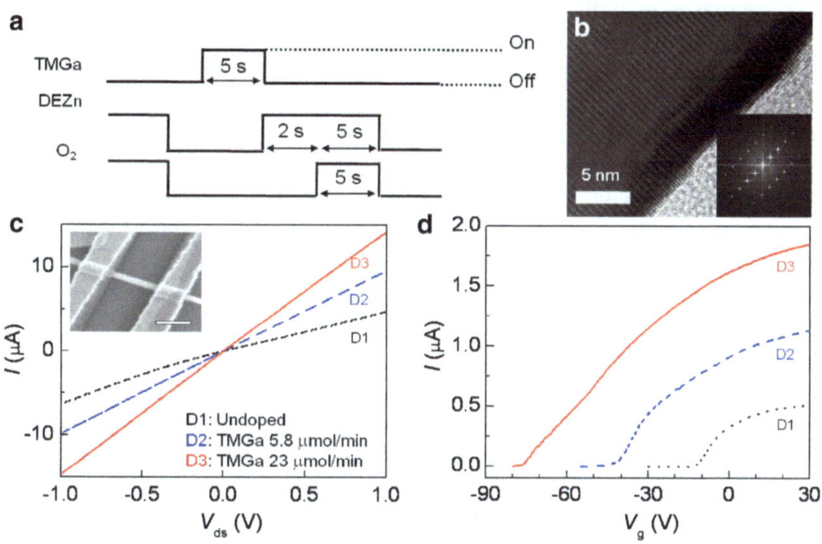

Fig. 2.5 Modulation doping of n-ZnO nanorods. (**a**) Schematic of the modulation doping process for growth of Ga-modulation-doped ZnO nanorods. Temporal operation of the metal-organic vapor and gas flow is depicted in sequence, from top to bottom. (**b**) Typical HR-TEM image of modulation-doped ZnO nanorods. The inset shows the electron diffraction pattern of a Ga-modulation-doped nanorod stem. (**c**) I_{ds}–V_{ds} characteristic curves and (**d**) I_{ds}–V_g transfer characteristic curves (at $V_{ds} = 0.1$ V) of individual ZnO nanorod devices with TMGa flow rates of 0, 5.8, and 23 μmol/min. The *inset* of (**c**) is the typical SEM image of the nanorod device where *scale bar* is 500 nm. Adapted with permission from [55]. Copyright 2009 American Institute of Physics

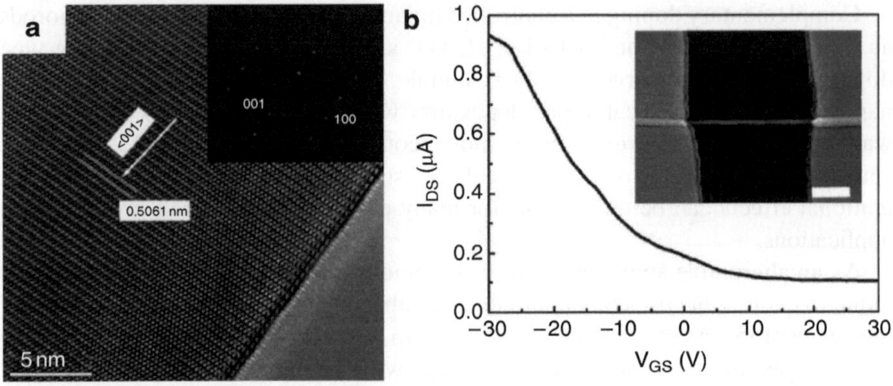

Fig. 2.6 Phosphorus-doped p-ZnO nanowires. (**a**) HR-TEM image of phosphorus-doped ZnO nanowires showing high-quality single crystal with extremely clean and smooth surface. The growth direction was along <0001> direction as indexed in the diffraction pattern. (**b**) I_{ds}–V_{gs} characteristic curve of p-ZnO nanowire transistor at $V_{ds} = 15\,V$. *Inset* is SEM image of the measured device. The *scale bar* is 500 nm. Adapted with permission from [56]. Copyright 2007 American Chemical Society

gradual increases in both threshold gate voltage and saturation current with Ga contents, indicating the increased carrier concentration in the modulation-doped ZnO nanorods. Thus, the modulation-doping technique provides a reliable route for controlling the conductivity of ZnO nanorods without negative effects on the electrical, optical, or structural characteristics.

P-type doping in ZnO nanorods is much more challenging, presumably due to high background n-type carrier concentration and self-compensation, caused by donor defects that form readily in undoped ZnO. However, preparation of p-type ZnO nanowires by in situ doping of phosphorus were recently reported [56–58]. For p-type ZnO nanowires, phosphorus pentoxide (P_2O_5) powder was used as a phosphorus dopant [56]. From the high-resolution TEM image in Fig. 2.6a, phosphorus-doped ZnO nanowires are shown to be single-crystalline. The transfer characteristics of phosphorus-doped ZnO nanowire FETs, measured at $V_{ds} = 15\,V$, demonstrated the increased conductance upon sweeping to negative gate voltage (Fig. 2.6b), strongly suggesting the p-type characteristic of ZnO nanowires by the phosphorus doping.

2.3 Position-Controlled Growth of ZnO and GaN Nanostructures

Despite the successful fabrication of high-quality semiconductor nanostructures on various substrates, practical applications of nanostructure LEDs have remained out of reach because of difficulties in manipulating and positioning individual nanostructures. Thus, demand has risen for the precise control of position and

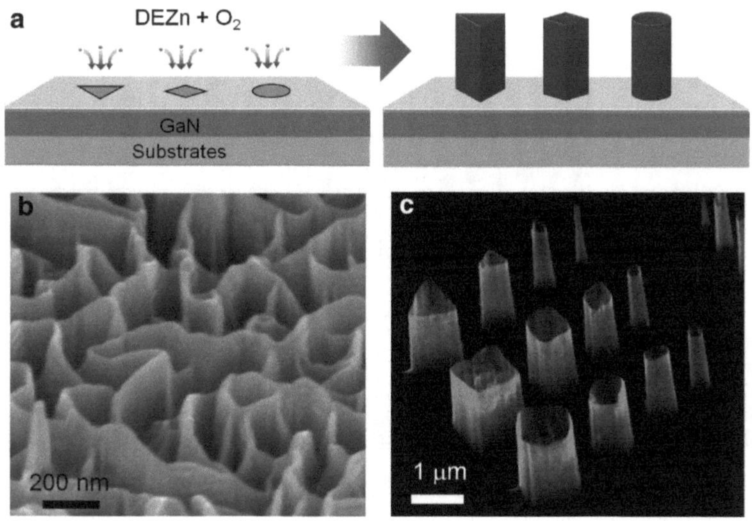

Fig. 2.7 (**a**) Schematics of the process for obtaining shape-controlled nanoarchitectures using nanowalls. (**b**) SEM image of ZnO nanowalls with a random network morphology on bare GaN/Si(111) substrates. (**c**) SEM image of shape-controlled ZnO nanotubes with triangular, square, and circular morphologies. The shapes of the nanowalls and nanotubes are determined by the pattern of the growth mask

dimension during nanostructure growth. However, this is a huge challenge for the bottom-up approach due to difficulties in controlling nucleation sites [48, 59, 60].

Shape- and position-controlled ZnO nanowalls were grown on patterned substrates with a GaN intermediate layer using catalyst-free MOVPE [59]. To control both the shape and position of ZnO nanowalls, a patterned SiO_2 growth-mask layer was prepared on the substrates using conventional lithography. ZnO nanowalls were then grown epitaxially on the substrates at 600°C using DEZn and oxygen reactant gases. Figure 2.7a shows schematics of the processes for shape- and position-controlled nanoarchitectures using nanowalls and the corresponding SEM images. Generally, when no growth mask is used, vertical ZnO nanowalls adopt the morphology of random networks (Fig. 2.7b). With a growth mask, the shape and position of the nanowalls are determined by the patterns formed on the substrate because the nanowalls grew selectively along the edges of the patterns. For example, as shown in Fig. 2.7c, diverse ZnO nanotubes with triangular, square, and cylindrical tube morphologies were grown selectively on the patterned substrate. This shape-controlled selective growth of ZnO nanowalls was attributed to the preferred nucleation and growth of ZnO at the pattern edges on GaN/Si substrates.

Furthermore, the diameters and positions of the ZnO nanotubes were controlled precisely by changing those of the hole patterns. Figure 2.8a shows SEM images of regular hexagonal arrays of vertical ZnO nanotubes with a uniform diameter of 400 nm at spacings of 0.8, 1.2, 1.6, and 2.4 μm. Figure 2.8b shows controlled nanotube diameters of 200, 400, 600, and 800 nm with a fixed 1.6μm spacing.

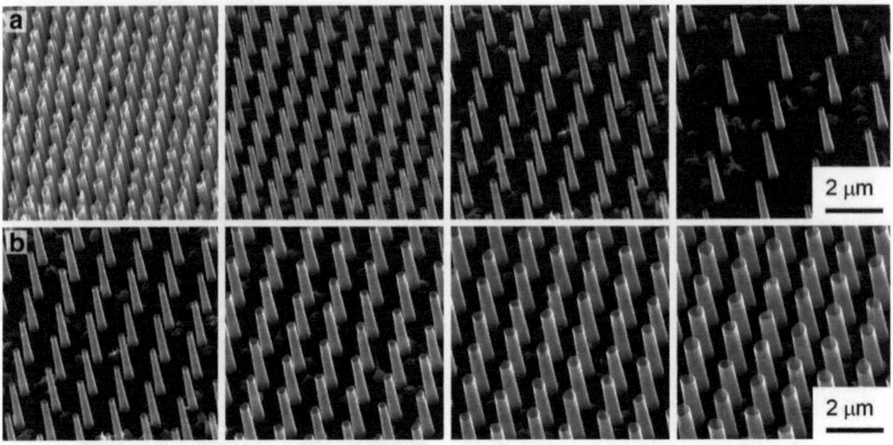

Fig. 2.8 SEM images of position- and diameter-controlled ZnO nanotube arrays. (**a**) Nanotube hexagonal array with spacings of 0.8, 1.2, 1.6, and 2.4 μm, left to right. (b) The nanotube diameters are 200, 400, 600, and 800 nm, left to right. Adapted with permission from [59]. Copyright 2009 John Wiley & Sons, Inc.

Smaller nanotube diameters resulted from smaller holes. Because the diameter and length of the ZnO tubes are determined by the hole diameters and the growth time, respectively, with a typical growth rate of approximately 34 nm/min, all dimension parameters and morphologies of ZnO nanotubes can be controlled simply by lithographic patterning and growth parameters of the MOVPE process.

The optical characteristics of the position-controlled ZnO nanotube arrays were investigated by PL spectroscopy at various temperatures between 10 K and room temperature [61]. As shown in Fig. 2.9a, the room temperature PL spectrum of ZnO nanotubes shows a dominant near band-edge (NBE) emission peak at 3.29 eV. No mid-gap emission, usually observed for PL spectra of bulk materials, was observed. The 10 K PL spectrum in Fig. 2.9b shows NBE emission peaks at 3.358 and 3.451 eV due to the ZnO nanotubes and GaN intermediate layer, respectively. The FWHM of the PL peaks was as small as 7 meV. This strong NBE emission with a small FWHM from the ZnO nanotubes suggests that the nanotube arrays were of high optical quality because of their high crystallinity and purity.

The optical characteristics of individual nanotubes were investigated using cathodoluminescence (CL) spectroscopy. Figure 2.9c shows CL monochromatic image of the ZnO nanotube array, which was measured at the photon energy of 3.36 ± 0.02 eV selected through a monochromator. Position-controlled ZnO nanotubes exhibited homogeneous and strong CL emission. Furthermore, the SEM and CL images are similar, implying that nonradiative transition or radiative transitions other than NBE due to defects did not occur. As shown in Fig. 2.9d, the 5 K CL spectrum of the nanotube array shows strong UV NBE emission peaks at 3.460, 3.361, 3.318, 3.310, 3.240, and 3.160 eV. Most of the CL peaks were

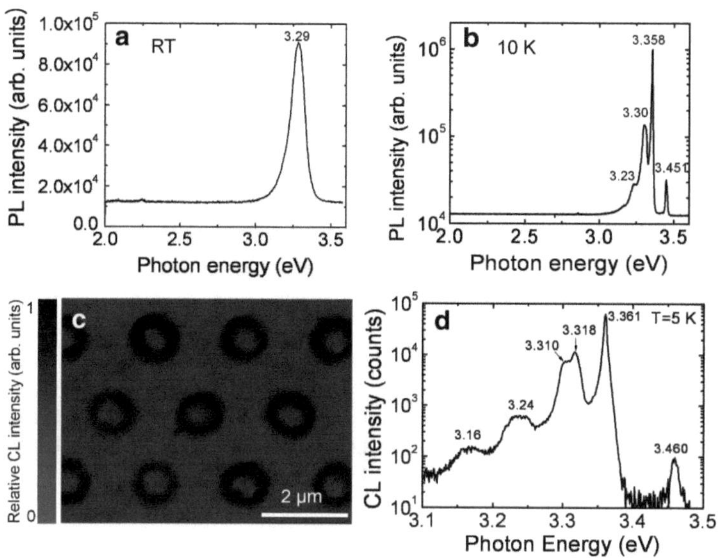

Fig. 2.9 PL spectra of ZnO nanotube arrays on patterned substrates at (**a**) room temperature and (**b**) 10 K. (**c**) CL monochromatic ($E = 3.36 \pm 0.02$ eV) image and (**d**) CL spectrum of ZnO nanotube arrays on patterned GaN/AlN/Si substrates. The CL image and spectrum were measured at 5 K. Adapted with permission from [61]. Copyright 2009 John Wiley & Sons, Inc.

tentatively ascribed to emissions from the ZnO nanotubes, except the peak at 3.460 eV from the GaN intermediate layer [62, 63].

The next challenge for constructing functional components of LEDs is to fabricate position-controlled nanomaterial heterostructures with compositional modulation since the heterostructure arrays with numerous quantum structures can be useful for many LED applications. In particular, coaxial nanorod/nanotube heterostructures that have QW layers on the cylindrical surfaces of nanorods or nanotubes can provide an ideal geometry for fabricating high-performance nanostructure-based LEDs with several advantages [64–68]. These nanoarchitecture arrays enable us to accurately control the position, thickness, and composition of quantum structures embedded in the nanoarchitectures. The controlled formation of quantum structures in nanoarchitectures will significantly enhance the optical and electrical characteristics of the LEDs. One strategy is the coaxial heteroepitaxial growth of quantum structures on the periphery of position-controlled nanotube arrays to increase the active volume of LEDs as well as continuous tuning of spectral wavelengths by changing the well thickness. The catalyst-free MOVPE allows the fabrication of coaxial nanotube heterostructure arrays with clean and abrupt interfaces by direct epitaxial growth of quantum structures over the entire surface of ZnO nanotubes.

Position-controlled nanotube quantum structures were fabricated by in situ heteroepitaxial growth of the QW and quantum barrier (QB) layers using catalyst-free

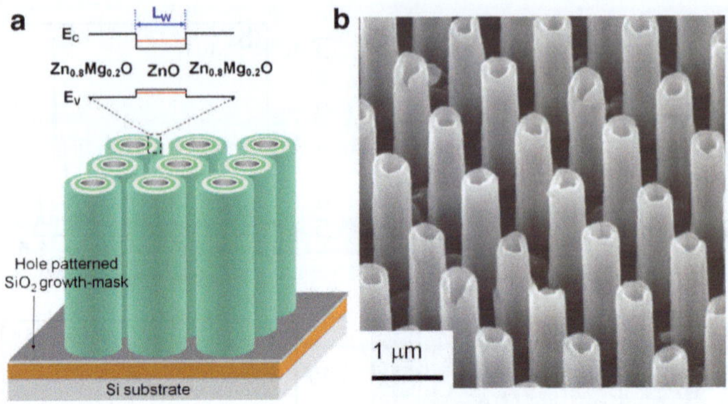

Fig. 2.10 (**a**) Schematic of a $ZnO/Zn_{0.8}Mg_{0.2}O$ coaxial nanotube SQW array and an energy band diagram of coaxial nanotube SQW. (**b**) SEM image of a $ZnO/Zn_{0.8}Mg_{0.2}O$ coaxial nanotube SQW array. The coaxial nanotube SQWs exhibited the mean diameter of 500 ± 30 nm

MOVPE [61]. Figure 2.10a shows a schematic of a $ZnO/Zn_{0.8}Mg_{0.2}O$ coaxial nanotube single-quantum-well (SQW) structure prepared along the periphery of the ZnO nanotube arrays. The Mg precursors for $Zn_{1-x}Mg_xO$ growth was *bis*-cyclopentadienyl-magnesium (cp_2Mg). The Mg content was controlled to be 20 at.% because $Zn_{1-x}Mg_xO$ with $x = 0.2$ is a good QB that confines carriers in a ZnO well layer [6]. Additionally, small lattice constants and thermal expansion coefficient mismatches between the ZnO and $Zn_{0.8}Mg_{0.2}O$ layers minimize the formation of interfacial defects in the nanotube quantum structures [69].

Figure 2.10b shows the SEM image of $Zn0.8Mg_{0.2}O/ZnO$ coaxial nanotube SQWs. At first, bare ZnO nanotubes with a mean diameter of 480 ± 20 nm and a length of 3.5 ± 0.1 μm were prepared. After the growth of the ZnO QW and $Zn_{0.8}Mg_{0.2}O$ QB shell layers for 90 and 120 s, respectively, the mean diameter of the coaxial nanotube heterostructures was increased to 500 ± 30 nm. By averaging the diameters of the individual coaxial nanotube heterostructures from microscopic images, the average growth rates of the ZnO QW and $Zn_{0.8}Mg_{0.2}O$ QB shells were estimated to be 0.30 and 0.33 Å/s, respectively.

The optical characteristics of position-controlled $ZnO/Zn_{0.8}Mg_{0.2}O$ coaxial nanotube SQW arrays were investigated using PL and CL spectroscopy. Figure 2.11a shows the 90 K CL spectrum and image of a single $ZnO/Zn_{0.8}Mg_{0.2}O$ coaxial nanotube SQW with a SQW width (L_W) of 4 nm. Two distinct CL emission peaks were observed at 3.35 and 3.41 eV, emitted from a core ZnO nanotube (I^{ZnO}) and SQW (I^{SQW}), respectively. Homogeneous and strong CL emissions were observed in spatially resolved, monochromatic CL images measured at photon energies of 3.35 ± 0.02 (red) and 3.41 ± 0.02 eV (green) corresponding to ZnO (I^{ZnO}) and SQW (I^{SQW}), respectively. In particular, I^{SQW} (green) was generally observed along the outer periphery of I^{ZnO} (red) due to the core ZnO nanotube, strongly suggesting that the ZnO SQW layer was conformally deposited on the outer surfaces of core ZnO nanotubes.

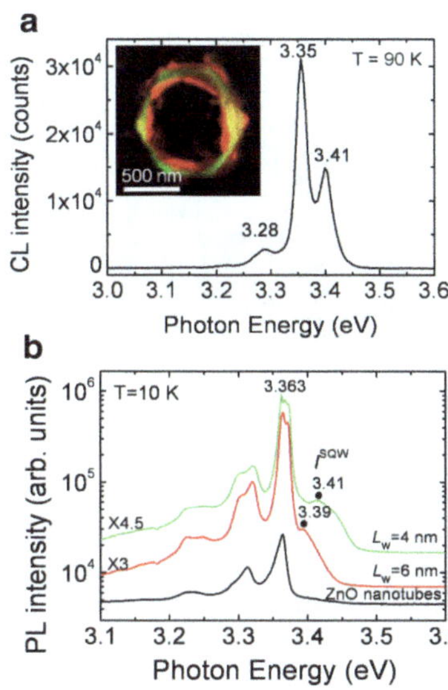

Fig. 2.11 (**a**) CL spectrum and CL image of individual $ZnO/Zn_{0.8}Mg_{0.2}O$ coaxial nanotubes. Monochromatic CL images measured at photon energies of 3.35 ± 0.02 (*red*) and 3.41 ± 0.02 eV (*green*) corresponding to ZnO (I^{ZnO}) and SQW (I^{SQW}), respectively. (**b**) Low-temperature PL spectra of ZnO nanotubes and $ZnO/Zn_{0.8}Mg_{0.2}O$ coaxial nanotube SQW arrays with well layer widths of 4 and 6 nm. Adapted with permission from [61]. Copyright 2009 John Wiley & Sons, Inc.

Quantum phenomena of the nanotube SQW arrays were further investigated by measuring PL spectra of the $ZnO/Zn_{0.8}Mg_{0.2}O$ nanotube SQWs with various QW widths. In addition to the PL peak at 3.363 eV ascribed to neutral-donor-bound excitons [28,62] in core ZnO nanotubes, as shown in Fig. 2.11b, a small PL peak was observed at 3.39 and 3.41 eV for 6- and 4-nm-thick SQWs, respectively. These blue shifts originated from quantized sublevel states created in coaxial nanotube SQWs due to the quantum-size effect. Simple theoretical calculations assuming a finite periodic square-well potential also confirmed that the systematic increase in PL emission energy with decreasing well width resulted from the quantum-confinement effect.

In addition to ZnO, GaN is another promising material for visible LED applications due to easy controls of n- and p-doping and band gap modulation over a wide wavelength range by alloying In and Al. Although most previous research on GaN nanorod growth has been performed using VLS growth methods [70–73], a few studies on catalyst-free growth of GaN nanorods have been reported [74–76]. In particular, GaN nanorod growth by catalyst-free MOVPE has been achieved only on substrates with both a GaN intermediate layer and a submicrometer hole-patterned mask layer [33, 77, 78].

Similar to the selective growth of ZnO nanotubes, position-controlled GaN nanorods have been grown on n^+-$GaN/Al_2O_3(0001)$ substrates with a hole-patterned SiO_2 mask layer using catalyst-free, selective MOVPE. To create

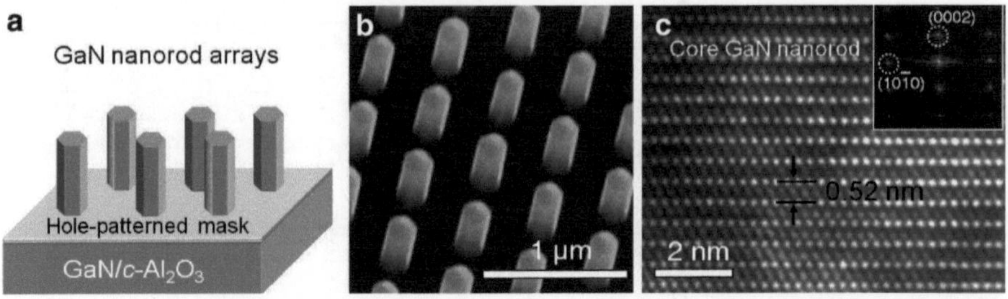

Fig. 2.12 (**a**) Schematic illustration of selective MOVPE of GaN nanorod arrays on n^+-GaN/Al_2O_3(0001) substrates. (**b**) SEM image of GaN nanorod arrays. (**c**) HR-TEM image of single-crystalline core n-GaN nanorods. The *inset* shows the diffraction pattern of the GaN nanorods. Adapted with permission from [33]. Copyright 2011 John Wiley & Sons, Inc.

hole-patterned arrays on a SiO_2 mask layer, various patterning techniques have been used, including *e*-beam or photolithography, laser interference lithography, nanoimprint lithography, and template-assisted lithography [33, 77, 78]. As shown in the SEM image of Fig. 2.12a, a vertically aligned GaN nanorod array exhibited excellent uniformity in length, diameter, and interdistance, all of which were controlled by changing the lithographic design and growth parameters.

The high-resolution TEM image in Fig. 2.12b clearly shows that the GaN nanorod was single-crystalline [33]. The lattice spacing between adjacent planes was ~0.52 nm, corresponding to the *d*-spacing of GaN(0001) planes. The diffraction pattern, obtained through fast Fourier transform (FFT) of the high-resolution TEM image, shows that the single-crystalline GaN nanorod grew along the *c*-axis of wurtzite. Furthermore, extended crystal defects, such as dislocations or stacking faults, were rarely observed in GaN nanorods, presumably due to nanoscale homoepitaxy in confined hole openings.

After the position-controlled growth of GaN nanorod arrays, $In_xGa_{1-x}N$/GaN multiple-quantum-well (MQW) layers were heteroepitaxially grown over the entire surface of each GaN nanorod array using selective MOVPE [33]. The spatial distributions of the thickness and composition of $In_xGa_{1-x}N$/GaN MQWs coated on GaN nanorod surfaces were investigated using cross-sectional scanning TEM. Figure 2.13a shows a low-magnification TEM image of the MQWs with a flat topmost plane, slants, and upright sidewalls, corresponding to the (0001), $\{10\bar{1}1\}$, and $\{10\bar{1}0\}$ planes of wurtzite GaN, respectively. In particular, the high-resolution TEM images in Figs. 2.13b, c clearly show significant large differences in the QW and QB thicknesses; the thicknesses of $In_xGa_{1-x}N$ QW and GaN QB topmost layers were 8 ± 2 and 22 ± 3 nm, respectively, while those on the sidewall facets were 1.4 ± 0.3 and 2.2 ± 0.3 nm, respectively. In addition to the two topmost $In_xGa_{1-x}N$/QW layers, another band-like $In_xGa_{1-x}N$ layer was also observed, denoted by the dotted line in Fig. 2.13b, which was attributed to buried $In_xGa_{1-x}N$/GaN MQWs grown along the nanorod slants. The formation of QW and QB layers with different thicknesses, depending on the GaN nanorod facets, may have resulted from the

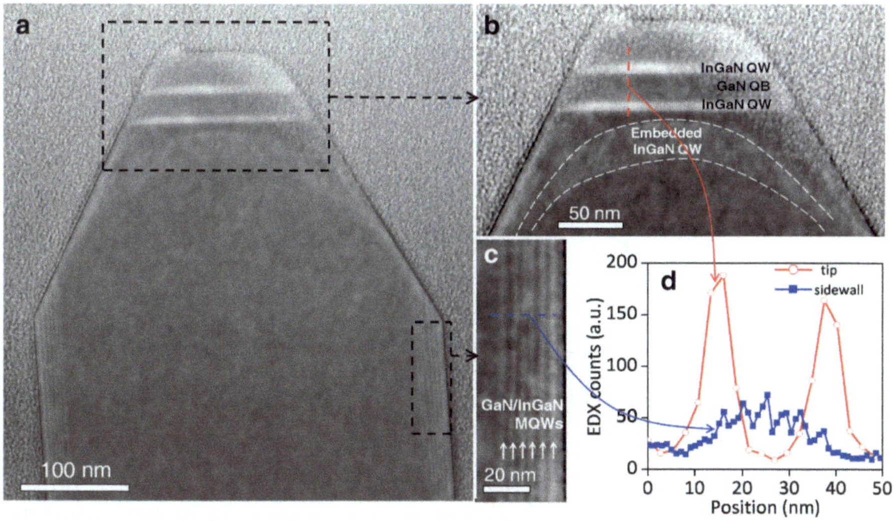

Fig. 2.13 (**a**) Low-magnification scanning TEM image of $In_xGa_{1-x}N/GaN$ nanorod MQWs composed of a flat topmost plane, slants, and upright sidewalls. (**b**)–(**c**) High-magnification scanning TEM images of the $In_xGa_{1-x}N/GaN$ MQWs formed on (**b**) topmost and (**c**) upright sidewall areas of GaN nanorods. (**d**) Energy-dispersive X-ray line profile of the indium L characteristic wavelength along the axial (*red dotted line* in (**b**)) and radial (*blue dotted line* in (**c**)) directions of the $In_xGa_{1-x}N/GaN$ nanorods. Adapted with permission from [33]. Copyright 2011 John Wiley & Sons, Inc.

anisotropic surface formation energies of GaN crystal planes that influence the diffusion of adatoms [48, 79]. In addition to thickness, the In concentration of $In_xGa_{1-x}N$ QWs formed at the tips of nanorods was four times higher than that on the sidewalls. The x values for $In_xGa_{1-x}N$ QWs on the topmost and sidewall facets were estimated to be ~0.6 and ~0.15, respectively. The $In_xGa_{1-x}N/GaN$ MQWs that formed anisotropically on the GaN nanorods play a critical role in the color tunability of the nanorod-embedded LEDs, as discussed in the next section.

2.4 Light-Emitting Device Applications

ZnO nanostructures offer useful functional components for diverse nanostructure-based LEDs [80, 81]. Recently, vertical ZnO nanostructure arrays have been widely used as active elements for light-emitting or lasing in nanophotonic devices and as passive elements for improving the light extraction of conventional thin-film LEDs [18, 42, 49].

A few papers reported fabrication of ZnO nanorod homojunction LEDs by an ion implantation, although ZnO p–n homojunction LEDs prepared using in situ doping methods have rarely been achieved presumably due to difficulty in

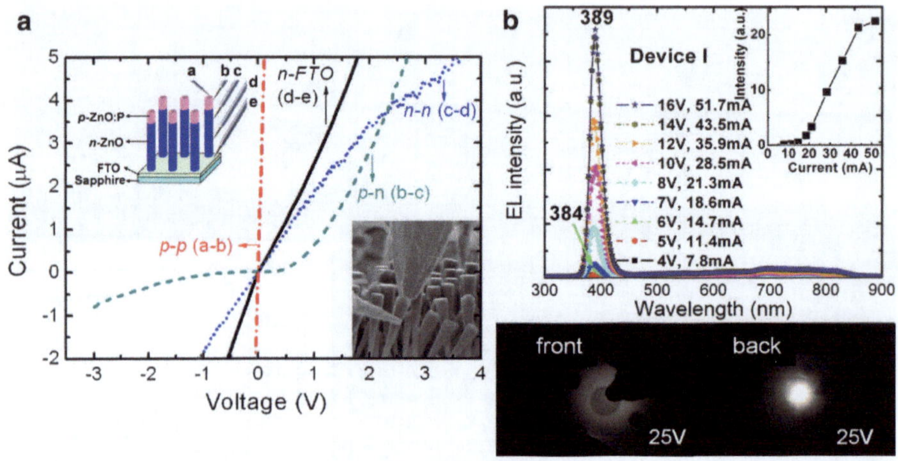

Fig. 2.14 ZnO p–n homojunction nanorod LEDs fabricated by an ion implantation. (**a**) $I-V$ characteristics of p–p, p–n, n–n, n–fluorine-doped tin oxide regions in a single ZnO nanorod. The *upper left and lower right insets* show the schematic diagram and a SEM image of proving by nanomanipulator, respectively. (**b**) Electroluminescence (EL) spectra at various applied current levels and the photographs of light emission from both front and back sides. The *inset* show the light output intensities as a function of forward applied current. Adapted with permission from [84]. Copyright 2009 American Institute of Physics

preparing p-type ZnO [82–84]. In particular, Sun et al. demonstrated UV emission from ZnO homojunction LEDs composed of p–n nanorod arrays, where p-type doping was achieved by phosphorous or arsenic ion implantation [82, 84]. The rectifying $I-V$ characteristic of individual p–n junction nanorods in Fig. 2.14a confirmed the formation of p-ZnO on the upper part of the nanorods. As shown in Fig. 2.14b, UV emission corresponding to NBE emission of ZnO was observed. Additionally, the LEDs exhibited a relatively weak and broad emission band in the visible range related to deep-level defects, which became stronger as implantation energy increased. Despite the successful demonstration of ZnO homojunction LEDs, ion implantation methods for p-type doping may not offer sophisticated device structures with thickness- and composition-controlled quantum structures required for high-efficiency LEDs.

To overcome the challenges associated with p-type doping, heterojunction LEDs, composed of n-type ZnO nanostructure arrays heteroepitaxially grown on a p-type thin film, have been suggested as an alternative to homojunctions for LED applications [49]. This nanostructure/film hybrid heterojunction device has several potential advantages over thin-film heterojunction devices. First, nanoepitaxy enables the growth of high-quality semiconductor nanostructures with a clean interface on a thin film, leading to high-efficiency LEDs by reducing nonradiative recombination at the junction. Second, carrier injection efficiency can be improved at the nanoscale junctions between individual nanostructures and underneath thin film because the carrier injection rate increases significantly for nanocontacts in Schottky diodes

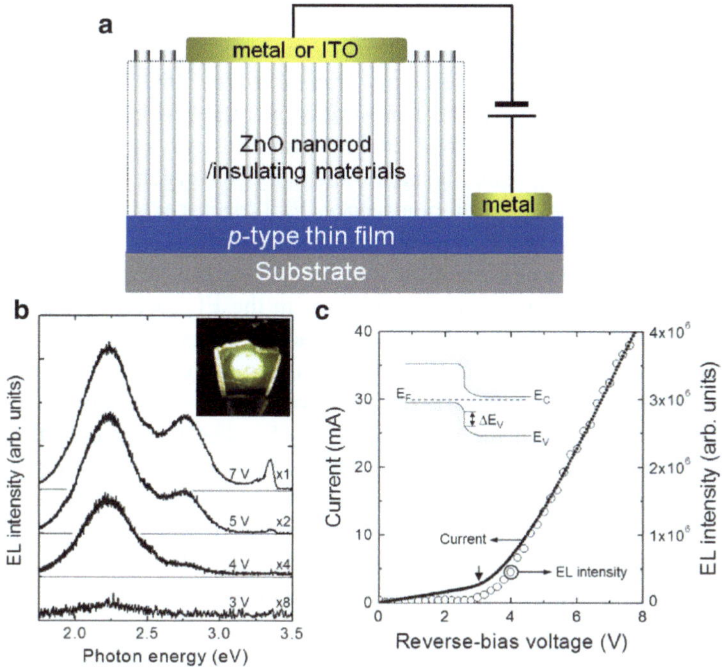

Fig. 2.15 (**a**) Schematic illustration of a general n-ZnO nanorods/p-type thin-film hybrid heterojunction LED. (**b**) EL spectra of an n-ZnO nanorods/p-GaN thin-film hybrid heterojunction LED. Under reverse bias, the EL spectra exhibited a yellow emission band centered at 2.2 eV (560 nm) and a weak blue emission band centered at 2.8 eV (450 nm). The *inset* is a photograph of light emission from the EL device at reverse-bias voltage of 5 V. (**c**) I–V characteristic and current–EL intensity measured at 2.2 eV (560 nm). The *inset* shows the band diagram of the p-GaN/n-ZnO heterojunction. Adapted with permission from [49]. Copyright 2004 John Wiley & Sons, Inc.

as reported previously [85, 86]. Third, light-extraction efficiency can be enhanced considerably due to reduced internal reflection on a rough surface composed of nanostructures [18].

Figure 2.15a shows a schematic diagram of a typical n-ZnO nanorod/p-type thin-film hybrid heterojunction LED. In this geometry, vertically aligned ZnO nanorod arrays were grown on p-type substrates via chemical and physical vapor-phase deposition or wet chemical methods, forming numerous nanoscale p–n junctions. For device fabrication, empty gaps between individual nanorods were filled with insulating supporting materials such as a polymer or spin-on glass (SOG) for electrical isolation, and then ohmic contacts on n-ZnO were formed separately with p-GaN contacts by depositing thin metal layers or transparent indium tin oxide (ITO). Among diverse p-type semiconductors available in industry, GaN thin film is the most widely used because it has a fundamental band gap energy (~3.4 eV) similar to ZnO, the same wurtzite crystal structure, and a low lattice constant misfit of 1.9% [49]. Other p-type thin films including SiC, Si, NiO, and $CuAlO_2$ have also been used [87–94]. However, their heterojunction devices exhibit poor performance because of a large energy barrier and/or many interface states at the junction.

For the first demonstration, vertically aligned n-type ZnO nanorod arrays were grown epitaxially on p-GaN (0001) substrates by catalyst-free MOVPE, and then Ti/Au ohmic contacts were formed on ZnO nanorod arrays after filling the gaps with a photoresist and exposing the nanorod tips by an oxygen plasma etching [49]. As shown in Fig. 2.15b, electroluminescence (EL) spectra exhibited a broad yellow emission band centered at 2.2 eV (560 nm), only under reverse-bias conditions and an additional weak blue emission at 2.8 eV (450 nm) as reverse bias increased. Based on PL data, yellow and blue emissions in the EL spectra were assigned to deep-level emission in ZnO and to the Mg-acceptor level in GaN, respectively. This unusual carrier transport under reverse bias may be explained in terms of type II band alignment of the p-GaN/n-ZnO heterojunction with a large valence band offset (ΔE_v), as shown in the inset of Fig. 2.15c. The large band offset formed at the heterojunction makes the tunneling barrier very thin by lowering the n-ZnO conduction band or raising the p-GaN valence band. Under a small reverse-bias voltage, the unoccupied conduction band minimum of n-ZnO is lower than the occupied valance band maximum of p-GaN. Thus, carrier transport by tunneling can occur even under small reverse-bias voltages and the tunneling probability increases with increasing reverse-bias voltage.

Many kinds of hybrid heterojunction LEDs with similar concepts and geometries have been reported by other groups [95–101]. For example, Zhang et al. recently demonstrated hybrid heterojunction LEDs using an n-ZnO nanowire array grown on p-GaN, exhibiting blue light emission under forward-bias conditions [99]. The device geometry is almost identical to the previous geometry except that a transparent ITO electrode was used instead of metal layers. As a result of using ITO contacts, the device exhibited resistive characteristics and emitted light under high forward-bias voltages. As a modified geometry, additionally, Jeong et al. reported the Al-doped ZnO film/n-ZnO nanowire array/p-GaN film structure for heterojunction devices by fabricating continuous ZnO thin films on top of ZnO nanowire arrays [102, 103], which also exhibited blue light emission under forward bias (Fig. 2.16a–c). This geometry has the advantage of simple device fabrication by depositing metal contacts on thin films without gap-filling or selective-etching processes. Furthermore, compared with film-based GaN/ZnO heterojunction LEDs, nanowire-inserted heterojunction LEDs showed stronger EL emission and enhanced electrical characteristics with smaller leakage current, lower turn-on voltage, and larger forward current (Fig. 2.16d). Such superior characteristics were explained by a low density of interfacial defects at the lattice-matched heterojunction interface and enhanced electron injection through nanoscale contacts.

Generally, p–n heterojunction devices are less efficient than homojunction devices because an energy barrier formed at the junction interface decreases carrier injection efficiency for heterojunction devices with a large band offset [104]. Accordingly, the nanorod/thin-film hybrid homojunction LEDs were fabricated using n-GaN/ZnO coaxial nanorod heterostructures. In this device as illustrated in Fig. 2.17a, n-GaN layers were grown epitaxially on the entire surface of ZnO

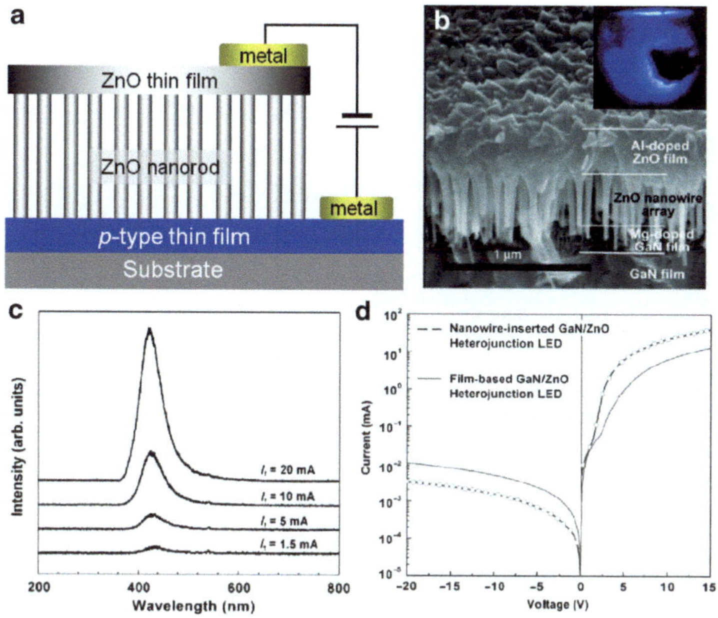

Fig. 2.16 N^+-film/n-nanowire/p-film hybrid heterojunction LED. (**a**) Schematic illustration, (**b**) SEM image, and (**c**) EL spectra of the ZnO film/ZnO nanowire array/GaN film hybrid heterojunction device. The *inset* of (**b**) is a photograph of blue light emission from the device at the forward current of 10 mA. (**d**) I–V curves of the film-based and ZnO-nanowire-inserted GaN/ZnO heterojunction devices. Adapted with permission from [103]. Copyright 2007 John Wiley & Sons, Inc.

nanoneedles using MOVPE [105]. Unlike previous devices using ZnO nanorods [53], dominant emission peaks were observed in the EL spectra (Fig. 2.17b) at 2.96 eV (430 nm) and 3.24 eV (382 nm), associated with deep acceptors in Mg-doped p-GaN and shallow donors in the n-GaN/ZnO nanostructures, respectively.

As mentioned in Sect. 2.3, the position-controlled nanorod/nanotube heterostructures can be the most promising material for the high-performance LEDs. The basic strategy for the fabrication of position-controlled nanoarchitecture LEDs is to use the well-developed growth techniques of GaN-based heterostructures on the surfaces of position-controlled ZnO nanotube arrays. The ability to control electrical conductivity amphoterically and to modulate the band gap from infrared to far UV by substitutional alloying of In and Al in III-nitrides enabled the fabrication of high-efficiency, full-color optoelectronic devices [106, 107]. Furthermore, because GaN has the same crystal structure as wurtzite ZnO with a small lattice misfit ($a = b = 3.249$ Å, $c = 5.206$ Å for ZnO and $a = b = 3.186$ Å, $c = 5.178$ Å for GaN), it enables high-quality GaN/ZnO heteroepitaxy with very few interfacial defects [66].

GaN/$In_xGa_{1-x}N$/GaN/ZnO nanoarchitecture LEDs were recently fabricated by epitaxial growth of the p- and n-GaN layer with $In_xGa_{1-x}N$/GaN MQW layers on

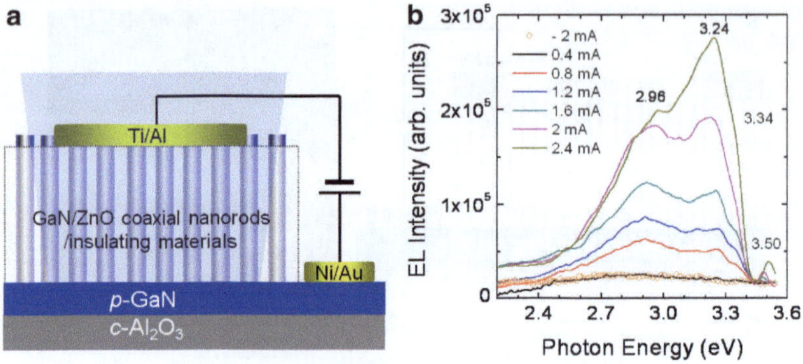

Fig. 2.17 Hybrid heterojunction LED using GaN/ZnO coaxial nanorod heterostructures. (**a**) Schematic illustration of the LED composed of n-GaN/ZnO coaxial nanorod heterostructures on the p-GaN substrate. (**b**) EL spectra at various applied current levels

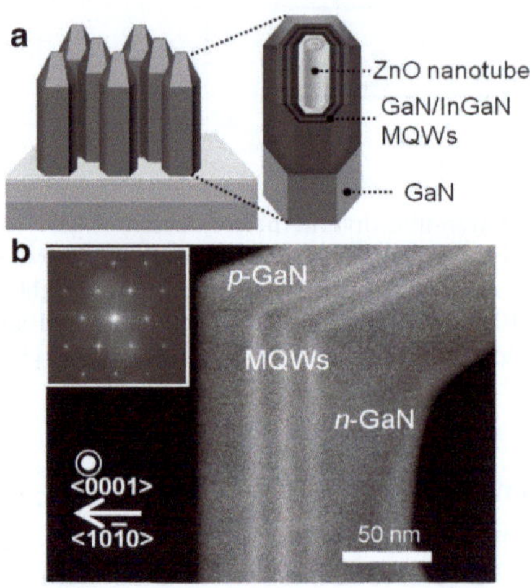

Fig. 2.18 Position-controlled GaN-based coaxial nanotube quantum structure arrays. (**a**) Schematic representation of position-controlled GaN/$In_{1-x}Ga_xN$/GaN/ZnO coaxial nanotube quantum structure arrays and their cross-sectional drawing. (b) Horizontal cross-sectional STEM image of the GaN/$In_{1-x}Ga_xN$/GaN/ZnO coaxial nanotube. *Inset* is the corresponding diffraction pattern obtained from a FFT process

the radial surfaces of position-controlled ZnO nanotube arrays [66,67]. Figure 2.18a shows the detailed structure of GaN/$In_xGa_{1-x}N$ coaxial nanotube MQWs. The horizontal cross-sectional scanning TEM image in Fig. 2.18b reveals that three bright lines corresponding to $In_xGa_{1-x}N$ QW layers alternated with clearly discriminated GaN QB layers, indicating that the three-period MQW layers were coaxially coated on the GaN/ZnO nanotube heterostructures. The corresponding FFT image of the coaxial nanotube quantum structure reveals the single crystallinity with sixfold rotational symmetry of the $\{10\bar{1}0\}$ planes.

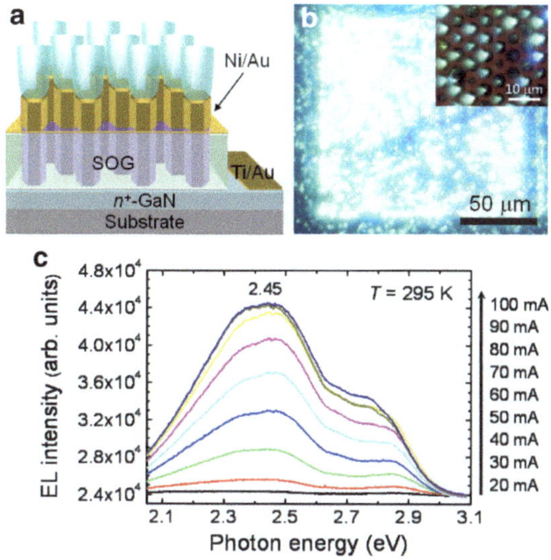

Fig. 2.19 $GaN/In_{1-x}Ga_xN/GaN/ZnO$ nanoarchitecture LED microarrays. (**a**) Schematic representation showing the device structure of position-controlled nanoarchitecture LED arrays. (**b**) Optical microscopic photographs of light emission from nanoarchitecture LED microarrays. The *inset* is the magnified image. (**c**) EL spectra of nanoarchitecture LED microarrays at various applied current levels of 20–100 mA

For the fabrication of vertical nanoarchitecture LED arrays, ohmic contacts were made on both the outermost p-GaN surface of nanoarchitectures and the n-GaN seed layer on a substrate. For electrical isolation between two metal electrodes, empty gaps between nanoarchitectures were filled with an insulating SOG layer (Fig. 2.19a). Figure 2.19b shows the hexagonal arrays of bluish-green light emission from nanoarchitecture LED microarrays under a forward current of 100 mA, where each spot corresponds to the light emission from individual nanoarchitecture LEDs. The EL spectra of nanoarchitecture LED microarrays, shown in Fig. 2.19c, exhibited a dominant peak centered at 2.45 eV and a shoulder around 2.85 eV, resulting from the $In_{1-x}Ga_xN/GaN$MQWs. In addition to the EL characteristics, the electrical characteristics were consistent with a typical rectifying behavior with a turn-on voltage of $\sim$3 V and a leakage current of $\sim 5 \times 10^{-4}$ A at –4 V. Above the turn-on voltage, the current increased rapidly with increasing bias voltage, increasing the light emission intensity.

This approach using position-controlled GaN/ZnO coaxial nanotube heterostructures has also been used to fabricate nanoarchitecture LEDs on Si substrates [108]. Because cracks form readily during the growth of a thick GaN film on Si, it is important to prepare crack-free thin GaN films for growing uniform ZnO nanotube arrays. Similarly, high-quality coaxial nanotube heterostructures were fabricated by heteroepitaxial growth of GaN-based LED structures on ZnO nanotube arrays. The LEDs emitted visible green light from individual nanoarchitecture LEDs,

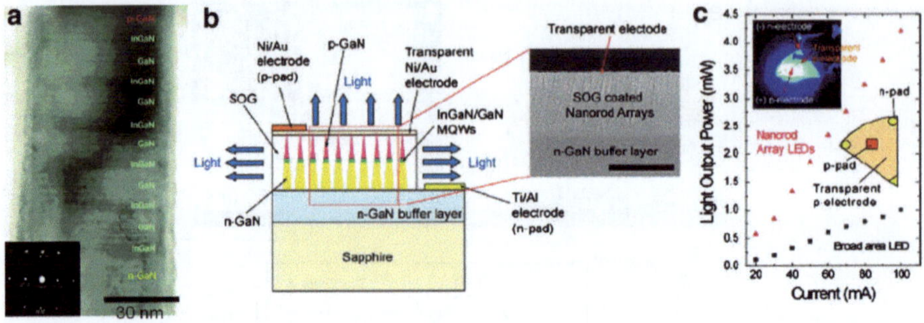

Fig. 2.20 (**a**) TEM image of $In_xGa_{1-x}N$/GaN MQWs in a nanorod. Inset shows the corresponding electron diffraction pattern. (**b**) Schematic diagram and SEM image of cross-sectional $In_xGa_{1-x}N/GaN$ MQW nanorod array LEDs. (**c**) Light output power-forward current $(L\text{–}I)$ characteristics of a six-period $In_xGa_{1-x}N$/GaN MQW nanorod LED using an on-wafer testing configuration, as compared to a conventional broad area LED. *Insets* show the top-view photograph image and schematic diagram of a blue emission from $In_{0.25}Ga_{0.75}N/GaN$ MQW nanorod LEDs at 20 mA. Adapted with permission from [9]. Copyright 2004 American Chemical Society

corresponding to a dominant EL peak at 2.35 eV. Additionally, the emission was bright enough to be seen with the naked eye under normal room illumination. The successful fabrication of nanoarchitecture LEDs on Si constitutes a promising strategy for resolving the problems associated with conventional thin-film LED fabrication on Si substrates for potential Si-compatible optoelectronics.

The combined approach of position-controlled growth of ZnO nanotube arrays and conventional III-nitride MOVPE enabled not only precise control of composition and layer thickness of the nanotube heterostructures but also fabrication of 3D nanoarchitecture LED microarrays. Furthermore, the ability to fabricate nanoarchitecture LED microarrays offers the potential for high-brightness LEDs as well as building blocks for integrated optoelectronic systems.

The early study on GaN nanostructure homojunction LEDs has been fabricated using axial p–n junction nanorod arrays because of the feasibility of both n- and p-type doping in GaN [9, 10, 13, 109]. Vertical GaN p–n junction nanorod arrays embedded with $In_{1-x}Ga_xN$/GaN MQWs were grown on sapphire substrates using metal-organic-hydride vapor-phase epitaxy (Fig. 2.20a) [9]. The arrays exhibited almost single-crystalline nature. For device fabrication, as schematically shown in Fig. 2.20b, nanorod arrays were buried in SOG to isolate individual nanorods and to make p-type electrodes on only the top p-type nanorod region. The electrical characteristics of nanorod LEDs were similar to those of thin-film LEDs. However, as shown in Fig. 2.20c, nanorod LEDs exhibited significant enhancement in light output despite a smaller active volume compared with thin-film LEDs. This was explained by an improvement in light-extraction efficiency due to reduced total internal reflection in nanorods as well as dislocation-free single crystallinity of nanorods. Similarly, GaN nanocolumn LEDs with $In_{1-x}Ga_xN$/GaN MQW layers were also fabricated on Si substrates using plasma-assisted molecular beam epitaxy

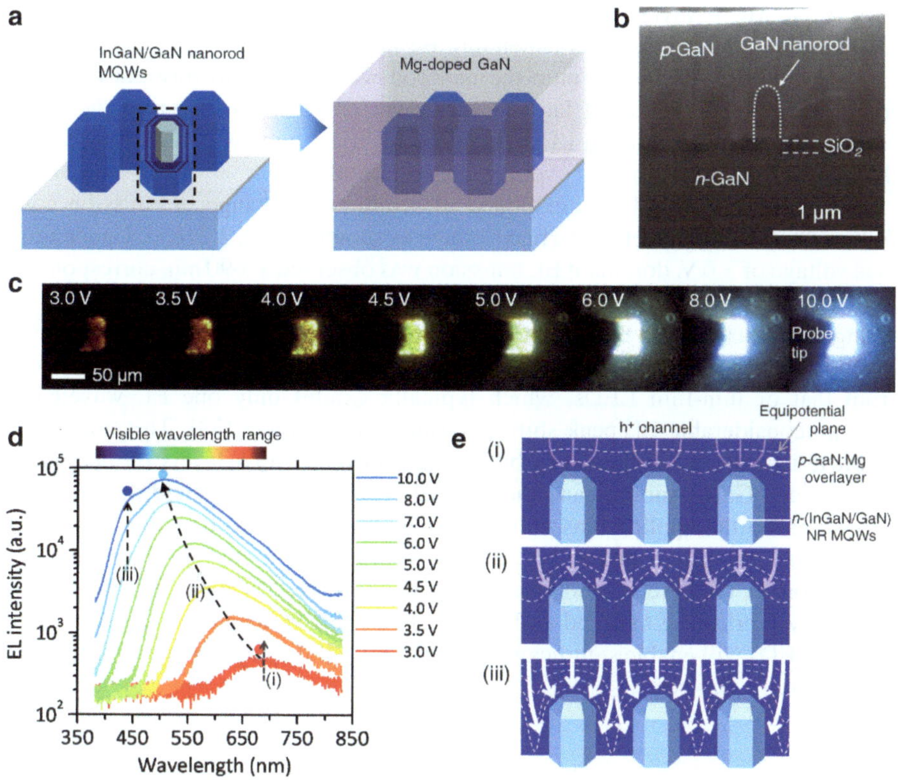

Fig. 2.21 (**a**) Schematic illustration for the fabrication of nanorod-embedded LEDs. (**b**) Cross-sectional SEM image of a Mg-doped GaN overlayer and $In_xGa_{1-x}N/GaN$ MQWs on GaN nanorod arrays. (**c**) Light emission photographs and (**d**) EL spectra of the nanostructured LEDs taken at various bias voltage levels from 3.0 to 10.0 V. (**e**) Schematic illustrations of the change of equipotential planes (*white dotted lines*) in the p-GaN overlayer of the nanostructured LEDs and paths of hole carriers (i) under a low electric field near the turn-on voltage, (ii) with increasing applied voltage, and (iii) at very high bias voltage. Adapted with permission from [33]. Copyright 2011 John Wiley & Sons, Inc.

(MBE) [10, 13]. The EL color of the nanocolumn LEDs was controlled in the visible range from green to red by adjusting the growth conditions of $In_{1-x}Ga_xN/GaN$ MQW layers.

In addition to axial nanostructure LEDs, thin-film LEDs embedded with nanorods were fabricated using position-controlled GaN nanorod arrays, which were color-tunable, depending on the applied electrical bias [33]. Figure 2.21a shows a schematic illustration of nanorod-embedded LEDs composed of $In_xGa_{1-x}N/GaN$MQW structures and a p-GaN overlayer on the GaN nanorod array. After selective growth of GaN nanorod arrays, $In_xGa_{1-x}N/GaN$ MQW layers were heteroepitaxially grown over the entire surface of each GaN nanorod array. Because of the anisotropic surface formation energies of GaN crystal planes, the anisotropic MQW layers were formed on the nanorod surfaces with different QW thicknesses and compositions depending on nanorod facets. Then, Mg-doped p-GaN was epitaxially coated onto the nanorod with MQWs to form an overlayer film, similar to the epitaxial lateral overgrowth process. As shown in the cross-sectional SEM image of the LED (Fig. 2.21b), a continuous p-GaN film, with a thickness of $\sim$700 nm formed over the GaN nanorods.

Nanorod-embedded LEDs exhibited the continuous change in emission color from red to blue with increasing bias voltage as shown in Fig. 2.21c, d. Under a low bias voltage of 3.0 V, dominant EL emission was observed at 690 nm, corresponding to red light. As the applied voltage increased to 10.0 V, the EL peak position changed gradually from 690 to 500 nm, with another peak centered at 440 nm. The visible color-tunable characteristic of nanorod-embedded LEDs was quite distinct from that of thin-film LEDs, which typically exhibit only one EL wavelength with no considerable EL peak shift at various bias voltages [33]. This unique EL characteristic can be explained by both the anisotropic MQW layers formed on the multifaceted GaN nanorods and the gradual change in electric field distributions, depending on the applied bias voltage in nanorod-embedded thin-film structures, as schematically illustrated in Fig. 2.21e. Furthermore, it enabled the monolithic integration of red, green, and blue LEDs that operated at a fixed drive current by simply fabricating nanorod-embedded LEDs with various device sizes. Color-tunable LEDs will enable significant advances in LED displays and many integrated optoelectronic devices.

2.5 Conclusions and Perspectives

This chapter provided an overview of catalyst-free MOVPE of ZnO and GaN nanostructures for visible LED applications. Recently, remarkable progress has been made in the growth, doping, and LED applications for ZnO nanostructures. A wide variety of growth methods for ZnO and GaN nanostructures have been developed and both n- and p-type doping in ZnO nanostructures have been demonstrated. In particular, vertically aligned high-density ZnO nanostructure arrays enabled the fabrication of diverse ZnO-nanostructure-based LEDs including nanostructure/film hybrid heterojunction LEDs and p–n homojunction nanorod LEDs. In contrast to high-density ZnO nanostructures, position-controlled growth of ZnO nanostructure arrays was extended to 3D nanoarchitectures based on coaxial heterostructures, including ZnO/MgZnO and InGaN/GaN/ZnO quantum structures, which are highly desirable for high-efficiency LED applications. However, there are challenges to achieving any ZnO-based 3D nanoarchitecture LED as an ideal system, including reliable p-type doping and controlled formation of quantum structures by band gap engineering. In our opinion, position-controlled 3D nanoarchitectures can be expanded to many other material systems for the fabrication of diverse optoelectronic devices, including LEDs, laser diodes, and solar cells.

Although this chapter focused primarily on ZnO-based nanostructures for LED applications, visible LEDs based on GaN nanorods may show more promising results for visible LED applications due to their easy p-doping control and $In_xGa_{1-x}N$/GaN MQW fabrication. In particular, vertical LED arrays based on radial nanorod quantum structures are interesting because they offer several advantages over axial nanorod LEDs. Despite recent progress in the fabrication of GaN-based coaxial nanorod heterostructures and their LED applications, several challenges remain. The first is to grow GaN nanorods with controlled positions and shapes on flexible, cheap, or large substrates. Large graphene substrates are potential candidates. The second challenge is to make good ohmic contacts on the nanostructures for high-performance LEDs. The nonflat morphology of the nanostructures makes it difficult to construct continuous and low-resistance metal contacts on the nanorod tips. Additionally, for device geometry with light emission through the nanorod tips, the metal layer should be thin enough to be transparent. Finally, previous reports on nanostructure-based LEDs indicate that the electrical characteristics are not yet sufficient for fabricating high-performance LEDs. Compared with thin-film LEDs, leakage current is very high, limiting device performance. Nevertheless, if these problems are resolved, coaxial nanorod heterostructures can offer new possibilities for the fabrication of solar cells and other sophisticated optoelectronic devices as well as visible LEDs.

Acknowledgements This work was financially supported by the National Creative Research Initiative Project (R16-2004-004-01001-0) of the Korea Science and Engineering Foundations (KOSEF).

References

1. X.F. Duan, Y. Huang, Y. Cui, J.F. Wang, C.M. Lieber, Nature **409**, 66 (2001)
2. P.J. Pauzauskie, P. Yang, Mater. Today **9**, 36 (2006)
3. T. Yatsui, S. Sangu, T. Kawazoe, M. Ohtsu, S.J. An, J. Yoo, G.-C. Yi, Appl. Phys. Lett. **90**, 223110 (2007)
4. W.I. Park, G.C. Yi, M. Kim, S.J. Pennycook, Adv. Mater. **15**, 526 (2003)
5. W.I. Park, S.J. An, J.L. Yang, G.C. Yi, S. Hong, T. Joo, M. Kim, J. Phys. Chem. B **108**, 15457 (2004)
6. E.S. Jang, J.Y. Bae, J. Yoo, W.I. Park, D.W. Kim, G.C. Yi, T. Yatsui, M. Ohtsu, Appl. Phys. Lett. **88**, 023102 (2006)
7. F. Qian, S. Gradečak, Y. Li, C.Y. Wen, C.M. Lieber, Nano Lett. **5**, 2287 (2005)
8. T. Yatsui, M. Ohtsu, J. Yoo, S.J. An, G.C. Yi, Appl. Phys. Lett. **87**, 033101 (2005)
9. H.-M. Kim, Y.-H. Cho, H. Lee, S.I. Kim, S.R. Ryu, D.Y. Kim, T.W. Kang, K.S. Chung, Nano Lett. **4**, 1059 (2004)
10. A. Kikuchi, M. Kawai, M. Tada, K. Kishino, Jpn. J. Appl. Phys. **43**, L1524 (2004)
11. Y. Sun, Y.H. Cho, H.M. Kim, T.W. Kang, Appl. Phys. Lett. **87**, (2005)
12. H.W. Lin, Y.J. Lu, H.Y. Chen, H.M. Lee, S. Gwo, Appl. Phys. Lett. **97**, 073101 (2010)
13. W. Guo, M. Zhang, A. Banerjee, P. Bhattacharya, Nano Lett. **10**, 3355 (2010)
14. C.-H. Lee, J. Yoo, Y.J. Doh, G.-C. Yi, Appl. Phys. Lett. **94**, 043504 (2009)
15. Y. Hu, J. Xiang, G. Liang, H. Yan, C.M. Lieber, Nano Lett. **8**, (2008)

16. Y. Huang, X.F. Duan, C.M. Lieber, Small **1**, 142 (2005)
17. D.J. Sirbuly, M. Law, P. Pauzauskie, H. Yan, A.V. Maslov, K. Knutsen, C.Z. Ning, R.J. Saykally, P. Yang, Proc. Natl. Acad. Sci. U.S.A. **102**, 7800 (2005)
18. S.J. An, J.H. Chae, G.-C. Yi, Appl. Phys. Lett. **92**, 121108 (2008)
19. M.T. Björk, B.J. Ohlsson, T. Sass, A.I. Persson, C. Thelander, M.H. Magnusson, K. Deppert, L.R. Wallenberg, L. Samuelson, Appl. Phys. Lett. **80**, 1058 (2002)
20. M.S. Gudiksen, L.J. Lauhon, J. Wang, D.C. Smith, C.M. Lieber, Nature **415**, 617 (2002)
21. R. Solanki, J. Huo, J.L. Freeouf, B. Miner, Appl. Phys. Lett. **81**, 3864 (2002)
22. W.I. Park, D.H. Kim, S.W. Jung, G.-C. Yi, Appl. Phys. Lett. **80**, 4232 (2002)
23. H.M. Kim, T.W. Kang, K.S. Chung, Adv. Mater. **15**, 567 (2003)
24. J. Noborisaka, J. Motohisa, T. Fukui, Appl. Phys. Lett. **86**, 213102 (2005)
25. X. Wang, X. Sun, M. Fairchild, S.D. Hersee, Appl. Phys. Lett. **89**, 233115 (2006)
26. W. Lee, M.C. Jeong, J.M. Myoung, Acta Mater. **52**, 3949 (2004)
27. W.I. Park, G.-C. Yi, M. Kim, S.J. Pennycook, Adv. Mater. **14**, 1841 (2002)
28. W.I. Park, Y.H. Jun, S.W. Jung, G.C. Yi, Appl. Phys. Lett. **82**, 964 (2003)
29. S.J. An, W.I. Park, G.-C. Yi, Y.J. Kim, H.B. Kang, M. Kim, Appl. Phys. Lett. **84**, 3612 (2004)
30. Y.-J. Kim, J.H. Lee, G.-C. Yi, Appl. Phys. Lett. **95**, 213101 (2009)
31. C.J. Park, D.K. Choi, J. Yoo, G.-C. Yi, C.J. Lee, Appl. Phys. Lett. **90**, 083107 (2007)
32. Y.J. Hong, S.J. An, H.S. Jung, C.-H. Lee, G.-C. Yi, Adv. Mater. **19**, 4416 (2007)
33. Y.J. Hong, C.-H. Lee, A. Yoon, M. Kim, H.-K. Seong, H.J. Chung, C. Sone, Y.J. Park, G.-C. Yi, Adv. Mater. **23**, 3284 (2011)
34. G.-C. Yi, C.R. Wang, W.I. Park, Semicond. Sci. Technol. **20**, S22 (2005)
35. U. Özgur, Y.I. Alivov, C. Liu, A. Teke, M.A. Reshchikov, S. Doğan, V. Avrutin, S.J. Cho, H. Morkoç, J. Appl. Phys. **98**, 041301 (2005)
36. E.V. Etzkorn, D.R. Clarke, J. Appl. Phys. **89**, 1025 (2001)
37. X.H. Wu, P. Fini, E.J. Tarsa, B. Heying, S. Keller, U.K. Mishra, S.P. DenBaars, J.S. Speck, J. Cryst. Growth **189**, 231 (1998)
38. A. Tsukazaki, A. Ohtomo, T. Kita, Y. Ohno, H. Ohno, M. Kawasaki, Science **315**, 1388 (2007)
39. L. Vayssieres, Adv. Mater. **15**, 464 (2003)
40. C.Y. Geng, Y. Jiang, Y. Yao, X.M. Meng, J.A. Zapien, C.S. Lee, Y. Lifshitz, S.T. Lee, Adv. Funct. Mater. **14**, 589 (2004)
41. L.E. Greene, M. Law, D.H. Tan, M. Montano, J. Goldberger, G. Somorjai, P. Yang, Nano Lett. **5**, 1231 (2005)
42. M.H. Huang, S. Mao, H. Feick, H.Q. Yan, Y.Y. Wu, H. Kind, E. Weber, R. Russo, P.D. Yang, Science **292**, 1897 (2001)
43. R.S. Wagner, W.C. Ellis, Appl. Phys. Lett. **4**, 89 (1964)
44. X.F. Duan, C.M. Lieber, Adv. Mater. **12**, 298 (2000)
45. H. Chik, J. Liang, S.G. Cloutier, N. Kouklin, J.M. Xu, Appl. Phys. Lett. **84**, 3376 (2004)
46. Y. Ding, P.X. Gao, Z.L. Wang, J. Am. Chem. Soc. **126**, 2066 (2004)
47. P.X. Gao, J.H. Song, J. Liu, Z.L. Wang, Adv. Mater. **19**, 67 (2007)
48. Y.J. Hong, J. Yoo, Y.-J. Doh, S.H. Kang, K.-j. Kong, M. Kim, D.R. Lee, K.H. Oh, G.-C. Yi, J. Mater. Chem. **19**, 941 (2009)
49. W.I. Park, G.-C. Yi, Adv. Mater. **16**, 87 (2004)
50. G. Kresse, J. Hafner, Phys. Rev. B **47**, 558 (1993)
51. D. Vanderbilt, Phys. Rev. B **41**, 7892 (1990)
52. J.P. Perdew, Y. Wang, Phys. Rev. B **45**, 13244 (1992)
53. M. Kim, Y.J. Hong, J. Yoo, G.-C. Yi, G. Park, K.-j. Kong, H. Chang, Phys. Status Solidi (RRL) **2**, 197 (2008)
54. G.-D. Yuan, W.-J. Zhang, J.-S. Jie, X. Fan, J.-X. Tang, I. Shafiq, Z.-Z. Ye, C.-S. Lee, S.-T. Lee, Adv. Mater. **20**, 168 (2008)
55. J. Yoo, C.-H. Lee, Y.-J. Doh, H.S. Jung, G.-C. Yi, Appl. Phys. Lett. **94**, 223117 (2009)
56. B. Xiang, P.W. Wang, X.Z. Zhang, S.A. Dayeh, D.P.R. Aplin, C. Soci, D.P. Yu, D.L. Wang, Nano Lett. **7**, 323 (2007)

57. G.D. Yuan, W.J. Zhang, J.S. Jie, X. Fan, J.A. Zapien, Y.H. Leung, L.B. Luo, P.F. Wang, C.S. Lee, S.T. Lee, Nano Lett. **8**, 2591 (2008)
58. S.S. Lin, J.I. Hong, J.H. Song, Y. Zhu, H.P. He, Z. Xu, Y.G. Wei, Y. Ding, R.L. Snyder, Z.L. Wang, Nano Lett. **9**, 3877 (2009)
59. Y.J. Hong, H.S. Jung, J. Yoo, Y.-J. Kim, C.-H. Lee, G.-C. Yi, M. Kim, Adv. Mater. **21**, 222 (2009)
60. B.J. Ohlsson, M.T. Björk, M.H. Magnusson, K. Deppert, L. Samuelson, L.R. Wallenberg, Appl. Phys. Lett. **79**, 3335 (2001)
61. J. Yoo, Y.J. Hong, H.S. Jung, Y.-J. Kim, C.-H. Lee, J. Cho, Y.-J. Doh, L.S. Dang, K.H. Park, G.-C. Yi, Adv. Funct. Mater. **19**, 1601 (2009)
62. B.K. Meyer, H. Alves, D.M. Hofmann, W. Kriegseis, D. Forster, F. Bertram, J. Christen, A. Hoffmann, M. Straßburg, M. Dworzak, U. Haboeck, A.V. Rodina, Phys. Status Solidi (b) **241**, 231 (2004)
63. A.K. Viswanath, J.I. Lee, S. Yu, D. Kim, Y. Choi, C.-H. Hong, J. Appl. Phys. **84**, 3848 (1998)
64. G.-C. Yi, S.J. An, Korea Patent No. 10–0644166 (November 2, 2006)
65. C.P.T. Svensson, T. Mårtensson, J. Trägårdh, C. Larsson, M. Rask, D. Hessman, L. Samuelson, J. Ohlsson, Nanotechnology **19**, 305201 (2008)
66. Y.J. Hong, J.-M. Jeon, M. Kim, S.-R. Jeon, K.H. Park, G.-C. Yi, New J. Phys. **11**, 125021 (2009)
67. C.-H. Lee, J. Yoo, Y.J. Hong, J. Cho, Y.-J. Kim, S.-R. Jeon, J.H. Baek, G.-C. Yi, Appl. Phys. Lett. **94**, 213101 (2009)
68. K. Tomioka, J. Motohisa, S. Hara, K. Hiruma, T. Fukui, Nano Lett. **10**, 1639 (2010)
69. W.I. Park, G.-C. Yi, H.M. Jang, Appl. Phys. Lett. **79**, 2022 (2001)
70. J.C. Johnson, H.J. Choi, K.P. Knutsen, R.D. Schaller, P.D. Yang, R.J. Saykally, Nature Mater. **1**, 106 (2002)
71. T. Kuykendall, P.J. Pauzauskie, Y.F. Zhang, J. Goldberger, D. Sirbuly, J. Denlinger, P.D. Yang, Nat. Mater. **3**, 524 (2004)
72. J. Yoo, Y.J. Hong, S.J. An, G.-C. Yi, B. Chon, T. Joo, J.W. Kim, J.S. Lee, Appl. Phys. Lett. **89**, 043124 (2006)
73. Y.B. Tang, X.H. Bo, C.S. Lee, H.T. Cong, H.M. Cheng, Z.H. Chen, W.J. Zhang, I. Bello, S.T. Lee, Adv. Funct. Mater. **18**, 3515 (2008)
74. H.M. Kim, D.S. Kim, Y.S. Park, D.Y. Kim, T.W. Kang, K.S. Chung, Adv. Mater. **14**, 991 (2002)
75. L.W. Tu, C.L. Hsiao, T.W. Chi, I. Lo, K.Y. Hsieh, Appl. Phys. Lett. **82**, 1601 (2003)
76. K. Kishino, H. Sekiguchia, A. Kikuchi, J. Cryst. Growth **311**, 2063 (2009)
77. P. Deb, H. Kim, V. Rawat, M. Oliver, S. Kim, M. Marshall, E. Stach, T. Sands, Nano Lett. **5**, 1847 (2005)
78. S.D. Hersee, X.Y. Sun, X. Wang, Nano Lett. **6**, 1808 (2006)
79. M. Funato, T. Kotani, T. Kondou, Y. Kawakami, Y. Narukawa, T. Mukai, Appl. Phys. Lett. **88**, 261920 (2006)
80. S.J. Pearton, W.T. Lim, J.S. Wright, L.C. Tien, H.S. Kim, D.P. Norton, H.T. Wang, B.S. Kang, F. Ren, J. Jun, J. Lin, A. Osinsky, J. Electron. Mater. **37**, 1426 (2008)
81. M. Willander, O. Nur, Q.X. Zhao, L.L. Yang, M. Lorenz, B.Q. Cao, J.Z. Perez, C. Czekalla, G. Zimmermann, M. Grundmann, A. Bakin, A. Behrends, M. Al-Suleiman, A. El-Shaer, A.C. Mofor, B. Postels, A. Waag, N. Boukos, A. Travlos, H.S. Kwack, J. Guinard, D.L. Dang, Nanotechnology **20**, 332001 (2009)
82. Y. Yang, X.W. Sun, B.K. Tay, G.F. You, S.T. Tan, K.L. Teo, Appl. Phys. Lett. **93**, 253107 (2008)
83. J.Y. Zhang, P.J. Li, H. Sun, X. Shen, T.S. Deng, K.T. Zhu, Q.F. Zhang, J.L. Wu, Appl. Phys. Lett. **93**, 021116 (2008)
84. X.W. Sun, B. Ling, J.L. Zhao, S.T. Tan, Y. Yang, Y.Q. Shen, Z.L. Dong, X.C. Li, Appl. Phys. Lett. **95**, 133124 (2009)
85. G.D.J. Smit, S. Rogge, T.M. Klapwijk, Appl. Phys. Lett. **81**, 3852 (2002)
86. W.I. Park, G.-C. Yi, J.W. Kim, S.M. Park, Appl. Phys. Lett. **82**, 4358 (2003)

87. H. Sun, Q.F. Zhang, J.L. Wu, Nanotechnology **17**, 2271 (2006)
88. A.C. Mofor, A. Bakin, U. Chejarla, E. Schlenker, A. El-Shaer, G. Wagner, N. Boukos, A. Travlos, A. Waag, Superlattices and Microstruct. **42**, 415 (2007)
89. Y.P. Hsieh, H.Y. Chen, M.Z. Lin, S.C. Shiu, M. Hofmann, M.Y. Chern, X.T. Jia, Y.J. Yang, H.J. Chang, H.M. Huang, S.C. Tseng, L.C. Chen, K.H. Chen, C.F. Lin, C.T. Liang, Y.F. Chen, Nano Lett. **9**, 1839 (2009)
90. M.H. Sun, Q.F. Zhang, H. Sun, J.Y. Zhang, J.L. Wu, J. Vac. Sci. Technol. B **27**, 618 (2009)
91. M. Willander, O. Nur, N. Bano, K. Sultana, New J. Phys. **11**, 125021 (2009)
92. J.-Y. Zhang, Q.-F. Zhang, T.-S. Deng, J.-L. Wu, Appl. Phys. Lett. **95**, 211107 (2009)
93. B. Ling, X.W. Sun, J.L. Zhao, S.T. Tan, Z.L. Dong, Y. Yang, H.Y. Yu, K.C. Qi, Physica E **41**, 635 (2009)
94. J.-Y. Wang, C.-Y. Lee, Y.-T. Chen, C.-T. Chen, Y.-L. Chen, C.-F. Lin, Y.-F. Chen, Appl. Phys. Lett. **95**, 131117 (2009)
95. E. Lai, W. Kim, P.D. Yang, Nano Res. **1**, 123 (2008)
96. C.H. Chen, S.J. Chang, S.P. Chang, M.J. Li, I.C. Chen, T.J. Hsueh, C.L. Hsu, Appl. Phys. Lett. **95**, 223101 (2009)
97. R. Guo, J. Nishimura, M. Matsumoto, M. Higashihata, D. Nakamura, T. Okada, Appl. Phys. B **94**, 33 (2009)
98. S.D. Lee, Y.S. Kim, M.S. Yi, J.Y. Choi, S.W. Kim, J. Phys. Chem. C **113**, 8954 (2009)
99. X.-M. Zhang, M.-Y. Lu, Y. Zhang, L.-J. Chen, Z.L. Wang, Adv. Mater. **21**, 2767 (2009)
100. X.Y. Chen, A.M.C. Ng, F. Fang, A.B. Djurisic, W.K. Chan, H.L. Tam, K.W. Cheah, P.W.K. Fong, H.F. Lui, C. Surya, J. Electrochem. Soc. **157**, H308 (2010)
101. J.T. Yan, C.H. Chen, S.F. Yen, C.T. Lee, IEEE Photonics Technol. Lett. **22**, 146 (2010)
102. M.-C. Jeong, B.-Y. Oh, M.-H. Ham, J.-M. Myoung, Appl. Phys. Lett. **88**, 202105 (2006)
103. M.-C. Jeong, B.-Y. Oh, M.-H. Ham, S.-W. Lee, J.-M. Myoung, Small **3**, 568 (2007)
104. M.W. Wang, J.O. McCaldin, J.F. Swenberg, T.C. McGill, R.J. Hauenstein, Appl. Phys. Lett. **66**, 1974 (1995)
105. S.J. An, G.-C. Yi, Appl. Phys. Lett. **91**, 123109 (2007)
106. S. Strite, H. Morkoç, J. Vac. Sci. Technol. B **10**, 1237 (1992)
107. F.A. Ponce, D.P. Bour, Nature **386**, 351 (1997)
108. C.-H. Lee, Y.J. Hong, J. Yoo, Y.-J. Kim, S.J. Lee, S.-R. Jeon, G.-C. Yi, IEEE J. Sel. Top. Quantum Electron. **17**, 966 (2011)
109. S.D. Hersee, M. Fairchild, A K. Rishinaramangalam, M.S. Ferdous, L. Zhang, P.M. Varangis, B.S. Swartzentruber, A.A. Talin, Electron. Lett. **45**, 75-U24 (2009)

Chapter 3
III–V Semiconductor Nanowires on Si by Selective-Area Metal-Organic Vapor Phase Epitaxy

Katsuhiro Tomioka and Takashi Fukui

Abstract The III–V nanowires (NWs) on Si are promising building blocks for future nanoscale electrical and optical devices on Si platforms. We review position-controlled growth of III–V NWs on Si substrate by selective-area growth and discuss how to control growth directions of III–V NW on Si. Finally, we demonstrate the integrations of III–V NW-based light-emitting diodes (LEDs) array on Si. These demonstrations should have broad applications in laser diodes and photodiodes with functionality not enabled by conventional NW.

3.1 Introduction

Tremendous advances in epitaxial technique have brought rapid progress to solid-state lighting (SSL) such as multicolor (from infrared to ultraviolet region) light-emitting diodes (LEDs) and high brightness LEDs. Blue LEDs are being commercialized in displays and lighting, and now the SSL requires more advanced materials with low cost, low power, and high brightness in the field of the epitaxial techniques. In planar LED architecture, the enlarged active layer requires high power for high brightness LEDs. Also, high brightness LEDs with low power requires integration of many chips because the intensity of each LED is low. This approach results in high cost. When many chips are integrated, the planar LED architecture will reach a limit because of these conflicting demands of epitaxial growth. Therefore, new architecture and nanomaterials are being explored for future low-cost, low-power, and high brightness SSL, because recent advances in epitaxial technique have enabled simple formation of low-dimensional semiconductor

---

K. Tomioka (✉)
Graduate School of Information Science and Technology, Research Center for Integrated Quantum Electronics (RCIQE), Hokkaido University, Kita 14, Nishi 9, Kita-ku, Sapporo, Japan
e-mail: tomioka@rciqe.hokudai.ac.jp

G.-C. Yi (ed.), *Semiconductor Nanostructures for Optoelectronic Devices*, NanoScience and Technology, DOI 10.1007/978-3-642-22480-5_3,

nanostructures. These semiconductor nanostructures achieve high-density SSL integration with low cost and low power. Moreover, the nanostructure achieves thresholdless lasing emission [1, 2].

Among the low-dimensional semiconductor nanostructures, semiconductor nanowires (nanowhiskers, nanorods, nanopillars, and nanocolumns: NWs) are strong candidates for future nanometer-scale electronic and optical devices [3–8], because they have small diameters and large surfaces area that enable high-density integration of active devices on various platforms and fabrication of various kinds of functional devices by using heterostructures. The surface area for the growth of the radial heterostructures enables the formation of core-shell and core-multishell (CMS) NWs. Moreover, a top surface with a small diameter can form axial heterostructures regardless of lattice mismatches. The use of the core-shell or axial NWs provide functionality to NW-based applications.

The conventional growth for NWs is that using vapor–liquid–solid (VLS) mechanism. The VLS mechanism uses catalysts and their liquid phase underneath metal particles for crystallization. In 1964, Wagner and Ellis reported the mechanism and formation of Si whiskers [9]. In their reports, sub-micrometer Si whiskers were grown with an Au catalyst. The VLS-grown Ge whiskers on corresponding bulk substrates were investigated in the 1970s by Givargicov [10]. In the early 1990s, Hiruma and his coworkers at Hitachi Laboratory investigated VLS-grown III–V compound semiconductor nanowhiskers [11], the forerunner of III–V NWs. After their pioneering works, the potential of NWs for future electronics and photonics was demonstrated in the early 2000s [3–8]. Since then, the VLS method has become exceedingly common because it can be used to synthesize almost all types of semiconductor NWs with a small size and high quality via rather simple procedures as mentioned in Chap. 1. The metal catalysts, however, often act as unintentional impurities inside the NWs and form deep levels [12] which would degrade performance of NW-based devices. Recently, a self-catalyzed method has been investigated in the NW growth [13, 14]. This method is categorized in the VLS method. The growth of radial heterostructures for VLS-grown NWs has been reported on Ge/Si [7], InGaN/GaN [15], and InP/InAs [16] core-shell NWs. The axial heterostructures have been investigated on Si/Ge [17], GaP/GaAs [6], and InAs/InP axial NWs [18].

Another approach for growing NWs is catalyst-free selective-area growth (SAG). This approach uses partially masked templates with lithographically defined opening patterns. The masked templates are made from amorphous or metal films, such as SiO_2, SiN_x, and Ti. Position-controlled polygonal nanostructures surrounded by several facets can be formed inside the openings, because the crystal growth proceeds under a faceting growth mechanism. The use of III–V (111)A or B substrates enables formation of vertically aligned hexagonal pillars surrounded by {-110} vertical sidewalls. These position-controlled NWs have been achieved in III–V compound semiconductors, such as GaAs [19–22], InP [23, 24], InAs [25], InGaAs [26–29], and GaAsP [30], as well as nitrides [31] and oxides [32]. As for core-shell NWs, position-controlled AlGaAs/GaAs [33], InAs/InP [34], and GaAsP/GaAs [35] core-shell NWs and InP/InAs/InP CMS NWs [36] have been reported by using selective-area metal-organic vapor phase epitaxy (SA-MOVPE). At the same time, position-controlled growth of InGaAs/GaAs [37], [38], InAsP/InP [39], and AlGaAs/GaAs [40] axial heterostructured NWs have been investigated.

As for the geometrical advantages of optical applications of NWs, this architecture allows light propagation along the length of the axis with light confined in the small diameter which closes to the limitation of length in light diffraction. This advantage has been used for fabricating NW Fabry–Perot cavities [41, 42] and optical pumped laser emissions [5, 43–45]. In addition, using this structure in optical devices could enable high-intensity optical devices due to the large surface area relative to the two-dimensional plane. Also, we can fabricate low-power optical devices using nanometer-scaled electrode structures.

The integrating of all LEDs with Si substrate will be effective for low-cost production of SSL. Using the heteroepitaxy of III–V compound semiconductors on Si could enable the less expensive production of light-emitting devices. For instance, replacing sapphire or gallium nitride (GaN) with Si substrate for fabricating blue LED would reduce the substrate cost to about a tenth. There are, however, several problems due to the mismatches in lattice and crystal structure between the group-IV materials and III–V compound semiconductors: (1) lattice mismatch; (2) difference in thermal expansion coefficient; (3) antiphase domain (or boundary) due to polarity. These mismatches form misfit dislocations and threading dislocations, which degrade device performance. Therefore, epitaxial techniques to overcome these mismatches have been investigated since the 1980s. So far, the use of buffer layer growth [46] to relax strains resulted from lattice mismatch; SAG, such as microchannel epitaxy (MCE) [47], to reduce the numbers of dislocations; and two-step growth [48] to suppress formation of antiphase domains have been proposed. There is no epitaxial technique that can completely overcome these dislocations, although the heteroepitaxial technique has been applied to GaN growth on Si and the blue LED [49]. However, it has become apparent that nanometer-scale epitaxial techniques of semiconductor NWs such as VLS and SAG are effective in overcoming these problems.

In this chapter, we will describe position-controlled growth of III–V NWs on Si substrate by SA-MOVPE and integration of III–V NW-based LEDs on Si substrate. In Sect. 3.2, optical applications of semiconductor NWs and their geometrical advantages in core-shell and CMS NWs are described. In Sect. 3.3, details of SA-MOVPE for growing NWs are explained. In Sect. 3.4, heteroepitaxial growth of III–V NWs on Si and effective methods to align vertical III–V NWs are described. Finally, fabrication of NW-based LEDs is discussed.

3.2 Optical Application of Semiconductor NWs

Light source using semiconductor NWs have been pioneered by Hiruma et al. [1]. They have fabricated near-infrared (NIR) LEDs by using VLS-grown GaAs NWs with axial p–n junctions. After their report, blue LEDs based on GaN nanocolumns were demonstrated by Kishino et al. in 1996 [50]. Since then, optically pumped stimulated emission from GaN NW [5] and multi-colored LEDs using GaN/InGaN CMS NWs were reported in the early 2000s [15]. The core-shell structure has one shell layer, and the CMS structure has multilayer shell on the sidewalls. Fabrication of the CMS NW-based LEDs has been investigated for NIR [50–52] and blue [53–56] regions.

There are two approaches to optical applications using semiconductor NWs, divided on the basis of the structures. One is to use the axial heterostructure for state-of-the art devices such as single photon emitters [57]. The other is to fabricate the core-shell structure for the light source and detectors. Applications based on the latter approach have become exceedingly due to increased research on optical application and photovoltaic devices using geometric characteristics of the NWs. We have also reported the fabrication of GaAs/AlGaAs CMS NW-based NIR LED [51] and InP core-shell NW-based solar cells using radial p–n junctions [58].

Next, we describe the geometrical advantages of semiconductor NWs for optical light sources and detectors. The NW advantages are obviously its large surface area and small diameter. Let us consider a core-shell NW with a diameter of 200 nm and a height of 3 μm as shown in Fig. 3.1a. The junction area will be 1.8 μm^2 when the NW has radial p–n junctions on the NW sidewalls. When we integrate the NWs with a pitch of 400 nm in a 50 μm^2 (the surface area is 2.5×10^3 μm^2), the total junction area of the NW array will be 2.8×10^4 μm^2. Thus, the junction area of the NW array will be approximately ten times larger than a planar LED [Fig. 3.1b]. In practice, it should be noted that this advantage will not be a simple comparison from problems such as contact resistance and surface states. When the device performance of a single NW LED is close to that of the planar LED, the LED chip area can be composed of a tenth of the area. Also, when these NWs are integrated on Si substrate, manufacturing costs will be ideally reduced to 1/100 of that of planar LEDs. In this manner, making the most of the geometrical advantages in NWs should be effective for making light sources with high brightness and low cost. Moreover, some functionality such as cavity effect as well as simple geometric advantage can be utilized on the NW because double-heterostructures (DH) can be formed on the sidewalls as shown in Fig. 3.1c.

The nanometer-scale footprint of the NW structure is feasible for Si photonics and monolithic integration of III–V NWs on Si platforms. Rapid progress has been made in the area of Si photonics, which are photonics made with Si-based

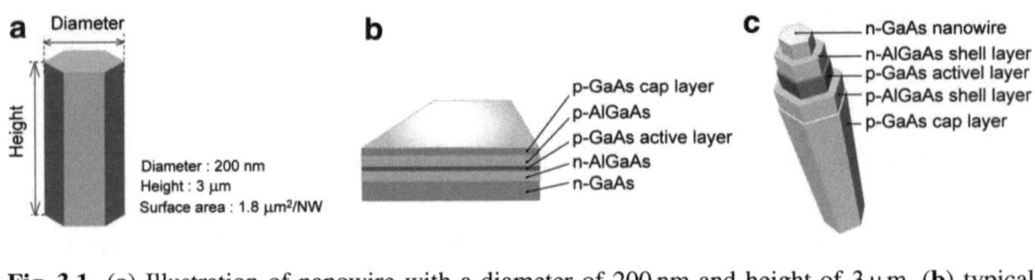

Fig. 3.1 (**a**) Illustration of nanowire with a diameter of 200 nm and height of 3 μm, (**b**) typical GaAs/AlGaAs-based light-emitting diodes (LEDs), and (**c**) core-multishell nanowire-based LED

materials [59–62]. The other major approach is the integration of the III–V compound semiconductor photonic devices on Si. Although optical interconnection and intraconnection with short-length distances can be used generally in the NIR region, it is difficult to fabricate Si-based light sources for the NIR region because Si is an indirect-band-gap semiconductor and the band gap is approximately 1,100 nm. On the other hand, almost all III–Vs have direct band gap and emit light at a wavelength sensitive to the Si-based avalanche photodiode (APD), and these III–Vs are easier to deploy on the Si platform. Therefore, heteroepitaxial technique of III–V compound semiconductor on Si is being used as an optical source of Si photonics which can replace Cu-based intraconnection of large-scale integration (LSI).

Let us recall the core-shell NW in Fig. 3.1a. When a single NW-light source (laser diodes) and APD can be realized on the Si platforms, the device area will be approximately 0.12 μm^2 per NW. This area is 1/5,600 that of a vertical cavity surface-emitting laser (VCSEL) with a diameter of 30 μm (the area is about 700 μm^2). The small footprint enables the integration of the other optical devices required for Si-photonics, such as waveguides and optical modulators, to the residual space.

We described the geometric advantages of the NW array and single NW architectures for optical applications. Applying these geometric advantages into optical devices integrated on Si platforms requires precise positioning for the NW growth. In the next section, we describe SA-MOVPE for growing position-controlled NWs.

3.3 Growth of NWs by Selective-Area Metal-Organic Vapor Phase Epitaxy

SA-MOVPE is a kind of template method that combines bottom-up (epitaxial growth) and top-down (lithography) approaches. This method enables the fabrication of nanostructures with lithographically defined positioning. The mask used in this method is usually amorphous film, such as SiO_2 and SiN_x. The SAG technique was developed in the early 1960s and was first used for Si-integrated circuits [63]. In 1965, Tausch et al. were the first to investigate the SAG of GaAs [64], and Rai-Choudhury investigated the SAG of GaAs by using metal-organic sources [65]. Then, Jones reported on the SAG of GaAs by using lithography-patterned native oxide mask [66]. The use of (111)B-oriented substrates enables the formation of inclined {110} facets and vertical {1–10} facets. The formation of the {1–10} vertical facets on the (111)B-oriented surface has been demonstrated by Fukui et al. [67] Tetrahedral GaAs have been fabricated by using this facet growth [68]. Moreover, quantum dots (QDs) have been selectively grown using lithographically patterned GaAs substrates [69–71]. Using the selectivity of the formation of the {1–10} facets, vertical hexagonal nanostructures could be formed, and these structures have been applied in laser and photonic crystals [72, 73]. Now, this approach is being used for growing NWs.

3.3.1 Process of SA-MOVPE for NW Growth

The growth process of the SA-MOVPE is illustrated in Fig. 3.2. After degreasing substrate (Fig. 3.2a), SiO_2 films with thicknesses of 10–30 nm are formed by RF sputtering, plasma-enhanced chemical vapor deposition (PECVD), or thermal oxidation. The thermal oxidation is usually used to form SiO_2 for III–V NW growth on Si because of the thermal tolerance of the film. Next, circular openings with regular pitch are formed on the amorphous films by using electron-beam (EB) lithography and wet chemical etching. Circular openings are arranged in a triangular lattice with a pitch of 0.4–3.0 μm. The opening diameter, d_0, ranges from 20 to 400 m. For the use of NWs in optical applications, d_0 of around 70–300 nm are used. Reactive-ion etching (RIE) is sometimes used for the etching. A scanning electron microscopy (SEM) image of the masked substrate is shown in Fig. 3.2b. Finally, NWs are grown by MOVPE. Figure 3.3 shows typical growth results of GaAs NWs and InAs NWs by SA-MOVPE. Vertical-aligned regular NW arrays are successfully fabricated. The (111)B- or (111)A-oriented surfaces are used for the NW growth by SA-MOVPE, because the III–V NWs are preferentially grown in <111>A or <111>B directions. The (111)A surface has the topmost layer arranged by group-III atoms, and the (111)B surface has the layer arranged by group-V atoms.

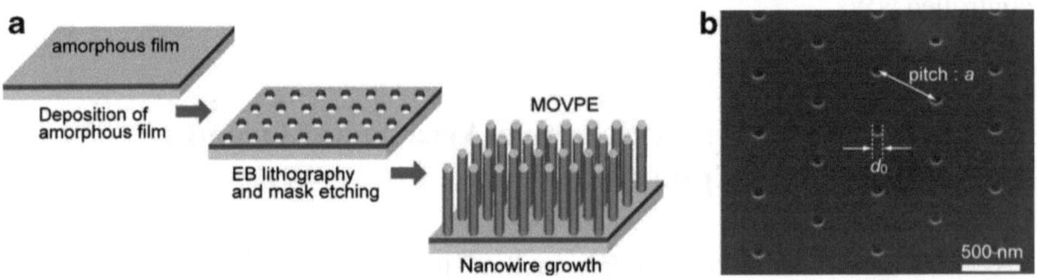

Fig. 3.2 (**a**) Fabrication processes for SA-MOVPE. After deposition of amorphous film, hole openings are formed by lithography and etching. NW growth is performed by MOVPE. (**b**) SEM image of masked substrate with hole openings

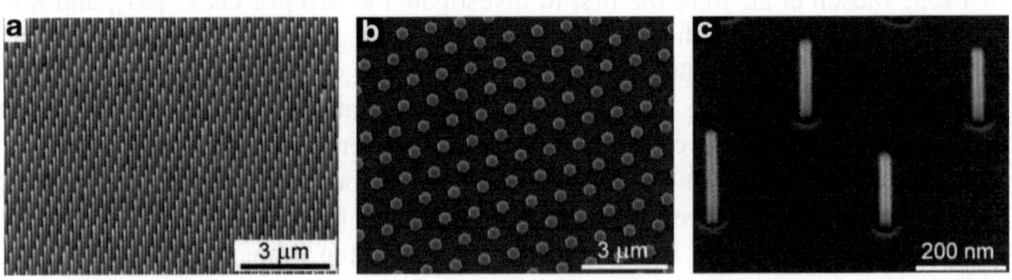

Fig. 3.3 Typical growth results of III–V NWs by SA-MOVPE. (**a**) GaAs NWs with diameter of 70 nm and height of 3 μm. The substrate is GaAs(111)B. (**b**) GaAs NWs with diameter of 400 nm and height of 400 nm. (**c**) InAs NWs with diameter of 20 nm and height of 400 nm. The substrate is InP(111)B

The NWs were grown using a low-pressure (0.1 atm) horizontal-reactor MOVPE system. The carrier gas used in this growth was pure hydrogen (H_2) that had been purified by passing it through a Pd film. The group-III-atom precursors used in this study were trimethylgallium (TMGa), trimethylindium (TMIn), and trimetylaluminium (TMAl), and the group-V-atom precursor was 5% arsine (AsH_3) in H_2. Silane (SiH_4) gas was used for n-type doping, and diethylzinc (DEZn) was used for p-type doping. Native oxide about 1-nm thick formed on the opened patterns and the oxide stem SAG.

3.3.2 *Crystal Shape in SA-MOVPE*

Equilibrium crystal shapes have polygonal morphology surrounded by close-packed planes with low surface energy. The low surface energy results in a stable surface with a slow growth rate. Thus, crystal shapes surrounded by stable crystal planes are formed by the faceting growth. For SAG, this trend appears prominently. The grown structure has low-index planes. The origin of the faceting in the SAG is explained simply by the difference in the surface chemical potential of the planes, i.e., the number of dangling bonds. This is because the growth rate roughly depends on the number of dangling bonds of the surface. Illustration of the atomic arrangement of GaAs from the <1–10> direction is shown in Fig. 3.4. Some low-index planes are arranged on the (001) planes. Interestingly, one of the family of {110} planes,{1–10} planes, are arranged normal to the (111)A or (111)B planes in this illustration. The

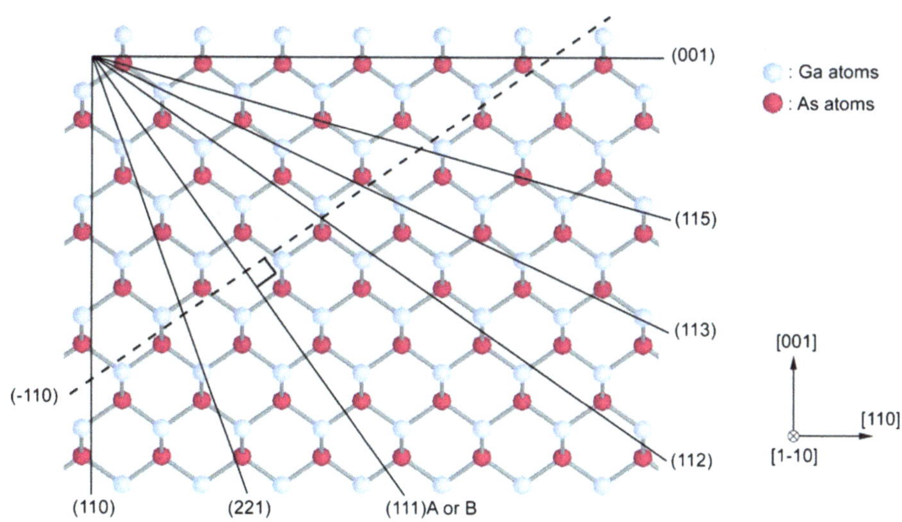

Fig. 3.4 Schematic of GaAs crystal structure with an injection of (1–10) direction. *Solid lines* show surface orientation of GaAs. The *dashed line* shows a family of {1–10} planes. These planes are normal to the (111)A or (111)B surface

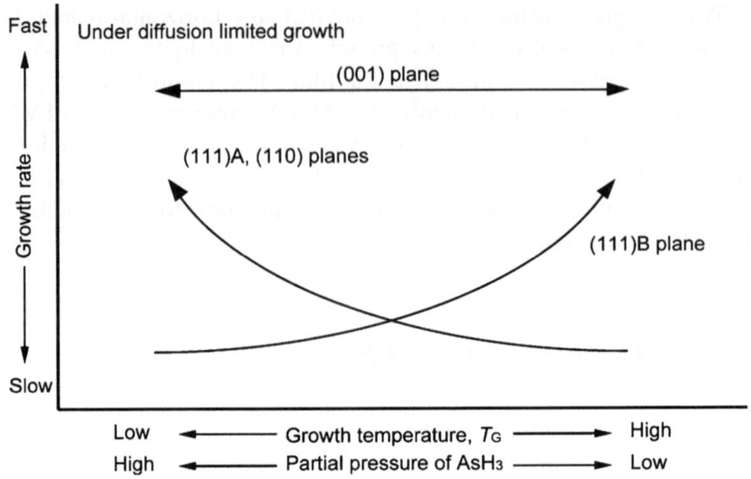

Fig. 3.5 Typical behavior in growth rate of GaAs with various growth parameters under diffusion-limited growth

planes are formed as facets normal to the vectors of the surface potential due to Wulff construction [74].

For instance, the growth rate of GaAs (001) surface is independent on the variation of growth temperature (T_G) and the partial pressure of AsH_3 ([AsH_3]) due to the diffusion-limited growth. On the other hand, the growth rates of the GaAs (110), (111)A, and (111)B surfaces tend to vary with the T_G and [AsH_3] as illustrated in Fig. 3.5. The growth rates of GaAs (111)A and (110) surfaces become faster than the (111)B planes under low T_G and high [AsH_3]. This is because As-trimers formed on the (111)B surface suppress the nucleation on GaAs (111)B surface. On the other hand, the growth rate of GaAs (001) and (111)B surfaces becomes faster than the GaAs (011) and (111)A surface under high T_G and low [AsH_3]. This is because the group-III adatoms evaporate easily from the (011) surface under high T_G. In this way, pyramidal nanostructures surrounded by (011) planes are formed on GaAs (001) [69] and tetrahedral structures surrounded by (111)B surface are formed under high T_G and low [AsH_3] [68]. Interestingly, the morphology of GaAs on (111)B-oriented surfaces becomes hexagonal pillar surrounded by vertical {1–10} facets. This is because the family of {-110} planes is arranged vertically from the (111)B surface. Therefore, hexagonal III–V NWs surrounded by {1–10} vertical facets can be formed on a (111)B-oriented surface, because these tendencies observed in SAG of GaAs commonly appear in other III–V compound semiconductors except for InP. It should be noted that vertical InP NWs are formed on (111)A-oriented surface in case of SA-MOVPE. This is because the preferential growth direction of InP NWs is the <111> A direction and phosphorous trimers can be formed on InP (111)A surface [75]. Next, typical optimum conditions for III–V NWs by SA-MOVPE are described.

3.3.2.1 GaAs NW

The optimum T_G window ranges from 700 to 750°C [19–22]. The MO sources are TMGa and AsH_3. Tertiarybutyl arsine (TBAs) is sometimes used for the growth [76]. The GaAs NWs are formed under a V/III ratio of 100–250. The optimum T_G window for GaAs NWs is a boundary between the formation of As-trimers and the desorption of adatoms on the (111)B surface. Formation of As-trimers on the (111)B surface becomes dominant below the optimum T_G, and the formation suppresses the GaAs growth on the (111)B surface. In this case, lateral overgrowth (LOG) along the <1–10> direction is dominant because the growth rate of {1–10} sidewalls is faster than that of the (111)B surface. On the other hand, desorption of adatoms is enhanced above the optimum T_G because this temperature range is desorption-limited growth. The LOG is mostly suppressed above the optimum T_G. In this case, the growth rate of GaAs NW is decreased because of desorption. Also, the growth of {1–10} sidewalls is suppressed due to desorption of group-V atoms.

The height of GaAs NWs is inversely proportional to the mask opening diameter [22]. This suggests that the migration of growth species on the NW sidewalls plays a major role in the SAG of GaAs NWs.

3.3.2.2 InAs NW

The optimum T_G window ranges from 540 to 580°C [25]. The MO sources are trimethylindium (TMIn) and AsH_3. The InAs NWs are formed under V/III ratio of 250. The LOG is enhanced below the optimum T_G, and the growth rate of InAs NW is reduced above the optimum T_G [25]. The mechanism of these behaviors is same as the GaAs NWs in SAG. The height of the InAs NWs is inversely proportional to the square of the mask opening diameter [25]. This means that the surface diffusion of growth species on the SiO_2 and NW sidewalls plays a major role in the NW growth.

3.3.2.3 InP NW

The optimum T_G window ranges from 600 to 660°C [23]. The MO sources are TMIn and tertiarybutyl-phosphine (TBP). The InP NWs are formed under a V/III ratio of 15–60 for SAG. The preferential growth direction is the <111>A direction. Thus, vertical InP can be formed on (111)A substrate. The LOG is enhanced below the optimum T_G and high [TBP]. The distinct feature of the InP NW is morphology and crystal structure difference. The morphology of the InP NWs becomes tapered under high T_G and low V/III ratio, and straight cylinder-shaped structures are formed under low T_G and high V/III. These differences have been found to be related to the crystal structures. The tapered InP NWs showed perfect wurtzite structure, and straight InP NWs showed twin-included zincblende structure [24]. The mechanism of the crystal structure difference is thought to be due to the difference in iconicity of InP.

3.3.2.4 GaP NW

The optimum T_G window ranges from 770 to 790°C. The MO sources are TMGa and TBP. The GaP NWs are formed under a V/III ratio of 200–250 for SAG. The preferential growth direction is the <111>B direction.

3.3.2.5 InGaAs NW

The optimum TG for InGaAs NWs on InP(111)B ranges from 630 to 670°C [26–28], and the T_G for Ga-rich InGaAs NWs on GaAs(111)B ranges from 650 to 700°C [29]. Moreover, the In/Ga composition will vary with the NW pitch due to the migration of growth species on SiO_2 and NW sidewalls. For instance, Ga composition of InGaAs NW will change from 62 to 80% with a pitch decrease from 6 to 0.6 μm [28]. The MO sources are TMIn, TMGa, and AsH_3. The height of InGaAs NWs was inversely proportional to the square of the opening diameter of the mask shown in Fig. 3.4c [28]. This was because the surface migration of In atoms was dominant during InGaAs NW growth.

3.3.3 Growth of Core-Shell Structures

The most distinctive feature of the SA-MOVPE for NW growth is isolation of LOG mode and NW growth mode by changing T_G or V/III ratio. For example, low T_G enhances the LOG of GaAs without NW growth after the GaAs NW growth under optimum T_G. [21] The diameter of the GaAs NW becomes larger than the mask opening without NW growth as shown in Fig. 3.6. This is because the growth rate

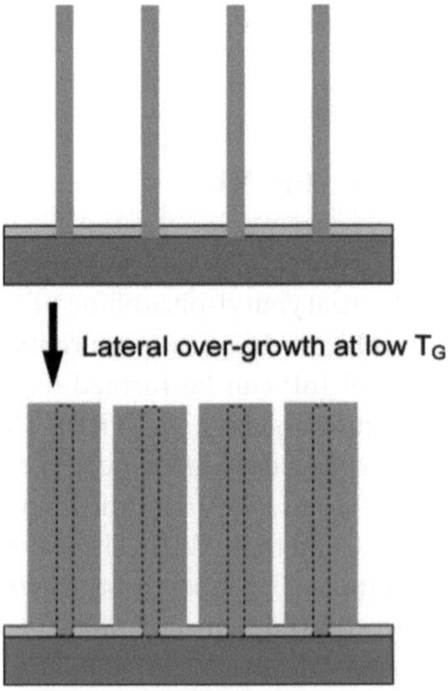

Fig. 3.6 Illustration of nanowire growth and LOG

of {1–10} sidewalls is increased due to the suppression of the desorption process on the {1–10} sidewalls. This characterization is similarly observed in other materials. The use of this property for the growth of heterostructures enables the formation of core-shell or CMS NWs such as GaAs/AlGaAs core-shell NWs [33, 51].

The surface states of NWs will be increased as compared with planar III–V because the NWs have large surface area. The surface states degrade the optical properties of III–V NWs and device performance through a nonradiative recombination process. The core-shell structures effectively passivate the surface states of the NWs. Also, the shell layer is stable at atmosphere as compared to a chemical treatment such as sulfur passivation [77]. GaAs/AlGaAs core-shell NWs had improved photoluminescence (PL) intensity up to 490 times due to the reduction of the surface states [78].

3.4 Heteroepitaxy of III–V NWs on Si Substrate

Next, we will describe the heteroepitaxy of III–V NWs on Si by SA-MOPVE. Recent advances in heteroepitaxial techniques such as the VLS [3–8] method and SAG [23–40] have enabled integration of III–V compound semiconductor (III–V) NWs with Si substrates. These materials integrated on Si have attracted much attention as building blocks for next-generation electronics and photonics because they can be used as fast channels in vertical nano-architectures, steep-slope switches, and vertical NW-based high-electron mobility transistors (HEMTs) on Si wafers. The III–V NWs can be used for optical circuits replacing Cu-based connections in devices such as nanometer (nm)-scale light source and detectors, as well as LSI chips.

Since Märtensson's pioneering work on As/P-related III–Vs NWs on Si was reported in 2004 [79], dozens of papers on III–V NW growth on Si substrates have been published [80–132]. It should be noted that the growth and applications of nitride-related NWs on Si were described by Kishino et al. in 1997 [133]. Almost all reports have focused on crystal growth of the III–V NWs on Si because we are only in the dawn of III–V NWs/Si integration. So far, the growth of GaAs [50–52, 78, 79, 81, 83, 89, 90, 92–96, 98, 99, 101, 104, 106–109, 111, 114–116, 118, 120–122, 125–128, 131], InP [80, 82, 83, 85, 87, 100, 102, 107, 109, 117], InAs [83, 88, 91, 93, 97, 105, 109, 110, 112, 113, 119, 123, 129, 130], GaSb [122, 124], and ternary-alloy [50, 51, 78, 85, 88, 106, 122, 127] NWs on Si have been reported, and these NWs were grown by the VLS, catalyst-free [88, 91, 96, 101, 105, 106, 110, 117, 120, 122–128, 130, 131], and SAG methods [51, 78, 97, 113, 119, 129, 132]. The III–V NW-based devices such as LEDs [50–52] and surrounding-gate FETs [105, 119] integrated on Si with these growth methods have been demonstrated, and unique devices using III–V NW/Si heterojunctions, such as solar cell [110], Esaki tunnel diodes [129], and tunnel FETs [132], have been reported.

In the heteroepitaxy of III–V NWs, one must consider not only the conventional problems in III–Vs/Si integration, but also the following four problems:

(a) positioning for NW sites, (b) polarity in III–V NWs (c) unintentional doping from Si substrate, (d) misfit dislocation at the heterointerface.

(a) *Positioning of NW sites*. Though positioning is a problem in other nanostructures, lithographic techniques such as positioning of metal catalysts in VLS or openings in SAG are effective for defining the NW sites.
(b) *Polarity in III–V NWs*. Conventional III–V NWs tend to grow in the <111>B or <111>A directions. For example, InAs NWs preferentially grow in the <111>B direction, so, vertically aligned InAs NWs can be grown on a III–V(111)B substrate. On the III–V(111)A surface, the InAs NWs are grown in three equivalent tilted <111>B directions. The <111> direction of group-IV semiconductors, on the other hand, does not show polarities different to III–Vs. Thus, in III–V/Si heteroepitaxy equivalent surface orientations and directions always occur on the Si(111) surface and these equivalencies form antiphase domains. Instead of the antiphase domain formation, equivalent growth directions always occur for III–V NW/Si integrations. That is, such III–V NWs on Si(111) grow to vertical <111> and three equivalent tilted <111> directions at the same time. The differences result either from the coexistence of (111)A and B surfaces that are formed when Si is eliminated by a metal catalyst during VLS growth, or from termination of group-III or V atoms on the Si(111) surface during SAG. For rational design of NW applications taking advantage of geometries, we have to control such equivalent growth directions into vertical the <111> direction.
(c) *Unintentional doping from Si substrate*. In VLS growth of NWs, metal catalysts corrosively etch the Si surface and release Si atoms that can diffuse into the NWs. In catalyst-free growth, Si atoms can easily diffuse into III–V NWs because of the high temperatures required for the catalyst-free growth. The unintentional doping from the Si substrate forms a gradual carrier distribution inside III–V NWs. In such cases, a highly doped n-type layer always forms close to the heterointerface resulting from the unintentional doping and degrades performance in III–V NW applications. To improve the performance, we have to suppress or control the doping. Effective ways to do that, however, have not been found, because there have been few investigations of this kind of doping. Low-temperature (LT) buffer growth could probably suppress the unintentional doping.
(d) *Misfit dislocation at heterointerface*. The heteroepitaxy of lattice-mismatched systems usually introduces misfit dislocation networks at the interface. The lattice mismatches are 4.1% for the GaAs/Si system, 8.1% for the InP/Si system, and 11.6% for the InAs/Si system. Generally, misfit dislocations are formed in the (111) plane with <2–1–1> directions. Misfit dislocations with a period corresponding to the value calculated from the lattice mismatch are observed in InAs NWs/Si systems [97]. These misfit dislocations could probably be avoided by reducing the diameter of the NWs, and Ertekin et al. and Flank have calculated the diameter below which the NW has coherent (without misfit dislocation) and plastic (with misfit dislocations) growth [134, 135]. Coherent

growth has recently been seen in experiments with thin GaAs NW/Si interfaces [78]. The effect of these misfit dislocations on NW-based applications on Si has not been investigated. These perceptions regarding dislocations will be important parameters for NW-based devices on Si and new concepts using III–V NW/Si heterojunctions [110, 129, 132].

In this section, we review the SAG of III–V NWs on Si substrate. We also explain key techniques for controlling the growth directions of III–V NWs on Si regardless of polarity. Also, we describe the results obtained using transmission electron microscopy (TEM) to investigate the heterointerface of III–V NWs on Si. We focus on InAs NWs and GaAs NWs.

3.4.1 Basic Concept for Selective-Area Growth of III–V NWs on Si

The SA-MOVPE initiates from an atomically flat surface without catalysts. For the SA-MOVPE of III–V NWs on Si, we should consider the surface reconstruction of Si, interface of SiO_2/Si, and incorporation process of group-III and group-V atoms on Si(111) surface at the same time. The methods for controlling these issues will be different for each growth material because the chemisorption processes of group-III atoms on the Si surface and the bonding strength between group-III atoms and Si are different at various temperatures. If we neglect these bases, vertically aligned III–V NWs will not be formed on the Si surface. In this section, we explain the fundamentals for SAG of III–V NWs on Si.

Various kinds of reconstruction of Si(111) surfaces have been observed under ultrahigh vacuum (UHV) as shown in Fig. 3.7. The c2 × 1, 7 × 7, and 1 × 1

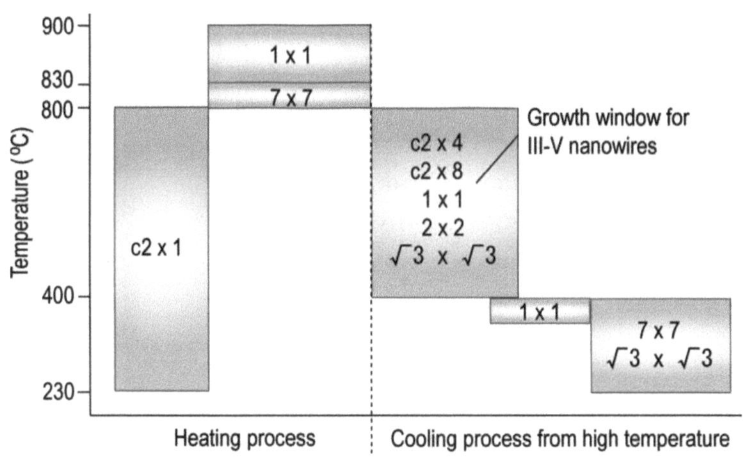

Fig. 3.7 Surface reconstructions of Si(111) during heating and cooling

reconstructed surfaces are formed under heating processes. Under cooling processes, the metastable reconstructed $n \times n$ ($n = 1, \sqrt{3}, 2$), c2×4, and c2×8 surfaces are formed at temperatures from 500 to 830°C and 1 × 1 reconstructed structure is formed at above 830°C and below 430°C [136–139]. We should note that the growth temperature windows for most III–V semiconductors are within the range in which metastable reconstructions form. These surface reconstructions suppress adsorption and nucleation processes of III–V materials because these surfaces are very stable and there are no dangling bonds for adsorption. Moreover, these metastable surface reconstructions are thought to randomize the orientation of dangling bonds and/or to disrupt the uniform nucleation of III–V growth on the Si surface. A promising way to avoid the formation of metastable surface reconstructions is to cool the Si(111) surface to 400°C in a H_2 ambient, because the 1 × 1 reconstructed surface that was formed at a higher temperature can regenerate at around 400°C.

Native oxides also suppress the nucleation process of III–Vs. Moreover, these oxides result in unexpected NW growth on Si[110]. The in situ thermal cleaning at above 900°C is effective for removing native oxides and forming the Si(111) 1 × 1 reconstructed surface because it decomposes SiO_2 and also alters the surface reconstruction of Si(111). These phenomena are basic processes for cleaning Si(111) to control the growth directions of III–V NWs on Si(111) by SA-MOVPE.

The SiO_2/Si interface has Si^{1+}, Si^{2+}, and Si^{3+} chemical structures [140, 141]. After the removal of native oxides under high temperature and the formation of the Si(111) 1 × 1 reconstructed surface, chemisorption of group-III/V atoms on the Si(111) 1 × 1 surface should be precisely modulated to align vertical III–V NWs. This is because the coexistence of vertical and tilted NWs shown in Fig. 3.8 will

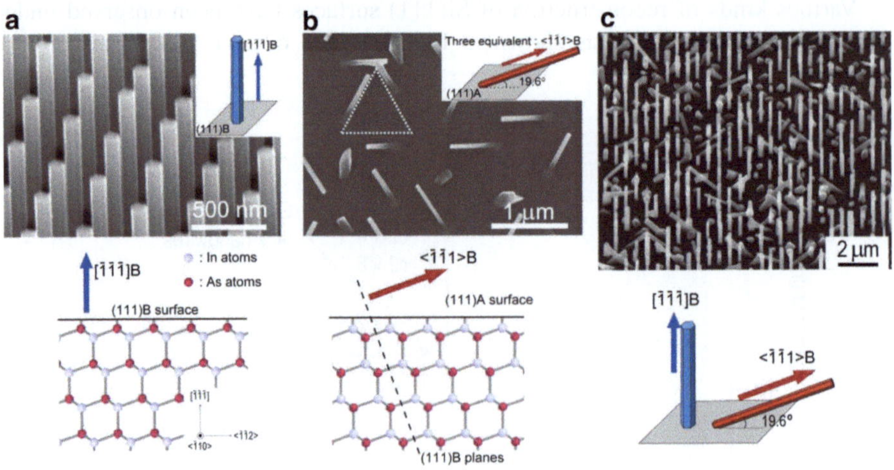

Fig. 3.8 (**a**) SEM image of InAs NW growth on InAs(111)B and schematics for InAs(111)B structure. *Blue arrow* shows growth direction of the InAs NWs. (**b**) SEM image of InAs NW growth on InAs(111)A and schematics of InAs(111)A structure. *Red arrow* shows growth direction of the InAs NWs. (**c**) Typical growth results of InAs NW on Si. Vertical NW and tilted NW are grown at the same time

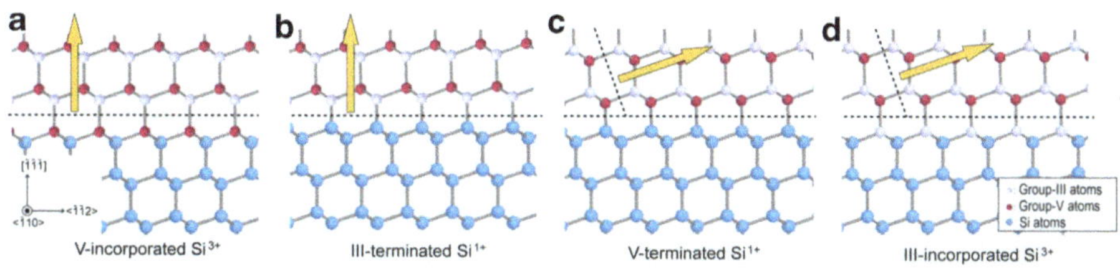

Fig. 3.9 Schematics of chemical structures: (**a**) Group-V-incorporating Si^{3+} structure, (**b**) group-III-terminated Si^{1+} surface, (**c**) group-V-terminated Si^{1+} surface, and (**d**) group-III-incorporating Si^{3+} structure. These are viewed from $< -110>$ direction. *Arrows* indicate III–VNW-growth direction

occur due to formation of the four types of chemical structures in Fig. 3.9a–d. Figure 3.9a, b depicts As-incorporating Si^{3+} and In-terminated Si^{1+}surfaces, which are equivalent to the (111)B-oriented surface. Figure 3.9c, d shows In-incorporating Si^{3+} and As-terminated Si^{1+} surfaces, which correspond to the (111)A-oriented surface. These structures are the results of termination of nonpolar Si by group-III and group-V atoms. Vertical NW growth can be obtained by simply forming an (111)B-oriented surface on the Si(111) surface. As shown in Fig. 3.9a, b, once As-incorporating Si^{3+} and/or In-terminated Si^{1+} has formed on the Si surface, only vertical III–V NWs should grow on the Si(111) substrates. Conversely, the growth directions of the III–V NWs can be controlled by optimizing the initial surface and growth conditions. To form an As-incorporating Si^{3+} surface, group-V atoms should be replaced with the outermost Si atoms of 1×1 reconstructed surface because it is equivalent to a V-atom-terminated Si^{3+} surface and a (111)B-oriented surface. Conveniently, the Si(111):As 1×1 reconstructed surface was found to be formed below 430°C in an As ambient [142]. The method of forming these (111)B-oriented surface is different for each III–V/Si system because these processes strongly depend on bonding strength between group-III atoms and Si atoms. Next, we describe typical growth results of InAs and GaAs NWs on Si(111) substrate.

3.4.2 Selective-Area Growth of InAs NWs on Si

Indium arsenide is a narrow-band-gap semiconductor and has high-electron mobility (at room temperature, 20 times higher than that of Si) because of its small electron effective mass. This material is less effective against surface depletion that results from surface states. This means that conductive InAs NWs can easily be obtained regardless of surface passivation. Although the lattice mismatch in InAs/Si heteroepitaxial system is approximately 11.6%, the VLS method and SA-MOVPE enabled epitaxy of InAs NWs on Si surface. This material is expected as to be a good building block for future vertically aligned channel on Si CMOS.

First, we studied the relation between the growth yields of InAs NWs and the AsH_3 supply conditions prior to the growth. In these studies, we usually used

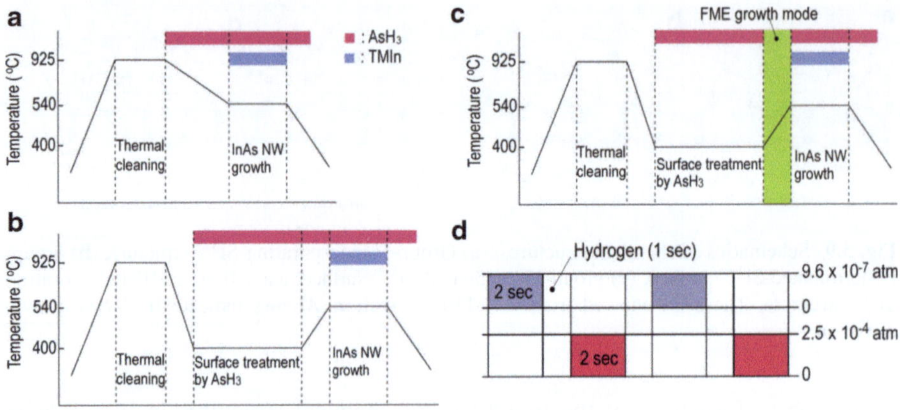

Fig. 3.10 (**a–c**) Schematic illustrations of gas-flow and temperature sequence for InAs NW growth on Si. (**d**) Schematic diagram of flow-rate modulation mode

Fig. 3.11 Percentage of growth results with a growth sequence of Fig. 3.10a as a function of [AsH_3]

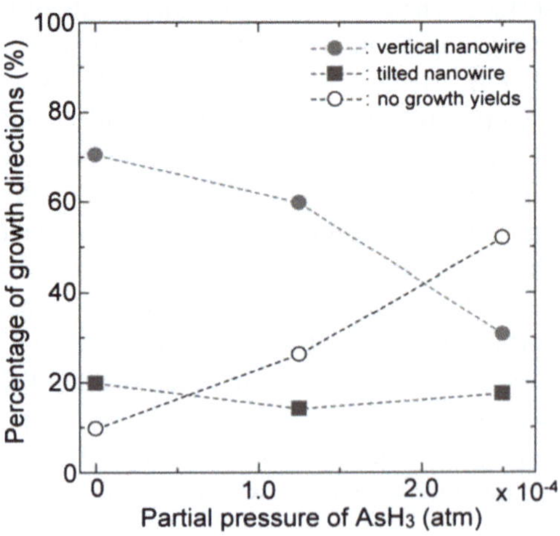

thermal cleaning at above 925°C for 5 min prior to the growth to remove native oxide from the opening area of patterns. The openings were 60 m in diameter and the partial pressure of AsH_3 prior to the growth was 2.5×10^{-4} atm. Figure 3.10a–d shows a series of growth sequences in which for each sequence we evaluated the percentage yields of vertical NWs, tilted NWs, and no growth. We confirmed the reproducibility of each percentage with 30 wafers: the standard deviation was within ±1%.

With the conventional sequence shown in Fig. 3.10a, the yield of vertical InAs NWs is approximately 31% and that of tilted NWs is 17%. As shown in Fig. 3.11, percentage of no growth yield could be as high as 52%. This is because desorption of In atoms from the Si substrate and complex Si reconstruction suppress the nucleation of InAs. It should be noted that the MOVPE system uses pure H_2 as the carrier gas, but no hydrogen terminates on the Si surface because the Si–H bond is thermally weak at 540–925°C. A complex As atom termination of the reconstructed Si(111) surface occurred prior to the growth, and this complex surface reconstruction blocked the SAG. Furthermore, In adatoms can easily be desorbed from the Si surface at high temperatures. The dependence of NW growth yields on the [AsH_3] during the cooling after thermal cleanings is shown in Fig. 3.11, where the vertical NW growth yield obtained without AsH_3 during the cooling after thermal cleaning was 70%. The percentage of no-growth yields increased with the increment of [AsH_3]. This indicated that formation of As-incorporated, reconstructed Si surface was increased and such reconstructed Si surface blocked the nucleation process. Such surface reconstructions and In atom desorption should, therefore, be controlled to ensure the formation of (111)B-oriented Si(111) surface.

Cooling the Si(111) surface to 400°C in H_2 ambient prevents problems due to metastable surface reconstructions because the 1×1 reconstructed surface can regenerate at 400°C. We therefore tried to convert the Si(111) 1×1 surface to a Si(111):As 1×1 reconstructed surface by treating it with AsH_3 at 400°C as shown in Fig. 3.10b. This sequence increased the vertical NW-growth yield to 67% and decreased the yield of 19.6°-tilted InAs NWs to 11%, and it had little effect on the no-growth yield (22%). The small fraction of 19.6°-tilted InAs NW indicates that As-terminated Si^{1+} surface was formed. Thus, 67% of the mask openings were changed to (111)B-oriented surface, but 11% of the openings formed (111)A-oriented surface in this case. Moreover, desorption of As adatoms occurred from the As-terminated Si^{1+} surface when the substrate temperature rose to the growth temperature. This desorption process is related to the no-growth yields in this case.

The percentage of the vertical-NW yield, tilted-NW yields, and no-growth yields were almost the same as the variation of [AsH_3] during the treatment of AsH_3 at 400°C as shown in Fig. 3.10b.

This indicates that formation of the As-terminated Si^{1+} surface and the desorption process of the As adatoms from the As-terminated Si^{1+} surface are not dependent on the [AsH_3]. When the In atoms adsorb onto the dangling bond of Si^{1+} after the desorption process, the surface will act as a (111)B-oriented surface. We therefore think that the control of desorption or decomposition of V atoms is important for suppressing the growth of tilted NWs.

Consequently, we used the growth sequence shown in Fig. 3.10c, d to avoid desorption of As adatoms. First, the substrate was cooled to 400°C after thermal cleaning. Next, AsH_3 was supplied at this temperature to form the As-incorporating Si^{3+} surface. Because As and In atoms should be efficiently supplied to the Si(111) surface in order to form a (111)B-oriented surface just before InAs NW growth, we used the flow-rate modulated epitaxy (FME) at 400°C [97]. FME is a method of alternately supplying group-III and group-V precursors during MOVPE. The purpose of the FME is to enhance the termination of As-incorporating Si^{3+} by In atoms and termination of bare Si^{1+} surfaces by In atoms because the termination of Si^{1+} surface by group-III atoms also forms (111)B-like surface. We also used a brief pulse of H_2 in the interval between the TMIn and AsH_3 supply to enhance the exchanges of supplied materials. Figure 3.10d outlines the optimum gas-flow

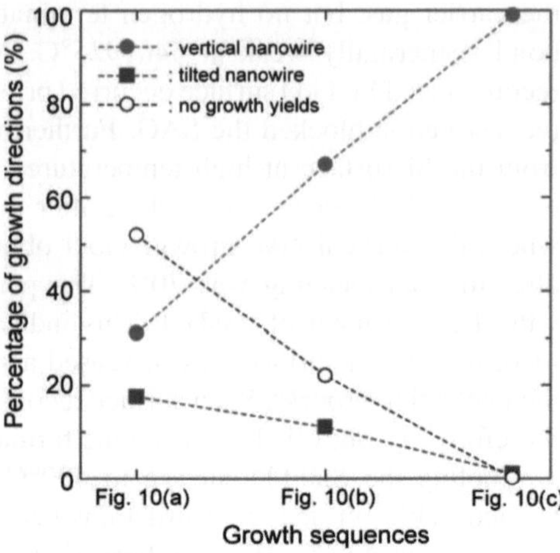

Fig. 3.12 Percentage of growth results with growth sequences shown in Fig. 3.10

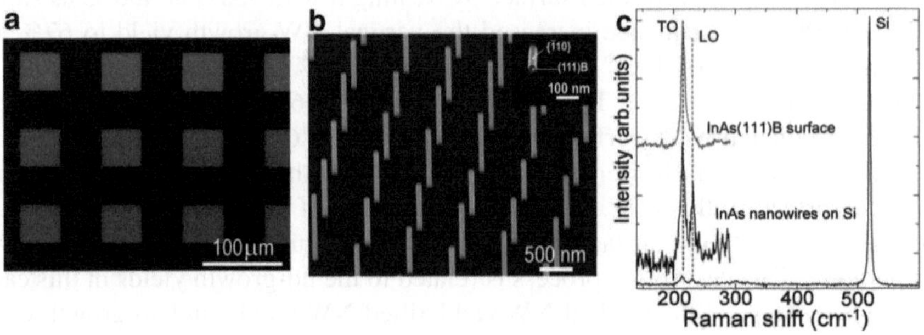

Fig. 3.13 (**a**) Overview of InAs NW arrays on patterned substrate. (**b**) 45° -tilted view showing vertical InAs NW array. *Inset* shows a plan-view of a InAs NW. Side facets are {−110} planes and hexagonal cross section is (111)B plane. (**c**) Raman spectra of the InAs NWs on Si

sequence of the FME. The FME mode was carried out for 20 cycles at 400°C, after which typical InAs NW growth was carried out at 540°C. As a result, 99% of the NWs were vertical, and 1% were tilted NWs, and the no-growth yield was zero as shown in Fig. 3.12. These results suggest that with this growth sequence, a (111) B-oriented surface was formed on Si(111).

The complete vertically aligned InAs NWs on Si(111) under optimized conditions are shown in Fig. 3.13a, b. Figure 3.13a is an overview of InAs NW arrays on Si(111) substrate. The position-controlled InAs NWs, shown in Fig. 3.13b, were formed within the gray square prepatterned regions (each $50 \times 50\,\mu m^2$) in Fig. 3.13a. The prepatterned regions were readily fabricated by using electron-beam lithography and wet chemical etching. Here, we used patterns with opening diameters of 70 nm and pitches ranging from 400 to 800 nm. The InAs NWs grew only in the openings and were oriented perpendicular to the surface. They were an average of 70 nm in diameter and average of 2 μm in height. Size fluctuation of NW diameters was within a standard deviation of ±4 nm. The length fluctuation of InAs

NWs in Fig. 3.13b was resulted from incubation time due to the initial nucleation on Si(111) surface. The inset of Fig. 3.13b shows a plan view of an SEM image. All the NWs had hexagonal cross sections with surrounding {−110} side facets. Raman scattering spectra of these NWs on Si and a reference InAs(111)B substrate are shown in Fig. 3.13c. TO and LO phonon spectra from the NWs and a strong Si LO phonon were observed. The TO and LO phonon spectra have no peak shifts as compared to those of the bulk InAs(111)B surface. This means that the strains generated from the lattice mismatch accommodate only at the interface, and these strains does not affect upper InAs NWs. The FME mode at low growth temperature thus seems to have accommodated the strains at the InAs/Si interface.

3.4.3 Selective-Area Growth of GaAs NWs on Si

Gallium arsenide (GaAs) was grown following the sequence summarized in Fig. 3.14. The basic concept for modifying the Si(111) surface into a (111)B-oriented surface is same as the InAs NW growth on Si. As described in Sect. 3.3.2, the optimum T_G for the GaAs NW is higher than that of InAs NWs. Therefore, Ga adatoms are frequently desorbed from the Si(111) surface under the high T_G, and nucleation will not proceed on the Si(111) surface. Figure 3.14a, b shows the typical growth sequence for GaAs NWs and the growth sequence with AsH_3 treatment and FME to produce Si(111) 1 × 1:As reconstructed surface. Details of the FME are shown in Fig. 3.14c. The use of these growth sequences disables the nucleation of GaAs on Si(111) surface as shown in Fig. 3.15a. This is because the Ga and As adatoms evaporated from the Si(111) surface at the high T_G. Therefore, we introduced low-temperature GaAs growth to avoid the desorption process of the adatoms after the formation of Si(111) 1 × 1:As reconstructed surface as shown in Fig. 3.14d.

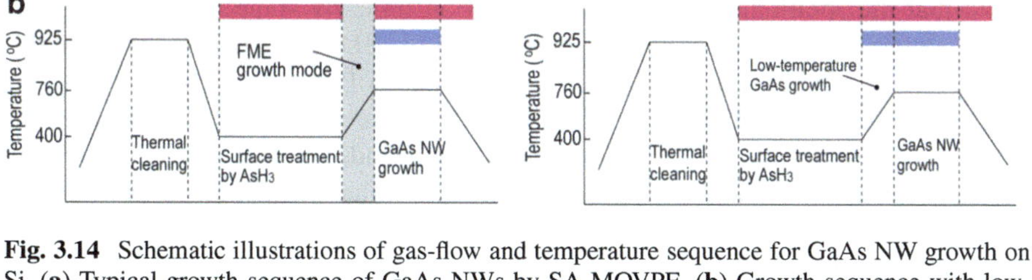

Fig. 3.14 Schematic illustrations of gas-flow and temperature sequence for GaAs NW growth on Si. (**a**) Typical growth sequence of GaAs NWs by SA-MOVPE. (**b**) Growth sequence with low-temperature FME. (**c**) Detail of the FME. (**d**) Growth sequence with low-temperature GaAs growth

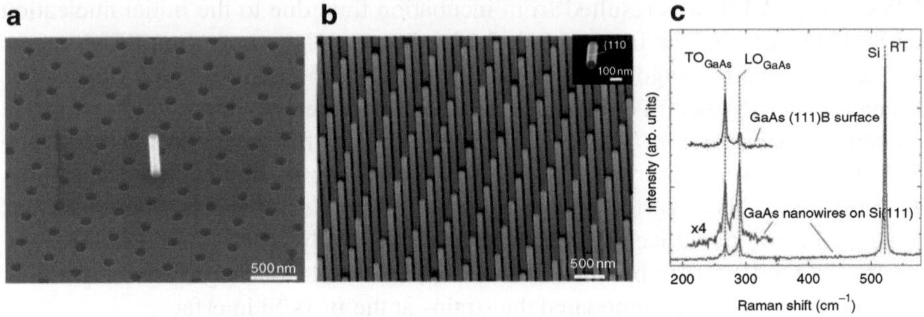

Fig. 3.15 (**a**) SEM image of GaAs with a sequence of Fig. 3.14b. (**b**) SEM image of GaAs NWs on Si substrate with a sequence of Fig. 3.14d. (**c**) Raman spectra of GaAs NWs on Si substrate

Following the steps, GaAs NWs could be grown in the vertical [111] direction on Si as shown in Fig. 3.15b. The yield of the vertically aligned GaAs NWs was 100%. The grown NWs had a hexagonal cross section having a (111)B top surface and {−110} sidewall facets. The GaAs NWs in Fig. 3.15b measured 70 nm in diameter and 1.7 μm in height. The standard deviation in diameter fluctuation was ±3 nm. Figure 3.15c shows the Raman spectra for GaAs NWs. LO and TO phonon peaks as well as a Si LO phonon peak can be observed. Neither LO or TO phonons indicates a peak shift to that of GaAs bulk. This means that the NWs grown on Si are pure GaAs without strain resulting from large lattice mismatch.

3.4.4 Size Dependence of the GaAs NW Growth on Si

The GaAs NW growth for different opening areas was investigated to obtain insights into the nucleation process of GaAs on Si(111). Figure 3.16a shows a plan view of an SEM image of planar GaAs growth on (111)B-oriented Si(111). Numerous triangular three-dimensional (3D) islands and their coalescence are formed on the surface. The Si(111) surface was modified to a (111)B-like surface using the sequence shown in Fig. 3.14d. The triangular shapes reflect a threefold symmetry for the (111)B surface and the orientation of the facets of the 3D islands is what one would expect from the symmetry. There are also 30°-rotated 3D islands, which mean that the rotational twin was introduced at an early stage of the nucleation, in Fig. 3.16a. The initial nucleation process for the 3D GaAs islands on the (111) B-oriented Si surface is assumed to be coalescence of triangular two-dimensional (2D) islands as illustrated in Fig. 3.16b. The coalescence of 2D islands with the rotational twins results in the generation of 3D islands with antiphase defects (or domains) at the boundary with coalescence. A similar nucleation process occurs when the openings are large, as shown in Fig. 3.16c, e. Here, d_0 was 600 nm in diameter. Hillock structures surrounded by complex facets were formed on the openings. This is because a large amount of coalescence occurred on the large

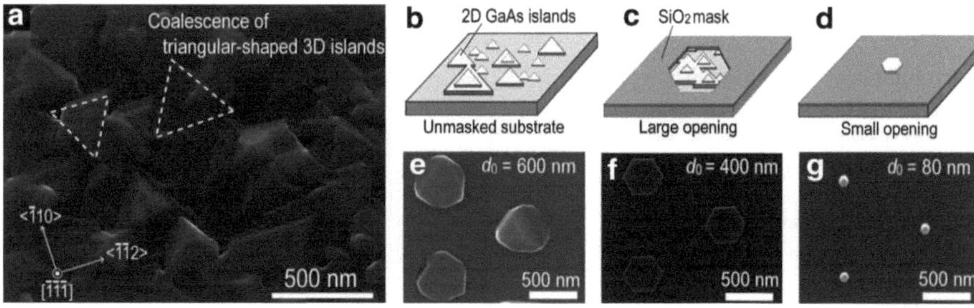

Fig. 3.16 (**a**) Plan view of GaAs planar growth on Si(111) substrate. *Dashed white triangles* represent two-dimensional (2D) islands. Plan view of SEM images of GaAs growth on masked Si(111). (**b**) Illustration of nucleation of 2D islands and their coalescence on unmasked Si(111) surface. Illustration of growth behavior of GaAs: (**c**) large opening and (**d**) small opening on masked Si(111) surface. (**e**) Diameter of openings, d_0, was 600 nm. (**f**) $d_0 = 400$ nm and (**g**) $d_0 = 80$ nm

surface areas, and formed many step and kinks within the openings. The roughness due to the steps and kinks enhanced the adsorption process of the growth species and formed large hillocks.

The hexagonal pillars surrounded with the $\{-110\}$ vertical sidewall facets and the (111)B top surface, on the other hand, could be formed on smaller openings, as shown in Fig. 3.16f, g corresponding to $d_0 = 400$ and 80 nm. For $d_0 = 400$ nm, there were partly hillock-shaped structures similar to those in Fig. 3.16g on some parts of the openings, but they are completely suppressed for $d_0 = 80$ nm. Therefore, we concluded that the sizes of the openings, which determined the number of 2D islands in the initial nucleation process, were important for growing GaAs NWs on the Si surface in SA-MOVPE.

The effect of the size of the openings was further confirmed with micro (μ)-PL measurements. Figure 3.17 shows the optical properties. The excitation laser spot for the μ-PL was 2 μm in diameter, and the wavelength was 632.8 nm. Because the pitch of the NWs is 600 nm, approximately ten NWs are included in the excitation area. Two main luminescence bands are observed from the GaAs grown on the planar region of Si as well as GaAs NWs on Si. The first is the near-band-edge emission at around 1.52 eV, and the second is the deep-level-related emission at around 1.33 eV. The emission is weak and dominated by deep-level-related recombination for planar GaAs growth on Si, while strong near-band-edge emissions were observed for GaAs NWs grown on Si, especially for $d = 70$ nm. This difference is thought to be due to the number of nonradiative recombination centers in grown materials. Nonradiative recombination centers were generated in planar GaAs presumably because of dislocations. Also, antiphase defects were formed in the boundary of coalescent 3D islands and lead to the deep-level-related emissions. The deep-level-related PL emissions are still dominant for $d_0 = 400$ nm as shown in Fig. 3.17, which implies that they still contain many antiphase defects originating from the coalescence of 2D islands, although their geometrical shape is in the form of a smooth top surface and facet sidewalls as already shown in

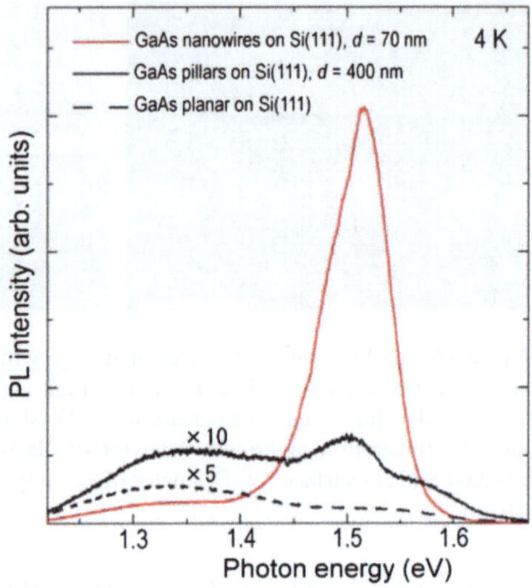

Fig. 3.17 PL spectra of GaAs NWs on Si(111). μ-PL spectra of GaAs NW array on Si. *Solid red line* represents GaAs NWs on Si(111) whose diameter (d) is 70 nm, *solid black line* describes GaAs pillar on Si(111) whose d is 400 nm, and *broken line* describes planar GaAs grown on Si(111) surface

Fig. 3.16f. On the other hand, the PL in GaAs NWs with diameters of 70 nm on Si improved because of a radiative recombination process due to the reduction of the antiphase defects.

Photoluminescence for GaAs NWs grown on Si(111) shows broad asymmetric spectra at around 1.52 eV, which means that the PL spectra are composed of several optical transitions as well as interband transition, such as free-exciton and carbon-related donor–acceptor pair transition. The full-width at half-maximum (FWHM) of all luminescence centers was several tens of meV. The slightly large FWHM in the PL spectra is commonly observed for homoepitaxial GaAs NWs [20]. [41]. Although the origin of the broadened PL bands in NWs is still unclear, the optical properties in thin GaAs NWs on Si are less effective from these defects due to lattice mismatch and coalescence.

3.4.5 Growth of GaAs/AlGaAs Core-Shell NWs on Si

The GaAs/AlGaAs core-shell NWs on the Si(111) surface were grown as schematically illustrated in Fig. 3.18a. Core-shell structures in NWs are useful for in situ passivation for applications to optical devices based on GaAs NWs. In addition, the growth of the shell layer is superior to the method of surface passivation using chemicals such as $(NH_4)_2S$ solution [77] because of its chemical stability in ambient. The growth direction with the SA-MOVPE method can be controlled radially or axially by controlling the growth temperature as explained in Sect. 3.3.3.

Figure 3.18b shows an SEM image of the vertical GaAs/AlGaAs core-shell NWs grown on Si(111). The AlGaAs layer was grown at 750°C for 10 min. The V/III ratio was 90. The diameter of the NWs during AlGaAs growth increased from

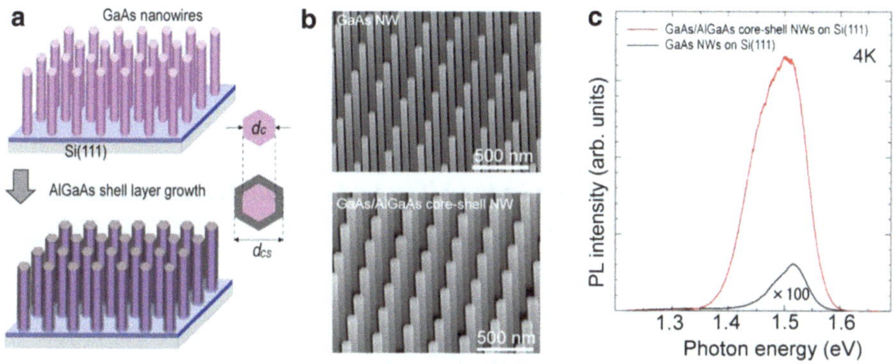

Fig. 3.18 (**a**) Illustration of GaAs/AlGaAs core-shell NW growth. (**b**) SEM image of GaAs NW and GaAs/AlGaAs CS NW on Si substrates. (**c**) PL spectra of GaAs NWs and GaAs/AlGaAs CS NWs at 4 K

75 to 150 nm, while the heights of the NWs remained constant at 1.75 μm. This means that the AlGaAs layer only grew radially, i.e., in the <−110> directions. The lateral growth rate of the AlGaAs was 7.5 nm/min. The μ-PL spectra of the GaAs/AlGaAs core/shell NWs were shown in Fig. 3.18c. The PL intensity of the core/shell NWs was dramatically enhanced (×490) compared to that of the GaAs NWs on Si. This indicates that the AlGaAs shell layer acted as a passivation layer to reduce the surface nonradiative recombination centers due to surface states in GaAs NWs. In the next section, we describe the fabrication of the GaAs/AlGaAs CMS NW-based LEDs.

3.5 Fabrication of III–V NW-based LEDs on Si Surface

One application of the III–V NWs on Si is in nanometer-scaled light sources and detectors for on-chip integration replacing Cu-based intrachip connections with high-performance optical interconnections in a small area. The large surface-to-volume ratio of the radial p–n junctions in CMS NWs can make the junction area larger than that of planar substrate with the same surface area [54, 143]. The vertical CMS NWs with radial p–n junctions are therefore desirable because of their area effectiveness and because they can improve the performance of LEDs, photodetectors, and solar cells. These CMS structures with radial p–n junctions also have potential for avalanche breakdown with low negative bias. The geometry of CMS NWs also makes them useful in free-standing NW-based APDs.

LEDs based on CMS NWs have been fabricated in wide band-gap semiconductors [53–56]. The NW-based light sources, which operate at wavelengths from 800 to 900 nm, are suitable for on-chip integrations with Si-photodiodes (PDs) and avalanche PD systems. However, there have been few investigations of III–V NW-based LEDs on Si for the NIR region [50–52]. In this section, we describe the integration of GaAs/AlGaAs CMS NW-based LEDs on Si substrate. This demonstration is a first step for electrically driven NW-based laser and free-standing NW APDs on Si platforms.

3.5.1 Growth of AlGaAs/GaAs/AlGaAs Double-Heterostructures in CMS NWs on Si

We designed double-heterostructure (DH) CMS NWs on Si to enhance the emission efficiencies in NW-based LEDs. The structure consists of an n-type GaAs NW as a core and n-AlGaAs, p-GaAs, p-AlGaAs, and p-GaAs shells for the innermost to outermost shells, which are sequentially grown on the sidewall of the GaAs core NW. The n-type and p-type AlGaAs layers are cladding layers for confinement of electrons and holes in the inner p-GaAs layer and also for photon generation in the p-GaAs layers. The p-GaAs wedged between the n- and p-AlGaAs layers is a tubular quantum well (QW), while the outer p-GaAs is a capping layer for Ohmic contacts.

After the growth of GaAs NWs, the n- and p-AlGaAs shell layers were formed at 700°C for 5 min in each. Also, the growth of p-GaAs well layer and capping layer was performed at 700°C for 3 min in each. Nominal carrier concentrations of planar GaAs and p-GaAs were 3.5×10^{17} and $4.0 \times 10^{18}\,\text{cm}^{-3}$. The donor and acceptor concentrations, N_D and N_A, of planar n-AlGaAs and p-AlGaAs were 9.0×10^{17} and $1.0 \times 10^{18}\,\text{cm}^{-3}$. The actual doping level in each layer of the CMS NWs is unclear because of difficulty in characterizing such thin layers. As mentioned in a previous report regarding core-shell InP NW solar cells [58], there is a possibility that these carrier concentrations for the CMS NWs were lower than those for the planar epitaxial layers.

Figure 3.19a, b shows the growth results of the DH structure CMS NW array on Si. Vertically aligned regular hexagonal NWs with diameters of 250 nm were fabricated on Si substrate. Unintentional kink and taper formations of the NWs resulting from high doping are not observed in these images. Thus, each NW had uniform shell layer thickness. The average heights were 3 μm. Figure 3.19c is an illustration of the CMS NW structure and Fig. 3.19d shows an SEM image taken from a cleaved and selectively etched CMS NW. In this image, the lateral thicknesses of both n- and p-AlGaAs layers are 25 nm, and the lateral growth rate was estimated to be 2.5 nm/min.

3.5.2 Fabrication of CMS NW-Based LEDs on Si

Figure 3.20 illustrates the NW-based LED structure. The detailed device process is explained in [51]. The Cr/Au thicknesses was Cr(10 nm)/Au(130 nm) metal with a 25-nm-thick n- and p-AlGaAs shell layer. In this LED structure, about 10^5 NWs are

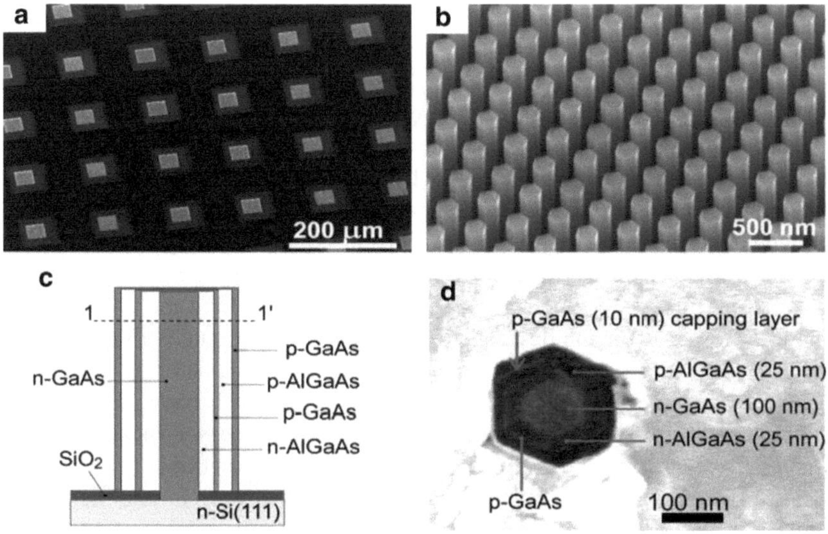

Fig. 3.19 (**a**) Overview of CMS-NW arrays on patterned substrate. (**b**) 30°-tilted view SEM image of CMS-NWs. (**c**) Illustration of CMS NWs. (**d**) SEM image of a CMS NW cross-section cut following *dashed line* (1–1′) in (**c**)

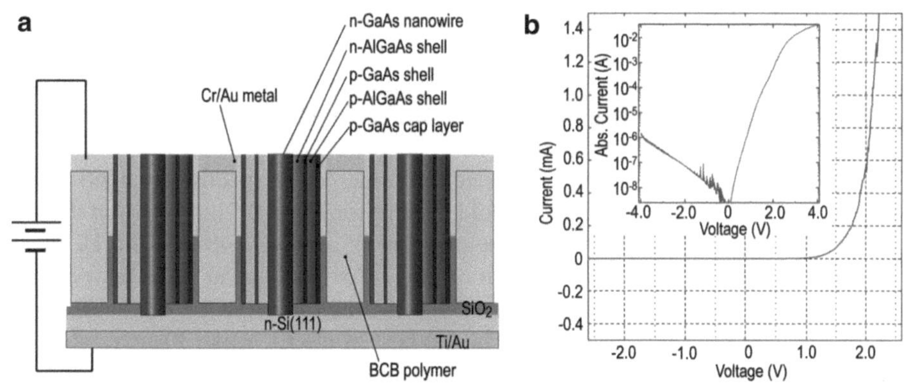

Fig. 3.20 (**a**) Schematics of GaAs/AlGaAs CMS NW-based LEDs on Si. About 10^5 NWs are connected in parallel. (**b**) Current–voltage (I–V) curve of the NW-based LED on Si

connected in parallel. The total junction area is estimated to be $1.3 \times 10^{-3}\,\text{cm}^2$. To determine recombination mechanism in the CMS NW-based LED on Si, we measured the current of the CMS NW devices at voltages from -4 to $4\,\text{V}$ using an HP 4156B parameter analyzer. In these measurements, the Cr/Au metal was positively biased, and the Ti/Au backside electrode was grounded. Figure 3.20b shows the current–voltage (I–V) curves for these structures. The I–V curves for all devices show moderate rectifying properties. The inset of Fig. 3.20b shows a semilogarithm plot of the I–V properties. In this structure, the depletion width is estimated to be 23 nm and the total thickness of this CMS layer was thicker than the depletion width.

The rectifying property for the CMS NW-based LED array with thick AlGaAs layers indicates gradual linear curves resulting from unexpected resistance. The resistances originated from series and contact resistance across the NWs and Si substrate. Also, premature turn-on behavior originating from surface states occurred. Considering these series resistance (R_S) with the CMS NW-based LED and shunt resistance to be infinity, the current equation is expressed as $J(n) = J_0 \exp[(V - R_S rm I)/n\mathrm{kT}]$ where n is the ideality factor of the p–n junction. From this equation, the n is estimated to be 3.8 for the device, and the R_S is calculated to be 45.3Ω. The n for the device is higher than conventional GaAs-based diodes ($n \sim 1.1$–1.5), and it is also higher than the thermal diffusion or recombination expected from the Sah–Noyce Schottky mode [144]. Such high ideal factors are usually observed for AlGaN-based diodes [145,146]. The reason for such high value of the n is thought to be the carrier tunneling across the junction.

The I–V curve shows slightly small leakage current in the negative-bias region. It should be noted that this behavior was not observed for CMS NWs on GaAs(111)B substrate (not shown here). This is because of the band discontinuity across the GaAs/Si junction. Heterojunction between III–V NW and Si forms band discontinuity. Several reports have investigated the valence and conduction band offset with photoemission spectroscopy [147, 148]. The potential barrier of the band offset leads to Schottky properties. The Schottky properties of III–V NW on Si have been reported. The discontinuity across the GaAs/Si junction would be affected by several factors, such as misfit dislocations, interface states, and the conductance of Si and III–V NWs. Further investigations are required to clarify this discontinuity across the III–V NWs/Si junctions.

Figure 3.21a shows the typical electroluminescence spectra obtained under several current conditions. The threshold current for EL is 0.5 mA (current density is $0.3\,\mathrm{A/cm^2}$) at 1.9 V. The EL peak position is around 1.48 eV. The EL peak position is shifted to 60 meV from that of the GaAs band gap at room temperature ($E_g = 1.42$ eV). This EL came from the DH structures, because the estimated width of the tubular GaAs QW was 7 nm. The FWHM of the EL was 130 meV, which is larger than the theoretical linewidth (1.8 kT) in LED spectra, and the PL spectra at room temperature also showed large FWHM. This is therefore because the EL spectra in Fig. 3.21a contain several luminescence centers resulting from donor–acceptor pair transitions from Si, C, and Zn impurities in the AlGaAs layers.

The EL intensity increased superlinearly with the current injection, and it was saturated from $I = 1.5$ mA ($1.2\,\mathrm{A/cm^2}$) as shown in Fig. 3.20b. The superlinear characteristic in the EL intensity at low current injection indicates that the CMS NW-based LED is similar to that of superluminescent LEDs [98]. This is because the AlGaAs shell layer surrounded by Cr/Au metal and oxides acts as a cavity because the reflectivity of the metal is above 95%. The origin of the saturation was the carrier overflow effect as observed in surface-emitting LEDs [149]. This is because the volume of the active region in the CMS NWs is very thin and parasitic resistance (contact resistance, series resistance, and so on) is high. We

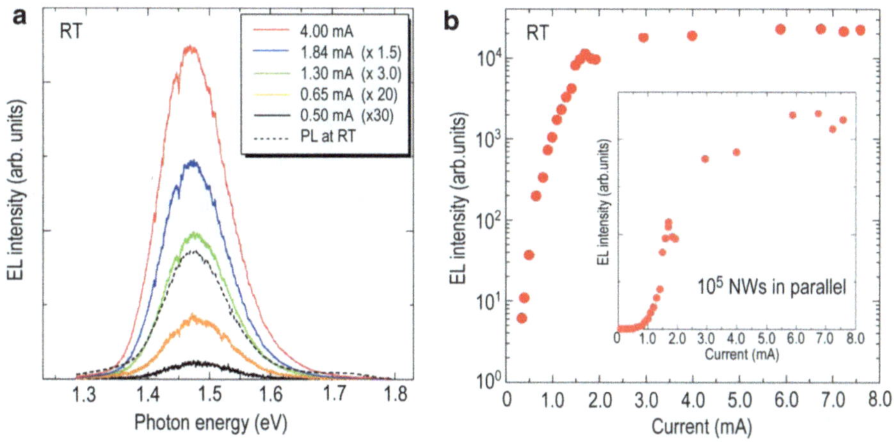

Fig. 3.21 (**a**) Electroluminescence (EL) spectra under several current injections at room temperature. *Dashed line* spectrum depicts the PL at room temperature. (**b**) Semilog plot of EL intensity as a function of injected current. *Inset* depicts linear-plot of EL intensity. About 10^5 NWs are connected in parallel

conclude that CMS NW-based LEDs are superluminescent surface-emitting LEDs. Moreover, shift of the EL peak position resulting from Joule heating on the junction temperature was not observed when the current injections were further increased. This is because the Si substrate acts as a heat sink due to its good thermal conductivity as compared to III–V compound semiconductors. This heterogeneous integration can, therefore, lead to thermally stable driving. Generally, GaAs-based LEDs fabricated on Si without buffering do not achieve such bright EL because of threading dislocations resulting from the thermal coefficient difference. Reduction of threading dislocations using SAG [47] has been reported, and we also found that InAs and GaAs NWs grown on Si contained no threading dislocations [78,97]. The nanometer-scaled selective-area technique, therefore, could control the generation of threading dislocations and produce better performing CMS NW-based LEDs directly grown on Si surfaces without buffering techniques.

3.5.3 GaAs/GaAsP CMS Structure and Multi-Quantum well Layers for Laser Diodes

One reason for using CMS NWs for optical devices is to make NW-based laser diodes (LDs) and APDs. As mentioned, a free-standing CMS NW has geometrical advantages with regard to junction area and self-cavity effects when we use a combination of metals. The AlGaAs/GaAs systems are not good materials for light sources because of the rapid degradation resulting from generation of defects such as dark-line dislocations and dark-spot defects [150–153]. There are several

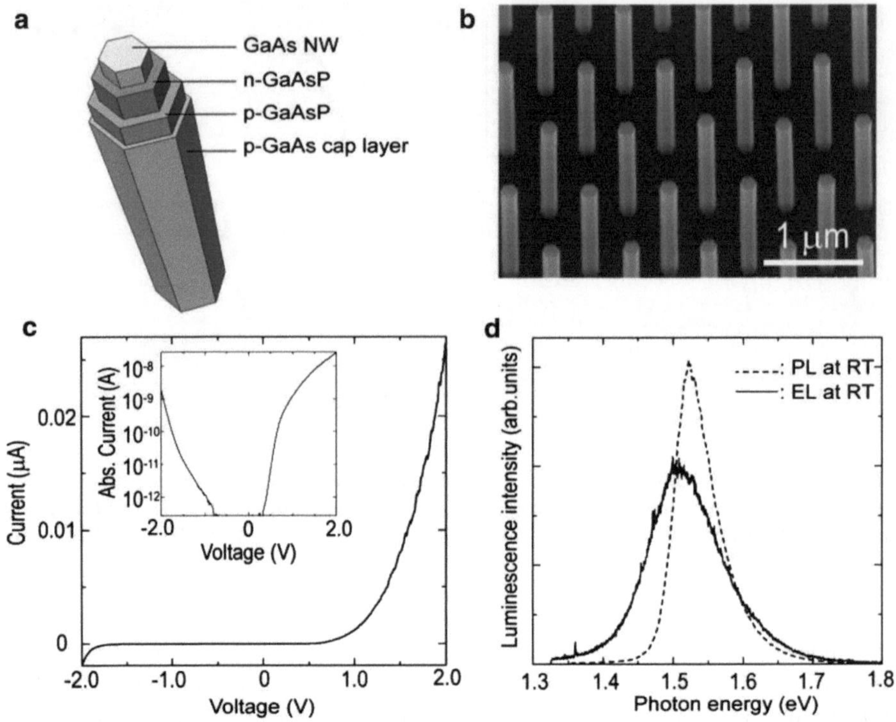

Fig. 3.22 (**a**) Illustration of cross section for GaAsP/GaAs CMS NW. (**b**) Growth results of the vertical GaAsP/GaAs CMS NWs. The total diameter is 180 nm. The diameter of the core GaAs NW is 80 nm. (**c**) I–V property for the CMS NWs. (**d**) EL (*solid line*) and PL (*dashed line*) spectra of the CMS NWs

approaches to making light-emitting devices without defects that degrade their performance. One is to use another material system, and another is to use multi-QW structures to enhance the optical gain. Here, we demonstrate the fabrication of GaAsP/GaAs CMS NW-based LEDs and an AlGaAs/20QWs/AlGaAs CMS NW-based LED array on Si as a first step for realizing electrically driven NW-based LDs.

As shown in Fig. 3.22a, we have designed and grown simple GaAsP/GaAs CMS NW-based LEDs on Si. Their structure consists of an n-type GaAs NW as a core and n-GaAsP, p-GaAsP, p-GaAs shells. TBP was used for the P source material. The n-GaAsP and p-GaAsP were grown at 650°C. The growth conditions for the AlGaAs/QWs/AlGaAs CMS NW-based LED array on Si were the same as those for the GaAs/AlGaAs CMS NW-based LED. For the QWs, we have designed 20-layer AlGaAs/GaAs multiple QWs (20-QWs) with the separation of 1-nm-thick $Al_{0.12}Ga_{0.88}$ As layers. The thickness of the GaAs well layers was constant at approximately 1 nm. The device processes were the same as those of GaAs/AlGaAs CMS NW-based LED arrays on Si.

Figure 3.22b depicts the GaAs/GaAsP CMS NWs on Si. Vertically aligned GaAsP/GaAs CMS NWs were successfully fabricated on Si(111). The total diameter was 180 nm, and the diameter of the core GaAs NW was 80 nm. The electrical properties and EL spectra of GaAsP/GaAs CMS NW-based LED on Si are shown in Fig. 3.22c, d. The $I-V$ curve showed Schottky-like behavior resulting from insufficient doping for GaAsP layers and some leakage. The $I-V$ curve shows the rectifying property with an n of 1.59 and R_s of 14.6 MΩ. Although the n is close to that of conventional planar LEDs, the $I-V$ curve showed poor rectifying properties. EL was observed at around 1.51 eV at room temperature. The dashed line in Fig. 3.22d shows the PL spectrum from the same sample at room temperature. The luminescent peaks were the same as that of the EL peak, which means current injections occurred across the p–n junction in the GaAsP regions. The threshold current for the EL was 0.08 μA, but the bias voltage was 2.5 V. This is because of high contact resistance.

We designed the DH structure with multi-QWs for the CMS structure shown in Fig. 3.23a. Vertically aligned CMS NWs with 20-QWs were successfully grown on Si(111). The total diameter of the NWs was 180 nm, and the core NW diameter was 50 nm. The total thickness of the AlGaAs shell layer was 25 nm. The total thickness of the 20-QWs was 40 nm. The moderate rectifying properties of the AlGaAs/20-QWs/AlGaAs CMS NW-based LED array on Si are evident in Fig. 3.23c. The n for the device was 1.93 and R_s was 50 kΩ. Such high resistance was thought to be from

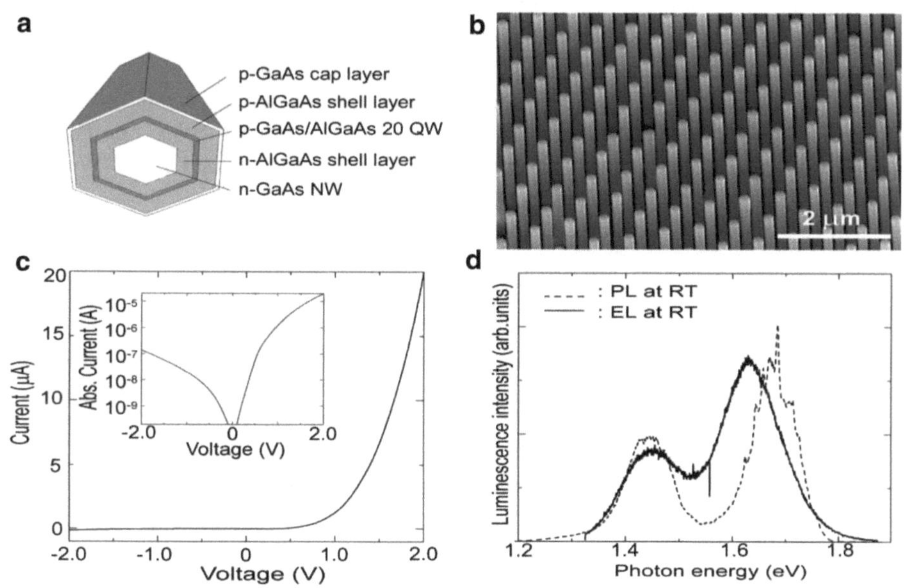

Fig. 3.23 (**a**) Illustration of cross section of AlGaAs/20-QWs/AlGaAs CMS NW. (**b**) Growth results of the vertical CMS NWs. The total diameter is 180 nm. The diameter of the core GaAs NW is 50 nm. The AlGaAs shell layer is 25 nm and the thickness of the QW is 40 nm. (**c**) $I-V$ property for the CMS NWs. (**d**) EL (*solid line*) and PL (*dashed line*) spectra of the CMS NWs

band offsets in GaAs/AlGaAs multi-QW layers. The EL peaks were observed at 1.45 and 1.65 eV. The former was due to the GaAs QW layers and the latter was due to the AlGaAs barriers inside the multi-QWs. The threshold current was 20 μA at 2.0 V. The dashed solid line in Fig. 3.23d shows the PL spectra at room temperature. The PL spectra showed small oscillation due to the cavity effect in multi-QW layers, like Bragg reflectors at around 1.65 eV, which means these structures have the potential to confine photons inside the NW. Further improvements in heat management, refractive index, and contact resistance are required for LD applications.

3.6 Summary

In this chapter, we reviewed SA-MOVPE for III–V nanowire (NW) growth and heteroepitaxy of III–V NW on Si substrate for optical applications. The crystal growth of SA-MOVPE is based on faceting growth without a catalyst. This growth technique enabled the position-controlled growth of vertically aligned III–V NWs on lithography-patterned substrates. Also, growth temperature altered axial NW growth direction and radial growth directions, and resulted in formation of CMS NWs. Moreover, the nanometer-scale footprint and precise control of the initial surface of the SA-MOVPE achieved growth of III–V NWs on Si substrate.

For the growth, modification of nonpolar Si(111) surface into (111)B-oriented surface and low-temperature growth were important for controlling the growth directions of the III–V NWs. The initial Si surface should be cleaned at high temperature to remove native oxides and to form a Si(111) 1×1 reconstructed surface. For controlling the growth directions of III–V NWs, flow-rate modulation epitaxy at low temperature was effective for aligning the InAs NWs. On the other hand, low-temperature growth was important for growing vertical GaAs NWs on Si substrate.

The use of the CMS structure and these III–V NWs growth on Si enabled the fabrication of CMS NW-based LEDs on Si substrate. Vertically aligned AlGaAs/GaAs/AlGaAs double-heterostructured CMS and GaAsP/GaAs CMS NWs with radial pn junctions were successfully fabricated on Si(111) substrate. Then, NW LEDs using these CMS NWs showed moderate rectifying properties and electroluminescence with wavelengths of 820–870 nm regardless of lattice mismatch in the GaAs/Si system. These NW LEDs at the NIR could enable new approaches to Si photonics because nanometer-scaled light sources at the NIR are feasible for Si-based APDs to replace Cu-based intrachip connections with high-performance optical interconnections in a small area.

We still have challenges for future Si photonics using III–V NW-based optical devices on Si. We have to develop the technologies in this section into electrically driven NW-based laser diodes (LDs) and III–V NW-based photodiodes and APDs. Device structures optimized for cavity effect of NWs, as following in Chap. 8, are required to realize NW-based LDs. Moreover, new concepts such as metal cavity and surface plasmonics [154, 155] should be considered for applications. After optimization and realization of NW-based LDs and APDs on Si platforms, these NW-based optical applications could open new fields in Si-CMOS technology with optical interconnection.

Acknowledgements This work was financially supported by a Grant-in-Aid for Scientific Research from the Ministry of Education, Culture, Sports, Science and Technology (MEXT). The authors thank Prof. Junichi Motohisa, Prof. Shinjiroh Hara, and Prof. Kenji Hiruma for fruitful discussions. Also, we would like to acknowledge Dr. Premila Mohan, Dr. Junichiro Takeda, Dr. Yang Lin, Dr. Ying Ding, Mr. Masatoshi Yoshimura, Mr. Yasunori Kobayashi, Mr. Tomotaka Tanaka, and Mr. Hajime Goto.

References

1. Y. Arakawa, H. Sakaki, Appl. Phys. Lett. **40**, 939 (1982)
2. Y. Arakawa, H. Sakaki, M. Nishioka, H. Okamoto, N. Miura, Jpn. J. Appl. Phys. **22**, L804 (1983)
3. K. Hiruma, M. Yazawa, T. Katsuyama, K. Ogawa, K. Haraguchi, M. Koguchi, H. Kakibayashi, J. Appl. Phys. **77**, 447 (1995)
4. Y. Huang, X. Duan, Y. Cui, L.J. Lauhon, K.-Ha Kim, C.M. Lieber, Science **294**, 1313 (2001)
5. M.H. Huang, S. Mao, H. Feick, H. Yan, Y. Wu, H. Kind, E. Weber, R. Russo, P. Yang, Science **292**, 1897 (2001)
6. M.S. Gudiksen, L.J. Lauhon, J. Wang, D.C. Smith, C.M. Lieber, Nature **415**, 617 (2002)
7. L.J. Lauhon, M.S. Gudiksen, D. Wang, C.M. Lieber, Nature **420**, 57 (2002)
8. J.C. Johnson, H-J. Choi, K.P. Knutsen, R.D. Schaller, P. Yang, R.J. Saykally, Nat. Mater. **1**, 106 (2002)
9. R.S. Wagner, W.C. Ellis, Appl. Phys. Lett. **4**, 89 (1964)
10. E.I. Givargizov, J. Cryst. Growth **31**, 20 (1975)
11. K. Hiruma, T. Katsuyama, K. Ogawa, M. Koguchi, H. Kakibayashi, G.P. Morgan, Appl. Phys. Lett. **59**, 431 (1991)
12. Z.X. Yan, A.G. Milnes, J. Electrochem. Soc. **129**, 1353 (1982)
13. A.F. Morral, C. Colombo, G. Abstreiter, J. Arbiol, J.R. Morante, Appl. Phys. Lett. **92**, 063112 (2008)
14. C. Colombo, D. Spirkoska, M. Frimmer, G. Abstreiter, A.F. i Morral, Phys. Rev. B **77**, 155326 (2008)
15. F. Qian, S. Gradeak, Y. Li, C-Y. Wen, C.M. Lieber, Nano Lett.. **5**, 2287 (2005)
16. X. Jiang, Q. Xiong, S. Nam, F. Qian, Y. Li, C.M. Lieber, Nano Lett.. **7**, 3214 (2007)
17. C.-Y. Wen, M.C. Reuter, J. Bruley, J. Tersoff, S. Kosambaka, E.A. Stach, F.M. Ross, Science **326**, 1247 (2009)
18. M.T. Björk, C. Thelander, A.E. Hansen, L.E. Jensen, M.W. Larsson, L.R. Wallenberg, L. Samuelson, Nano Lett. **4**, 1621 (2004)
19. J. Motohisa, J. Noborisaka, J. Takeda, M. Inari, T. Fukui, J. Cryst. Growth **272**, 180 (2004)
20. J. Noborisaka, J. Motohisa, T. Fukui, Appl. Phys. Lett. **86**, 213102 (2005)
21. K. Ikejiri, J. Noborisaka, S. Hara, J. Motohisa, T. Fukui, J. Cryst. Growth **298**, 616 (2007)
22. K. Ikejiri, T. Sato, H. Yoshida, K. Hiruma, J. Motohisa, S. Hara, T. Fukui, Nanotechnology **19**, 265604 (2008)
23. P. Mohan, J. Motohisa, T. Fukui, Nanotechnology **16**, 2903 (2005)
24. Y. Kitauchi, Y. Kobayashi, K. Tomioka, S. Hara, K. Hiruma, T. Fukui, J. Motohisa, Nano Lett. **10**, 1699 (2010)
25. K. Tomioka, P. Mohan, J. Noborisaka, S. Hara, J. Motohisa, T. Fukui, J. Cryst. Growth **298**, 644 (2007)

26. M. Akabori, J. Takeda, J. Motohisa, T. Fukui, Nanotechnology **14**, 1071 (2003)
27. T. Sato, J. Motohisa, J. Noborisaka, S. Hara, T. Fukui, J. Cryst. Growth **310**, 2359 (2008)
28. T. Sato, Y. Kobayashi, J. Motohisa, S. Hara, T. Fukui, J. Cryst. Growth **310**, 5111 (2008)
29. M. Yoshimura, K. Tomioka, K. Hiruma, S. Hara, J. Motohisa, T. Fukui, Jpn. J. Appl. Phys. **49**, 04DH08 (2010)
30. S. Fujisawa, T. Sato, S. Hara, J. Motohisa, K. Hiruma, T. Fukui, Jpn. J. Appl. Phys. (in press)
31. H. Sekiguchi, K. Kishino, A. Kikuchi, Appl. Phys. Lett. **96**, 231104 (2010)
32. Y-J. Kim, C-H. Lee, Y.J. Hong, G-C. Yi, S.S. Kim, H. Cheong, Appl. Phys. Lett. **89**, 163128 (2006)
33. J. Noborisaka, J. Motohisa, S. Hara, T. Fukui, Appl. Phys. Lett. **87**, 093109 (2005)
34. P. Mohan, J. Motohisa, T. Fukui, Appl. Phys. Lett. **88**, 013110 (2006)
35. B. Hua, J. Motohisa, Y. Kobayashi, S. Hara, T. Fukui, Nano Lett. **9**, 112 (2009)
36. P. Mohan, J. Motohisa, T. Fukui, Appl. Phys. Lett. **88**, 133105 (2006)
37. L. Yang, J. Motohisa, J. Takeda, K. Tomioka, T. Fukui, Nanotechnology **18**, 105302 (2007)
38. J.N. Shapiro, A. Lin, P.S. Wong, A.C. Scofield, C. Tu, P.N. Senanayake, G. Mariani, B.L. Liang, D.L. Huffaker, Appl. Phys. Lett. **97**, 243102 (2010)
39. H. Sasakura, K. Humano, I. Suemune, J. Motohisa, Y. Kobayashi, M. van Kouwen, K. Tomioka, T. Fukui, N. Akopian, V. Zwiller, J. Phys. Conf. Ser. **193**, 012132 (2009)
40. A. Hayashida, T. Sato, S. Hara, J. Motohisa, K. Hiruma, T. Fukui, J. Cryst. Growth **312**, (2010) 3592
41. B. Hua, J. Motohisa, Y. Ding, S. Hara, T. Fukui, Appl. Phys. Lett. **91**, 131112 (2007)
42. Y. Ding, J. Motohisa, B. Hua, S. Hara, T. Fukui, Nano Lett. **7**, 3598 (2007)
43. S.F. Yu, C. Yuen, S.P. Lau, W.I. Park, G-C. Yi, Appl. Phys. Lett. **84**, 3241 (2004)
44. L.K. van Vugt, S. Ruhle, D. Vanmaekelbergh, Nano Lett. **6**, 2707 (2006)
45. B. Hua, J. Motohisa, Y. Kobayashi, S. Hara, T. Fukui, Nano Lett. **9**, 112 (2009)
46. Y. Takagi, H. Yonezu, K. Samonji, T. Tsuji, N. Ohshima, J. Cryst. Growth **187**, 42 (1998)
47. T. Nishinaga, T. Nakano, S. Zhang, Jpn.J. Appl. Phys. **27**, L964 (1988)
48. M. Akiyama, Y. Kawarada, K. Kaminishi, Jpn. J. Appl. Phys. 23, L843 (1984)
49. T. Egawa, T. mooku, H. Ishikawa, K. Ohtsuka, T. Jimbo, Jpn. J. Appl. Phys. **41**, L663 (2002)
50. C.P.T. Svensson, T. Martensson, J. Tragardh, C. Larsson, M. Rask, D. Hessman, L. Samuelson, J. Ohlsson, Nanotechology **19**, 305201 (2008)
51. K. Tomioka, J. Motohisa, S. Hara, K. Hiruma, T. Fukui, Nano Lett. **10**, 1639 (2010)
52. L.C. Chuang, F.G. Sedgwick, R. Chen, W.S. Ko, M. Moewe, K.W. Ng, T.-T.D. Tran, C. C-Hasnain, Nano Lett. **11**, 385 (2011)
53. S.J. An, J.H. Chae, G-C. Yi, G.H. Park, Appl. Phys. Lett. **92**, 121108 (2008)
54. C.-H. Lee, J. Yoo, Y.J. Hong, J. Cho, Y.-J. Kim, S.-R. Jeon, J.H. Baek, G-C. Yi, Appl. Phys. Lett. **94**, 213101 (2009)
55. E. Lai, W. Kim, P. Yang, Nano Res. **1**, 123 (2008)
56. H.-W. Lin, Y.-J. Lu, H.-Y. Chen, H.-M. Lee, S. Gwo, Appl. Phys. Lett.**97**, 073101 (2010)
57. M.T. Borgström, V. Zwiller, E. Muller, A. Imamoglu, Nano Lett. **5**, 1439 (2005)
58. H. Goto, K. Nosaki, K. Tomioka, S. Hara, K. Hiruma, J. Motohisa, T. Fukui, Appl. Phys. Exp. **2**, 035004 (2009)
59. B. Jalali, S. Fathpour, J. Light. Tech. **24**, 4600 (2006)
60. H. Rong, S. Xu, Y.-H. Kuo, V. Sih, O. Cohen, O. Raday, M. Panicca, Nat Photonics **1**, 232 (2007)
61. Y. kang, H.-D. Liu, M. Morse, M.J. Paniccia, M. Zadka, S. Litski, G. Sarid, A. Pauchard, Y.-H. Kuo, H.-W. Chen, W.S. Zaoui, J.E. Bowers, A. Beling, D.C. McIntosh, X. Zheng, J.C. Campbell, Nat Photonics **3**, 59 (2009)
62. B. Corcoran, C. Monat, C. Grillet, D.J. Moss, B.J. Eggleton, T.P. White, L. O'Faolain, T.F. Krauss, Nat Photonics **3**, 206 (2009)
63. B.D. Joyce, J.A. Baldrey, Nature **195**, 485 (1962)
64. F.W. Tausch, A.G. Lapierre, J. Electrochem. Soc. **112**, 706 (1965)
65. P. Rai-Choudhury, J. Electrochem. Soc. **116**, 1745 (1969)

66. S.H. Jones, K.M. Lau, J. Electrochem. Soc. **134**, 3149 (1987)
67. T. Fukui, S. Ando, Y.K. Fukai, Appl. Phys. Lett. **57**, 1209 (1990)
68. T. Fukui, S. Ando, Y. Tokura, T. Toriyama, Appl. Phys. Lett. **58**, 2018 (1991)
69. K. Kumakura, K. Nakakoshi, J. Motohisa, T. Fukui, H. Hasegawa, Jpn. J. Appl. Phys. **34**, 4387 (1995)
70. F. Nakajima, Y. Miyoshi, J. Motohisa, T. Fukui, Appl. Phys. Lett. **83**, 2680 (2003)
71. Y. Miyoshi, F. Nakajima, J. Motohisa, T. Fukui, Appl. Phys. Lett. **87**, 033501 (2005)
72. S. Ando, N. Kobayashi, H. Ando, J. Cryst. Growth **145**, 302 (1994)
73. T. Hamano, H. Hirayama, Y. Aoyagi, Jpn. J. Appl. Phys. **36**, L286 (1997)
74. G. Wulff, Z. Kristallogr **34**, 449 (1901)
75. C.H. Li, Y. Sun, D.C. Law, S.B. Visbeck, R.F. Hicks, Phys. Rev. B **68**, 085320 (2003)
76. K. Sladek, V. Klinger, J. Wensorra, M. Akabori, H. Hardtdegen, D. Grützmacher, J. Cryst. Growth **312**, 65 (2010)
77. B.J. Skromme, C.J. Sandroff, E. Yablonovitch, T. Gmitter, Appl. Phys. Lett. **51**, 2022 (1987)
78. K. Tomioka, Y. Kobayashi, J. Motohisa, S. Hara, T. Fukui, Nanotechnology **20**, 145302 (2009)
79. T. Mårtensson, C. Patrik, T. Svensson, B.A. Wacasar, M.W. Larsson, W. Seifert, K. Deppert, A. Gustafsson, L.R. Wallenberg, L. Samuelson, Nano. Lett **4**, 1987 (2004)
80. Y. Watanabe, H. Hibino, S. Bhunia, K. Tateno, T. Sekiguchi, Physica E **24**, 133 (2004)
81. V. Khorenko, I. Regolin, S. Neuman, W. Prost, F.-J. Tegude, H. Wiggers, Appl. Phys. Lett. **85**, 6407 (2004)
82. E.P.A.M. Bakkers, J.A. van Dam, S.D. Franceschi, L.P. Kouwenhoven, M. Kaiser, M. Verheijen, H. Wondergem, P. van der Sluis, Nat Mater. **3**, 769 (2004)
83. A. L Roest, M. A Verheijen, O. Wunnicke, S. Serafin, H. Wondergem, E.P.A.M. Bakkers, Nanotechnology **17**, S271 (2006)
84. K. Tateno, H. Hibino, H. Gotoh, H. Nakano, Appl. Phys. Lett. **89**, 033114 (2006)
85. M. Mattila, T. Hakkatainen, H. Lipsanen, Appl. Phys. Lett. **89**, 063119 (2006)
86. B. Mandl, J. Stangl, T. Mårtensson, A. Mikkelsen, J. Eriksson, L.S. Karlsson, G. Bauer, L. Samuelson, W. Seifert, Nano Lett. **6**, 1817 (2006)
87. S.S. Yi, G. Girolami, J. Amano, M.S. Islam, I. Kimukin, Appl. Phys. Lett. **89**, 133121 (2006)
88. H.D. Park, S.M. Prokes, M.W. Twigg, R.C. Cammarata, A.-C. Gaillot, Appl. Phys. Lett. **89**, 223125 (2006)
89. P.K. Mohseni, C. Maunders, G.A. Botton, R.R. LaPierre, Nanotechnology **18**, 445304 (2006)
90. S.-G. Ihn, J.-I. Song, Y.-H. Kim, J.Y. Lee, I.-H. Ahn, IEEE Trans. Nanotech. **6**, 384 (2007)
91. T. Mårtensson, J.B. Wagner, E. Hilner, A. Mikkelsen, C. Thelander, J. Stangl, B.J. Ohlsson, A. Gustafsson, E. Lundgren, L. Samuelson, W. Seifert, Adv. Mat. **19**, 1801 (2007)
92. S.-G. Ihn, J.-I. Song, Nanotechnology **18**, 355603 (2007)
93. L.C. Chuang, M. Moewe, C. Chase, N.P. Kobayashi, C. Chang-Hasnain, S. Crankshaw, Appl. Phys. Lett. **90**, 043115 (2007)
94. S.-G. Ihn J.-I. Song, Nano Lett. **7**, 39 (2007)
95. J.H. Paek, T. Nishiwaki, M. Yamaguchi, N. Sawaki, Phys. Status Solidi C **5**, 2740 (2008)
96. F. Jabeen, V. Grillo, S. Rubini, F. Martelli, Nanotechnology **19**, 27511 (2008)
97. K. Tomioka, j. Motohisa, S. Hara, T. Fukui, *Nano Lett.* **8**, 3475 (2008)
98. X.-Y. Bao, C. Soci, D. Susac, J. Bratvold, D.P.R. Aplin, W. Wei, C.-Y. Chen, S.A. Dayeh, K.L. Kavanagh, D. Wang, Nano Lett. **8**, 3755 (2008)
99. K.A. Dick, K. Deppert, L. Samuelson, L.R. Wallenberg, F.M. Ross, Nano Lett. **8**, 4087 (2008)
100. R.L. Woo, R. Xiao, Y. Kobayashi, L. Gao, N. Goel, M.K. Hudait, T.E. Mallouk, R.F. Hicks, Nano Lett. **8**, 4664 (2008)
101. F. Jabeen, V. Grillo, S. Rubini, F. Martelli, Nanotechnology **19**, 27511 (2008)
102. L.C. Chuang, M. Moewe, S. Crankhaw, C. Chang-Hasnain, Appl. Phys. Lett. **92**, 013121 (2008)
103. G. Zhang, K. Tateno, T. Sogawa, H. Nakano, J. Appl. Phys. **103**, 014301 (2008)
104. G. Zhang, K. Tateno, T. Sogawa, H. Nakano, Appl. Phys. Exp. **1**, 064003 (2008)

105. C. Rehnstedt, T. Mårtensson, C. Thelander, L. Samuelson, L.-E. Wernersson, IEEE Trans. Electron Dev. **55**, 3037 (2008)
106. M. Moewe, L.C. Chuang, S. Cranlshaw, C. Chase, C. Chang-Hasnain, Appl. Phys. Lett. **93**, 023116 (2008)
107. S.T. Boles, C.V. Thompson, E.A. Fitzgerald, J. Cryst. Growth **311**, 1446 (2009)
108. H. Detz, P. Klang, A.M. Adnrews, A. Lugstein, M. Steinmair, Y.J. Hyun, E. Bertagnolli, W. Schrenk, G. Strasser, J. Cryst. Growth **311**, 1859 (2009)
109. E. Cirlin, V.G. Dubrovskii, I.P. Soshnikov, N.V. Sibirev, Y.B. Sammsonenko, A.D. Bouravleuv, J.C. Harmand, F. Glas, Phys. Status Solidi RRL **3**, 112 (2009)
110. W. Wei, X.-Y. Bao, C. Soci, Y. Ding, Z.L. Wang, D. Wang, Nano Lett. 9, 2926 (2009)
111. Z. Zhao, K. Yadavalli, Z. Hao, K.L. Wang, Nanotechnology **20**, 035304 (2009)
112. S. Roddaro, P. Caroff, G. Bisasiol, F. Rossi, C. Bocci, K. Nilsson, L. Fröberg, J.B. Wagner, L. Samuelson, L.-E. Wernersson, L. Sorba, Nanotechnology **20**, 285303 (2009)
113. M. Cantoro, G. Brammertz, O. Richard, H. Bender, F. Clemente, M. Lays, S. Degroode, M. Caymax, M. Hwyns, S.D. Gendt, J. Electrochem. Soc. **156**, H860 (2009)
114. J.H. Kang, Q. Gao, H.J. Joyce, H.H. Tan, C. Jagadish, Y. Kim, D.Y. Choi, Y. Guo, H. Xu, J. Zou, M.A. Fickenscher, L.M. Smith, H.E. Jackson, J.M. Yarrison-Rice, Nanotechnology **21**, 035604 (2010)
115. H. Huang, X. Ren, X. Ye, J. Guo, Q. Wang, Y. Yang, S. Cai, Y. Huang, Nano Lett. **10**, 64 (2010)
116. C. Kallesoe, K. Molhave, K.F. Larsen, D. Engstrom, T.M. Hansen, P. Boggld, T. Mårtensson, M. Borgström, L. Samuelson, J. Vac. Sci. Technol. B **28**, 21 (2010)
117. S. Yu, G. Miao, Y. Jin, m L. Zhang, H. Song, H. Jiang, Z. Li, D. Li, X. Sun, Physica E **42**, 1540 (2010)
118. G. Zhang, K. Tateno, H. Gotoh, T. Sogawa, H. Nakano, Jpn. J. Appl. Phys. **49**, 015001 (2010)
119. T. Tanaka, K. Tomioka, S. Hara, J. Motohisa, E. Sano, T. Fukui, Appl. Phys. Exp. **3**, 025003 (2010)
120. G. Statkute, A.G. Nasibulin, M. Sopanen, T. Hakkarainen, E. Kauppinen, H. Lipsanen, Jpn. J. Appl. Phys. **49**, 020213 (2010)
121. S.-G. Ihn, M.-Y. Ryu, J.-I. Song, Solid State Com. **150**, 729 (2010)
122. S. Pissard, K.A. Dick, X. Wallart, P. Caroff, Appl. Phys. Lett. **96**, 121901 (2010)
123. G. Koblmüller, S. Hertenberger, K. Vizbaras, M. Bichler, F. Bao, J.-P. Zhang, G. Abstreiter, Nanotechnology **21**, 365602 (2010)
124. S. Schulz, M. Schwartz, A. Kuczkowski, W. Assenmacher, J. Cryst. Growth **312**, 1475 (2010)
125. G.E. Cirlin, V.G. Dubrovskii, Y.B. Samsonenko, A.D. Bouravleuv, K. Durose, Y.Y. Proskuryakov, B. Mendes, L. Bowen, M.A. Kaliteevski, R.A. Abram, D. Zeze, Phys. Rev. B **82**, 035302 (2010)
126. S. Breuer, M. Hilse, A. Trampert, L. Geelhaar, H. Riechert, Phys. Rev. B **82**, 0775406 (2010)
127. J. Tatebayashi, A. Lin, P.S. Wong, R.F. Hick, D.L. Huffaker, J. Appl. Phys. **108**, 034315 (2010)
128. S. Plissard, K.A. Dick, G. Larrieu, S. Godey, A. Addad, X. Wallart, P. Caroff, Nanotechnology **21**, 385602 (2010)
129. M.T. Björk, H. Schmid, C.D. Bessire, K.E. Moselund, H. Ghoneim, S. Karg, E. Lörtscher, H. Riel, Appl. Phys. Lett. **97**, 163501 (2010)
130. B. Mandl, J. Stangl, E. Hilner, A.A. Zakharov, K. Hillerich, A.W. Dey, L. Samuelson, G. Bauer, K. Deppert, A. Mikkelsen, Nano Lett. **10**, 4443 (2010)
131. P. Krogstrunp, R. P-Biro, E. Johnson, M.H. Madsen, J. Nygård, H. Shtrinkman, Nano Lett. **10**, 4475 (2010)
132. K. Tomioka, T. Fukui, Appl. Phys. Lett. **98**, 083114 (2011)
133. A. Kikuchi, M. Kawai, M. Tada, K. Kishino, Jpn. J. Appl. Phys. **43**, L1524 (2004)
134. E. Ertekin, P.A. Greaney, D.C. Chrzan, T.D. Sands, J. Appl. Phys. **97**, 114325 (2005)
135. F. Glas, Phys. Rev. B **74**, 121302 (2006)
136. K. Takayanagi, Y. Tanishiro, S. Takahashii, M. Takahashi, Surf. Sci. **164**, 367 (1985)

137. T. Hoshino, K. Kokubun, H. Fujiwara, K. Kumamoto, T. Ishimaru, I. ohdomari, Phys. Rev. Lett. **75**, 2372 (1995)
138. M. Hoshino, Y. Shigeta, K. Ogawa, Y. Homma, Surf. Sci. **365**, 29 (1996)
139. M. Chiba, Y. Tanishiro, H. Minoda, K. Yagi, Surf. Sci. **441**, 179 (1999)
140. F.J. Himpsel, F.R. McFeely, A. Taleb-Ibrahimi, J.A. Yarmoff, G. Hollinger, Phys. Rev. B **38**, 6084 (1988)
141. T. Hattori, T. Aiba, E. Iijima, Y. Okube, H. Nohira, N. Tate, M. Katayama, Appl. Surf. Sci. **104–105**, 323 (1996)
142. M.A. Olmstead, R.D. Bringans, R.I.G. Uhrberg, R.Z. Bachrach, Phys. Rev. B **34**, 6041 (1986)
143. E.C. Garnett P. Yang, J. Am. Chem. Soc. **130**, 9224 (2008)
144. C.T. Sah, R.N. Noyce, W. Shockley, Proc. IRE **45**, 1228 (1957)
145. H.C. Casey, J. Muth, S. Krishnankutty, J.M. Zavada, Appl. Phys. Lett. **68**, 2867 (1996)
146. P. Perlin, M. Osinski, P.G. Eliseev, V.A. Smagley, J. Mu, M. Banas, P. Sartori, Appl. Phys. Lett. **69**, 1680 (1996)
147. A.D. Katnani G. Margaritondo, Phys. Rev. B **28**, 1944 (1983)
148. Y.C. Zhou, Z.H. Zhu, D. Crouse, Y.H. Lo, Appl. Phys. Lett **73**, 2337 (1998)
149. S.K. Ray, K.M. Groom, H.Y. Liu, M. Hopkinson, R.A. Hogg, Jpn. J. Appl. Phys. **45**, 2542 (2006)
150. R.L. Hartman A.R. Hartman, Appl. Phys. Lett. **23**, 147 (1973)
151. H. Yoneze, M. Ueno, T. Kamejima, I. Sakuma, Jpn. J. Appl. Phys. **13**, 835 (1974)
152. R. Ito, H. Nakashima, O. Nakada, Jpn. J. Appl. Phys. **13**, 1321 (1974)
153. S. Yamakoshi, O. Hasegawa, H. Hamaguchi, M. Abe, T. Yamaoka, Appl. Phys. Lett. **31**, 627 (1977)
154. S.-H. Kwon, J.-H. Kang, C. Seassal, S.-K. Kim, P. Regreny, Y.-H. Lee, C.M. Lieber, H.-G. Park, Nano Lett. **10**, 3679 (2010)
155. H. Im, N.C. Lindquist, A. Lesuffleur, S.-H. Oh, ACS Nano **4**, 947 (2010)

Chapter 4
Synthesis and Properties of Aluminum Nitride Nanostructures

Daniel S.P. Lau and X.H. Ji

Abstract Aluminum nitride (AlN) is a wide band gap semiconductor that has potential applications in both deep ultraviolet (UV) optical devices and high-power electronics devices. We review the recent development of one-dimensional (1D) AlN nanostructures. Various synthesis strategies for AlN nanostructures are presented. We pay particular attention in the doping of AlN nanostructures. The magnetic and optical properties of the AlN nanostructures are studied in detail. The future prospect of 1D AlN nanostructures for deep UV light-emitting diodes is presented.

4.1 Introduction

4.1.1 Overview

Since the discovery of carbon nanotubes (CNTs) [1], one-dimensional (1D) semiconductor nanomaterials with controlled dimensions and morphology have attracted great interest from chemists, because of their unique optical, electrical, and magnetic properties and the possibility of fabricating nanodevices. The tremendous interest in nanoscale structures stems from their size-dependent properties. Nanostructured semiconductor materials are of particular interest because of their wide range of

D.S.P. Lau (✉)
Department of Applied Physics, The Hong Kong Polytechnic University, Hung Hom, Kowloon, Hong Kong, P.R. China
e-mail: apsplau@polyu.edu.hk

X.H. Ji
School of Materials Science and Engineering, South China University of Technology, Wushan RD., Tianhe District, Guangzhou, P.R.China, 510641
e-mail: jxhong@scut.edu.cn

G.-C. Yi (ed.), *Semiconductor Nanostructures for Optoelectronic Devices*, NanoScience and Technology, DOI 10.1007/978-3-642-22480-5_4,

applications in areas such as photonics [2], electronics [3], nanoelectromechanical systems (NEMS) [4, 5], and the life science [6]. For 1D semiconductor nanomaterials, their peculiar and unique physical properties arising from quantum confinement, including electronic quantum transport and enhanced radiative recombination of carriers, excite their promising potentials in extensive applications and for fabricating short-wavelength nanolasers, field-effect transistors, ultrasensitive nanosized gas sensors, nanoresonators, transducers, actuators, nanocantilevers, and field emitters (FEs).

Aluminum nitride (AlN) is a III–V compound with a hexagonal wurtzite crystal structure that has gained increasing interest for electronics, piezoelectric, and photoelectric applications. It has the largest band gap in group III nitrides, which makes it an essential material when pursuing even shorter wavelength of light emission devices. AlN-based light-emitting diodes (LEDs) with a PIN (p-type/intrinsic/n-type) and MIS (metal–insulator–semiconductor) structure at an emission wavelength of 210 nm have been demonstrated recently [7]. AlN-based AlN/GaN/InN compounds allow covering the UV band from UVA (320–400 nm) and UVB (280–320 nm) all the way into the UVC range (100–280 nm), which makes it promising for the application in deep UV LEDs, UV lasers, etc. [8–10]. AlN is also attractive for field emission applications because of its low-electron affinity [11] and thermal stability [12]. In addition, AlN is desirable for its chemical and thermal stability, mechanical strength, and compatibility to silicon and other group III nitrides, which makes it adaptable to a large variety of environments and enables large freedom in its device fabrication, such as in electrical packaging and in composites [13].

This chapter reviews current research activities that focus on the controlled growth of 1D AlN nanostructures by various methods. The morphologies, sizes, compositions, and microstructures controlled by catalysis, doping, reactant, and temperature are discussed. Moreover, structural properties in terms of Raman shifting, optical properties, and ferromagnetic properties of AlN nanostructure are also discussed. We conclude this chapter with some perspectives on the future research directions in this promising material.

4.1.2 Properties of AlN

Aluminum nitride is one of the most versatile III–V compounds. It has a hexagonal crystal structure and is a covalently bonded material with space grouping of P63mc, consisting of two interpenetrating hexagonal closely packed sublattices, each with one type of atoms (Al and N), offset along the c-axis by 0.385 of the cell height (0.385c), as shown in Fig. 4.1 [14]. The wurtzite AlN crystal can be regarded as stacked [AlN4] tetrahedral units. The N and Al atoms are stacked in ABABAB sequence, which leads to a high polarity along the [001] direction, making AlN possible to form 1D nanostructures [Fig. 4.1b]. AlN possesses a large energy band gap of about 6.2 eV, high thermal conductivity

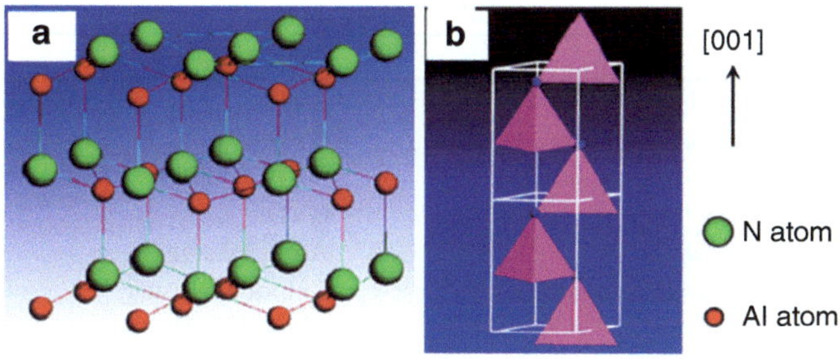

Fig. 4.1 (**a**) Hexagonal closely-packed structure of wurtzite AlN. (**b**) [AlN4] tetrahedral units stacking along the [001] direction

[$\sim$280 W/mK at 300 K], high breakdown voltage (15 kV/mm), high electrical resistivity ($10^9 \sim 10^{11}\Omega$m), high mechanical properties (hardness of 20 GPa), good thermal expansion coefficient [$(2.9 \sim 3.4) \times 10^{-6}\,K^{-1}$], fast sound propagation ($\sim$9 km/s), relatively high piezoelectric constants, and very low values of electron affinity (0.25 eV) [15–20]. It is stable to very high temperatures in inert atmospheres. But surface oxidation occurs at above 700°C in air. A layer of aluminum oxide protects the material from $\sim$700°C up to 1,370°C, at which bulk oxidation occurs. Furthermore, AlN keeps its mechanical and piezoelectric properties at temperatures above 1,000°C [21–24]. Therefore, AlN is widely used in many applications [25–28].

4.2 Synthesis of AlN Nanostructures

During the growth of nanostructured materials, the evolution of a solid from a vapor, a liquid, or a solid phase involves nucleation and growth. As the concentration of the building units (atoms, ions, or molecules) of a solid becomes sufficiently high, they aggregate into small nuclei or clusters through homogeneous nucleation. These clusters serve as seeds for further growth to form larger clusters. Vapor-phase growth is a considerably popular and extensively used method for the synthesis of 1D nanostructures, such as nanowires, nanobelts, nanorods, nanoneedles, and whiskers. Control of the supersaturation level determines the structural growth morphology. Preparation of AlN nanostructures mainly relies on two synthetic processes: a vapor-assisted route and a template-originated method. Among the vapor-phase growth methods, vapor–liquid–solid (VLS) and vapor–solid (VS) growth are discussed in the text. Table 4.1 gives selected AlN nanostructures synthesized by various methods.

As one of the most popular growth methods, chemical vapor deposition (CVD) is a well-established and reliable method for the synthesis of AlN nanostructures. In principle, the vapor-phase deposition technique is a simple process in which condensed or powder materials are vaporized at an elevated temperature and the

Table 4.1 Selected AlN nanostructures synthesized by various methods

	Morphology	Fabrication method	References
Undoped AlN	Nanotips	CVD	[29]
	Nanoneedle, nanorods	CVD	[30]
	Nanowires, nanorods	HVPE	[31]
	Nanowisker	CVD	[32]
	Nanowires	Arc discharge	[33, 34]
	Nanorods, nanoneedles	Chloride-assisted	[35, 36]
	Nanowire	Chloride-assisted	[37]
	Nanotips	Metal-catalyzed CVD	[38, 39]
	Nanocones	Ni-catalyzed chloride-assisted	[40]
	Nanonecklace	Metal-catalyzed CVD	[41]
	Hierarchical nanostructures	Ni-catalyzed CVD	[42]
	Nanowires	PVD	[43]
Doped AlN	Tb-doped nanorods	RF sputtering	[44]
	Co-doped nanorod arrays	Arc discharge	[45]
	Sc-doped sixfold-symmetrical Hierarchical nanostructures	Arc discharge	[46]
	Y-doped nanorods	Arc discharge	[47]
	Mn-doped nanowires	Au-catalyzed CVD	[48]
	Fe/Cu-doped nanorods	Chloride-assisted CVD	[49, 50]
	Si-doped nanoneedles	Co-catalyzed chloride-assisted	[51]

resultant vapor phases condense under certain conditions (temperature, pressure, atmosphere, substrate, etc.) to form the desired products. In the following section, the AlN nanostructures synthesized via VLS and VS growth are discussed in detail.

4.2.1 Vapor–Liquid–Solid Growth

The growth of AlN nanostructures via a gas-phase reaction involving the VLS process has been widely studied. According to this mechanism, the anisotropic crystal growth is promoted by the presence of the liquid alloy/solid interface.

4.2.1.1 Catalytic VLS

Shi and coworkers synthesized uniform AlN nanotips (AlNNTs) by CVD using silicon substrate covered with a metal layer (Au, Pt, and Al) as a catalyst [39]. The following is the experimental process in brief: an aluminum oxide boat, carrying the metal-coated silicon substrate and pure aluminum powder, was placed inside a quartz tube. The Al powder was kept upstream and the metal-coated Si substrate was placed upside down and downstream with respect to the flow of the nitriding

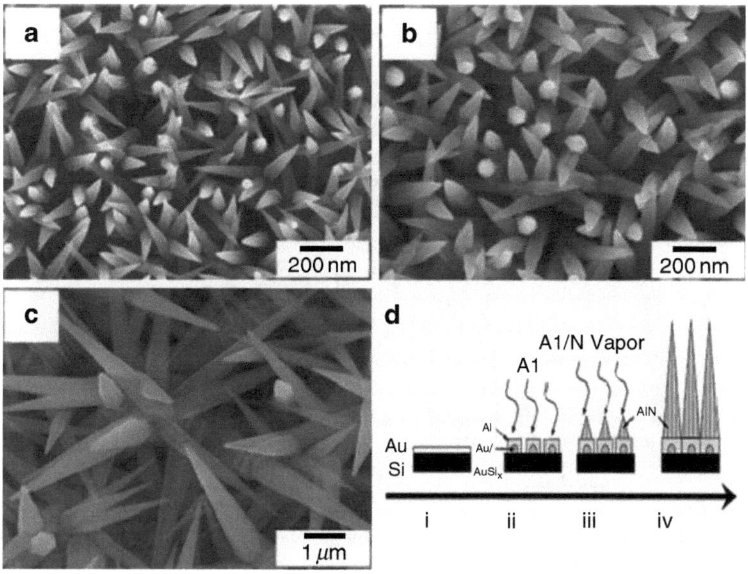

Fig. 4.2 Typical SEM images of the AlN nanotips grown with an Au layer of (**a**) 7 nm, (**b**) 15 nm, and (**c**) 50 nm thickness of Si. (**d**) Schematic diagram of the growth mechanism of AlN nanotips. (i) Au layer coated on the Si substrate. (ii) Gold or gold silicide nanoparticles form as the nucleation sites for the subsequent aluminum deposition. (iii) Al and N are absorbed on the nucleation sites bringing about the initial growth of AlN nanotips. (iv) AlN nanotips elongate with time when the reaction temperature is kept at 950°C [39]

gas, NH_3. During the growth process, the furnace temperature was ramped up to 950°C and maintained for 30 min with NH_3 flow rate of 30 sccm. Figure 4.2a shows the SEM images of AlNNTs on Si substrate with Au layer of various thicknesses. The morphology (base diameter, apex, and length) of the AlNNTs depended on the thickness of the metal layer. The growth mechanism of these AlNNTs was proposed in Fig. 4.2b. As the reaction temperature was ramped up from room temperature to 950°C, droplets of catalyst metals or their respective silicide phases were formed, resulting in the subsequent AlNNTs growth following VLS mechanism.

Besides Au, Ni was frequently used as catalytic metal as well [41]. It was demonstrated that at elevated temperature, Ni atoms on the silicon wafer aggregate to form small clusters/droplets, and Al vapor and N species continually dissolve into these Ni droplets to produce AlN species. When AlN species reach a supersaturated level in a droplet, AlN crystal grows out from the supersaturated alloy to form 1D nanowire along the direction having the lowest interface energy followed by VLS mechanism. Wu's group [52] synthesized aligned AlN nanowires under the confinement of anodic porous alumina template (APAT) reactions. The $Ni(NO_3)_2$ was adsorbed on the pore wall of APAT by dipping into the ethanol solution of $Ni(NO_3)_2$ in advance. The trace of Ni species as catalyst was determined in favor of reaction of the Al vapor and NH_3/N_2 diffused into the channels of APAT to form

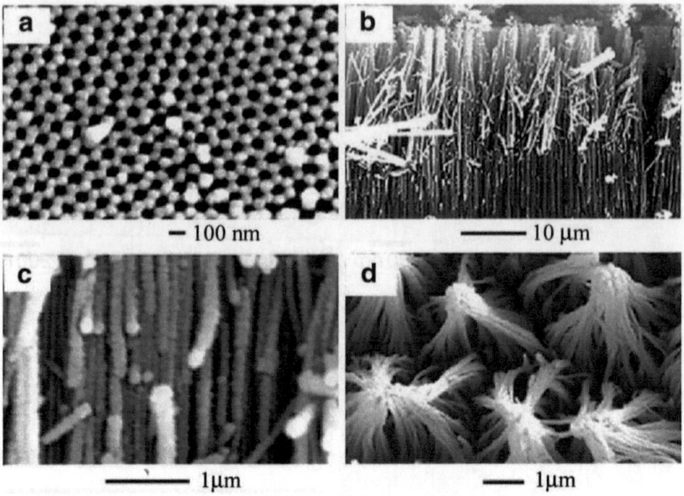

Fig. 4.3 (**a**) SEM image of the APAT fabricated by the two-step anodization (pore size ca.75 nm); (**b**) side view (SEM) of the as-prepared h-AlN nanowire array (length 30 μm); (**c**) enlargement of (**b**) (diameter ca.100 nm); and (**d**) top view of h-AlN nanowire array [52]

AlN species, as shown in Fig. 4.3. It was found that the aligned h-AlN nanowire arrays obtained were longer than 30 μm in length and approximately 100 nm in diameter, which reflects the pore diameter.

4.2.1.2 Self-Catalytic VLS

Nanostructured AlN can be grown using self-catalytic VLS mechanism; it is possible for Al or other element to serve as catalyst [39]. Chloride-assisted method is one of the examples. Haber et al. [53] firstly fabricated AlN nanowhiskers using $AlCl_3$ as a promoter; some improvements have been made to reduce the synthesis temperature by replacing Al with $AlCl_3$ as a precursor [54–56]. Yu et al. [57] prepared AlN nanostructures on silicon substrates by evaporating the mixture of Al powders and $NiCl_2$ in the temperature range of 680–800°C under NH_3/N_2 atmosphere. The involved chemistry reactions could be understood as follows:

$$3NiCl_2(s) + 2Al(s\ or\ l) \rightleftharpoons 2AlCl_3(g) + 3Ni(s)$$

$$AlCl_3(g) + 2Al(l) \rightleftharpoons 3AlCl(g)$$

It was demonstrated that the formation of the AlN nanostructures should start from the nitridation of Al droplets. When Ar was switched to NH_3/N_2 at reaction temperature, the Al droplets were nitrified to AlN particles, acting as the seeds for the subsequent epitaxial growth of AlN nanostructures. In their experiments, the same procedure but without the addition of $NiCl_2$ into Al powder was also performed for comparison. It was noted that nothing could deposit on the Si substrate in the low-temperature range of 680–800°C. As a result, the addition of $NiCl_2$ into Al powder plays an important role for the generation of gaseous Al precursor for the synthesis of AlN nanostructures.

4.2.2 Vapor–Solid Growth

Vapor–solid (VS) growth has proved to be an effective catalyst-free method for synthesizing nanostructured materials. The key feature in VS process is the existence of a large difference in growth rate along various crystal orientations. In a typical process, the vapor species is firstly generated by evaporation, chemical reduction, and other kinds of gaseous reactions. These species are subsequently transported and condensed onto the surface of a solid substrate. Researchers developed various techniques to synthesize AlN nanostructures via VS growth, including (1) direct reaction of Al with N_2/NH_3; (2) template-confined reaction; (3) chloride-assisted method; and (4) arc-discharge method.

4.2.2.1 Direct Reaction

Different AlN nanostructures were synthesized by directly nitrifying Al powder [58–61]. The synthesis of these AlN nanostructures was regarded as a vapor–solid reaction of vaporized aluminum with NH_3/N_2. High-purity straight and stacked sheet AlN nanowires were fabricated by Lei and coworkers by the direct nitridation method at high temperature [60]. The synthesis of AlN nanowires was carried out in a horizontal tubular furnace with two open-end straight alumina tubes. The furnace temperature was heated to 900°C under ammonia/nitrogen flow at 80 sccm. Subsequently, the alumina boat was put in the center of the furnace. Then, the furnace was heated to 1,350°C under NH_3/N_2 flow at 60 sccm and was kept at 1,350°C for half an hour. After the furnace temperature was cooled to the room temperature in the flow of ammonia/nitrogen atmosphere, the gray white product was obtained from the surface of the alumina boat. Figure 4.4 gives SEM images of the large-scale AlN nanowires with high density. The nanowires are composed of straight nanowires and a small quantity of hexagonal stacked nanosheets along the growth direction. The structural evolution between nanorods and nanotips were studied by vaporizing Al powders under ammonia (NH_3) ambient in a tubular furnace [61]. The shape evolution of the AlN nanostructures as a function of growth temperature is shown in Fig. 4.5. The growth of AlN nanostructures proceeds by the transport of Al vapor to the growth region as the temperature inside the quartz tube is ramped up. However, it should be noted that although there was a thin Au layer on Si, it was not detected at the tip of the nanostructures or in its body. The body of the nanostructure is purely hexagonal AlN growing along [001] and

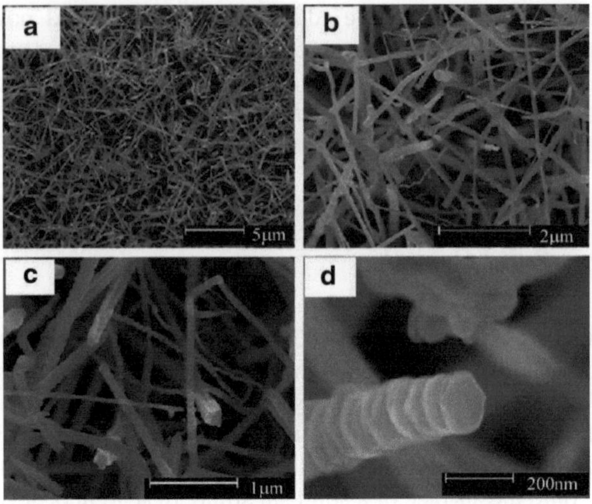

Fig. 4.4 SEM images of the as-prepared AlN nanowires. (**a**)–(**d**) are SEM images with different magnification showing the configuration of the as-prepared AlN nanowires [60]

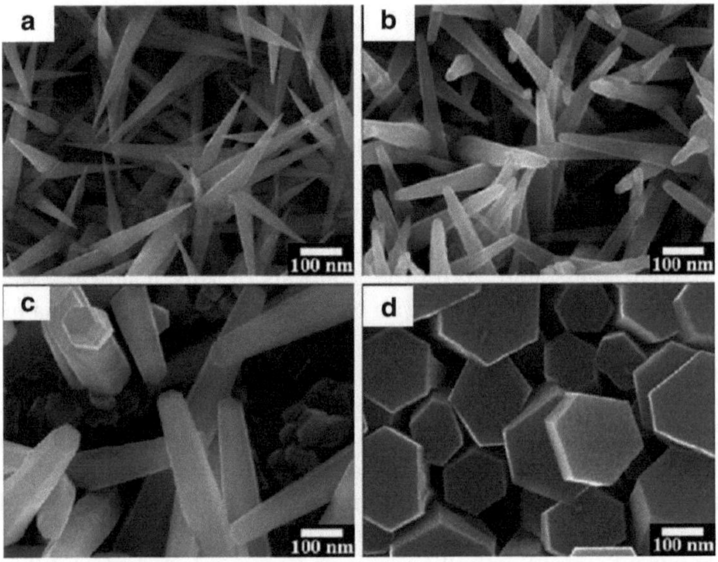

Fig. 4.5 Typical SEM images of the AlN nanostructures on silicon substrates (coated with 15 nm of gold) grown under (**a**) 950°C, (**b**) 1,000°C, (**c**) 1,100°C, and (**d**) 1,200°C, respectively [61]

no metallic Al phase is present. Therefore, VS growth mechanism was concluded. A diffusion-mediated growth model incorporating stacking of hexagonal platelets as growth units and an Ehrlich–Schwoebel barrier at the step edge has been proposed to explain the formation of high aspect ratio nanotips with substantial microscopic and structural evidence.

4.2.2.2 Template-Confined Reaction

Template-directed synthesis represents a straightforward route to 1D nanostructures. In this approach, the template simply serves as a scaffold within (or around) which a different material is generated in situ and shaped into a nanostructure with its morphology complementary to that of the template. Channels in porous membranes provide one class of the templates in the synthesis of 1D nanostructures. Anodic aluminum oxide (AAO) [62] and CNTs [63,64] are commonly used in the synthesis of AlN. Yuan et al. [62] used chemical method to convert alumina nanowires into AlN nanowires; growth temperature was at 1,300–1,400°C for 2 h in ammonia atmosphere with a flow of ~150 sccm. Figure 4.6a, b displays the FE-SEM images of densely packed alumina nanowires produced by chemical etching of the porous alumina film in diluted NaOH solution and the corresponding AlN nanowires obtained. The following reactions are involved:

$$Al_2O_3(\text{nanowire}) + 4Al \rightarrow 3Al_2O(\text{nanowire})$$

$$Al_2O(\text{nanowire}) + 2NH_3 \rightarrow 2AlN(\text{nanowire}) + H_2O + H_2$$

Zhang et al. [65] reported AlN nanowires synthesized in bulk from CNTs at relatively low temperatures. Figure 4.7 gives example of SEM images. This method

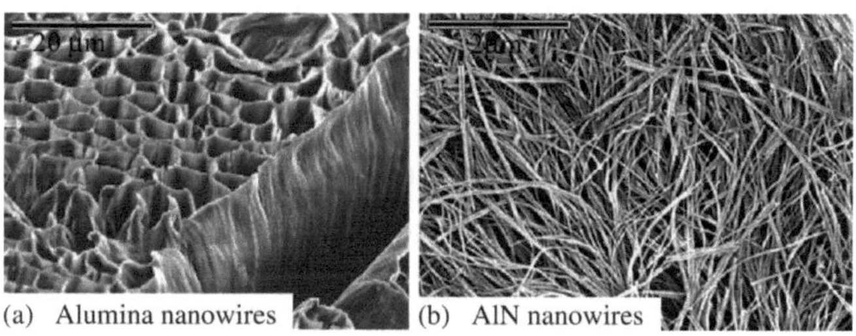

Fig. 4.6 (**a**) FE-SEM images of densely packed alumina nanowires produced by chemical etching of the porous alumina film in diluted NaOH solution. (**b**) Close-packed AlN nanowires obtained by chemical conversion of the alumina nanowires [62]

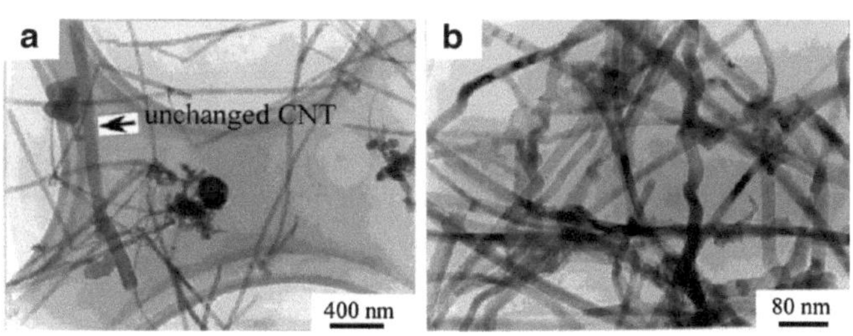

Fig. 4.7 AlN nanowires synthesized by using the carbon nanotubes with different average diameters: (**a**) 35 nm; (**b**) 15 nm. A thick CNT (>100 nm) remains unchanged after reaction (marked by an *arrow*) [65]

produced AlN nanowires through the reaction of the CNTs, Al, and Al_2O_3 in a flowing NH_3 atmosphere. AlN nanowires were produced according to the following reactions:

$$2Al_2O(g) + C(s) + 4NH_3 \rightarrow 4AlN(s) + CO + 5H_2 + H_2O(g)$$

CNTs acted as the template and play a key role in controlling the size of the AlN nanowires.

In this technique, the generated nanostructure has complementary morphologies to that of the template in which the template contains very small holes within the host material and these empty spaces are filled with a chosen material to form nanowires.

4.2.2.3 Chloride-Assisted Method

Owing to the fact that no catalyst was involved, vapor–solid growth was proposed for the synthesis of AlN nanostructures through chloride-assisted method [42, 66–68]. In He and coworkers' study, $AlCl_3$ and $(NH_4)_2CO_3$ were used as aluminum and nitrogen sources [66]. In brief, the experiment was carried out in a horizontal tube furnace. $AlCl_3$ was placed at the center of the tube furnace, where the temperature was the highest. $(NH_4)_2CO_3$ powder was positioned at the front part of the furnace upstream of $AlCl_3$. The substrates, silicon wafer with an ultrathin Ni film, were placed in a lower temperature zone, downstream from $AlCl_3$. The typical SEM images of the morphologies of AlN are displayed in Fig. 4.8; the inset is the cross-section image. The chemical reactions involved in the process include

1. The decomposition of $(NH_4)_2CO_3$: $(NH4)_2CO_3(s) \rightarrow 2NH_3(g) + CO_2(g) + H_2O(g)$
2. AlN formed via the gas-phase reaction of $AlCl_3$ and NH_3:$AlCl_3(g)+4NH_3(g) \rightarrow AlN(s) + 3NH_4Cl(g)$

It was suggested that the mechanism for the formation of the AlN tips on the surface of columnar AlN nanorods forms an entirely nanotip-covered structure.

Fig. 4.8 Typical SEM images of the morphologies of the as-synthesized AlN product. The *inset* is a cross-sectional image [66]

The formation of the AlN nanostructure consisted of three steps: (1) Al and N vapors increased from the thermal decomposition of $AlCl_3$ and $(NH_4)_2CO_3$. The Ni catalyst provides an energetically favored site for the absorption of incoming AlN vapor (or clusters), resulting in the formation of the solid-solution compound, yielding the 1D AlN nuclei at the interface between the substrate and the supersaturated compound solution, and resulting in the continuing growth of 1D columnar AlN nanorods. (2) The columnar AlN nanorods could also be used as template for the adsorption of AlN vapors during the process, resulting in the formation of AlN nuclei on the side surfaces of growing AlN columnar nanorods. Finally, the AlN hierarchical nanostructures were obtained.

Large-scale single-crystalline AlNNT arrays have been reported by Ji and coworkers [29] via a facile catalysis-free approach using $AlCl_3$ powder and NH_3 as source materials. In a typical process, a clean Si(100) substrate was placed in a quartz tube at 4–5 cm away from the source materials. The $AlCl_3$ was loaded into a ceramic boat and placed into the center of the quartz tube. After the furnace was pumped down to 10^{-3} Torr, NH_3 at 50 sccm was introduced into the tube. When the temperature of the furnace reached 1, 000°C, the quartz tube was pushed into the hot zone of the furnace where the source material was located at the center of the furnace. The growth was maintained for 2 h at 1, 000°C. Shown in the inset of Fig. 4.9a is the typical X-ray diffraction (XRD) pattern of the obtained AlN nanostructures. The XRD pattern agrees well with the standard wurtzite AlN, revealing a wurtzite AlN structure. Figure 4.9a shows the plan- and tilted-view SEM image of AlNNTs. Hexagonal cross sections and layer-stacked structure along the growth direction are clearly observed from the typical low-magnification TEM images of the AlNNTs, as shown in Fig. 4.9b, c. The growth of the AlNNTs revealed that the self-catalytic Al, resulting from the decomposition of $AlCl_3$ at the very beginning of the reaction, acts as nucleation sites, and then forms AlN with subsequently absorbed N ions resulting from the decomposition of NH_3.

Zhang et al. [67] reported the synthesis of the new sixfold-symmetrical AlN hierarchical nanostructures assembled by AlN nanoneedles through the chemical reaction between $AlCl_3$ and NH_3 with the vaporization temperature (VT) of $AlCl_3$ between 120 and 165°C and the reaction temperature (RT) higher than 1,050°C.

Fig. 4.9 (**a**) The typical low-magnification SEM image of the large-scale AlN nanotips; *inset* in (**a**) shows XRD pattern of AlN nanotips. (**b**) and (**c**) Typical low-magnification TEM image of the AlN nanotips; *inset* in (**b**) shows a SAED pattern of the single-crystalline AlN nanotips [29]

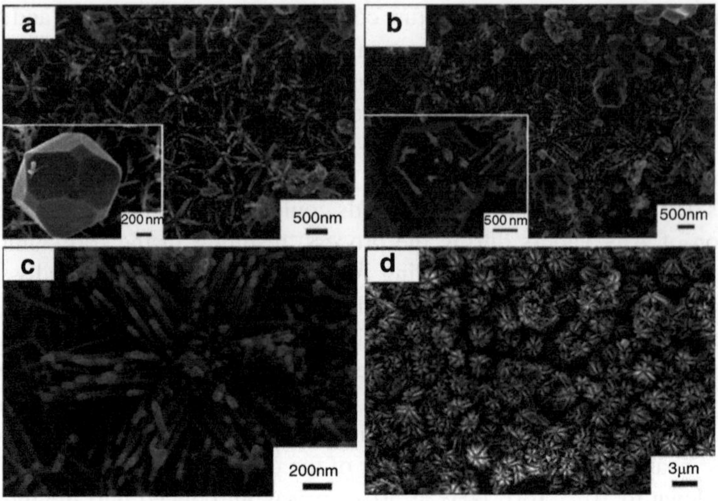

Fig. 4.10 Typical SEM images of the AlN products obtained by successive regulation of the RT and VT. (**a**) RT 1,100°C and VT 140°C. (**b**) RT 1,100°C and VT 165°C. (**c**) Enlargement of an individual nanoflower in image (**b**). (**d**) RT 1,200°C and VT 165°C. *Insets* in images (**a**) and (**b**) are enlargements of an individual microcrystal [67]

It should be noted that the morphologies, sizes, and densities of the AlN nanostructures could be controllably modulated by changing the reaction temperature and the vapor pressure of $AlCl_3$, as shown in Fig. 4.10. In the report of Song et al. [68], hexagonal AlN nanorod and nanoneedle arrays were synthesized through the direct reaction of $AlCl_3$ and NH_3 at about 750°C. Growth temperature was much lower in comparison with Zhang et al. [67]. The evolution from nanorods to nanoneedles was found to be affected by the ratio of NH_3 to Ar, as shown in Fig. 4.11. It was found that the diameter of the AlN nanorods and nanoneedles varied significantly as the NH_3 flow rate increased. The growth mechanism was deemed as surface diffusion-governed nucleation mechanism.

4.2.2.4 Arc-Discharge Method

Arc-discharge method is a common technique to fabricate nanostructured materials since Iijima synthesized CNTs in 1991 [1] by using this method. Subsequently, AlN nanotubes were synthesized by Tondare et al. [69] using arc plasma technique. Lei and coworkers [46, 70–74] reported different AlN nanostructures through direct reaction between Al vapor and nitrogen gas in direct current (DC) arc-discharge plasma. Shen et al. [33] also reported AlN nanowires prepared by this method. Their experimental processes were similar. In brief, the pure metal Mo rod was used as the cathode. The Al (purity 99.99%) column as raw material was tightly inserted into a water-cooled anode copper crucible. A certain distance was set between the tip of the Mo cathode and the Al column. Before arc discharging,

Fig. 4.11 Typical SEM images of the AlNnanorods and nanoneedles grown with NH_3/Ar flow ratios at (**a**) 3/80 (sccm), (**b**) 12/80 (sccm), (**c**) 20/80 (sccm), (**d**) 30/70 (sccm) [68]

the reaction chamber was evacuated and flushed with pure Ar gas (purity 99.99%) to remove the residual oxygen, and then pure N_2 was filled into the chamber as reaction gas. When the arc was ignited, the discharge voltage and current was kept constant. After a few minutes, a layer of white product was suspended around Mo cathode. During the arc-discharge process, the Al vapors are generated from the Al anode under accelerated electron bombardment and then react with the nitrogen radicals from the N_2 atmosphere parallel to the high-temperature plasma processes. The Al–N species are formed in plasma, and then Al–N species were transported by the thermal convection to a certain temperature zone to form into AlN nuclei. The AlN nuclei absorb continually the Al and N atom, and then the AlN crystal grew along a certain direction and formed AlN nanostructures. Normally, there was no catalyst used in the arc-discharge method. The thermal convection produced in arc-discharge process can automatically provide a vapor transport and condensation process, which was responsible for the formation of AlN nanostructures. Figure 4.12 gives examples of AlN nanostructures synthesized by the arc-discharge method [33, 70].

4.2.2.5 Other Vapor-Assisted Route

Besides the CVD technique, physical vapor transport growth has also been exploited, including molecular beam technique, sputtering, and sublimation epitaxy [43]. Highly oriented AlN single crystal nanowires with aspect ratio up to 600, diameter in the range of 40–500 nm, and 100 μm lengths, have been synthesized via a VS growth mechanism. The AlN nanowires were grown at 1,750°C. Figure 4.13

Fig. 4.12 SEM image of arc-discharged method. (**a**) and (**b**) are from [33], (**c**) and (**d**) are from [70]

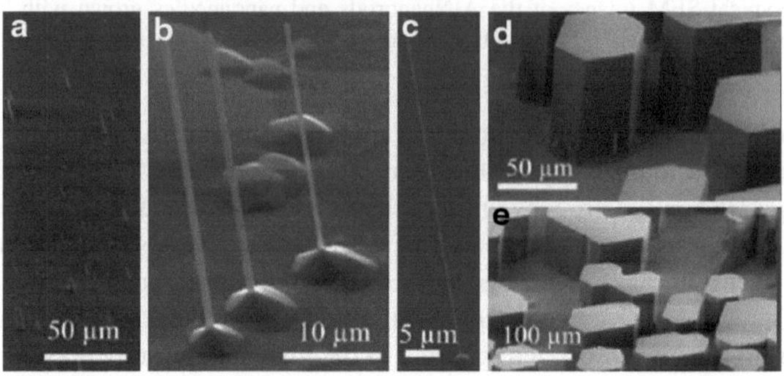

Fig. 4.13 Morphology evolution of nanowires grown at 1,750°C with different steady-state growth time and temperature rise, respectively: (**a**) 0 min and $10^{\circ}\mathrm{C\,min}^{-1}$; (**b**) 5 min and $5^{\circ}\mathrm{C\,min}^{-1}$; (**c**) 5 min and $30^{\circ}\mathrm{C\,min}^{-1}$. (**d**) SEM images of microrods with a length of ~100 μm each. (**e**) Coalescence events [43]

shows a typical SEM image of the AlN nanowires. It was found that the length and density of the nanowires varied with growth time and temperature. The AlN nanowires with an aspect ratio of 600 were obtained in the work. The apparent growth rate in this vapor-phase growth process is high, nearly 800μm h^{-1}. It should be noted that the growth rate along the c-direction is not linear; it saturates with time and the crystal feeding switches from normal to lateral. Nanowires converted to microrods when the length of the nanowires reached ~100 μm followed by lateral thickening and proceeded to coalesce owing to the longer growth time. It is believed that switching from normal to lateral growth occurs due to a decrease of the nitrogen concentration in a quasi-closed growth cell.

Carbothermal procedures provide convenient means of preparing AlN nanowires. One of the examples, an intimate mixture of Al or Al_2O_3 powder and carbon

Fig. 4.14 TEM image of the AlN nanowires. *Inset* of top-right is HRTEM of the apex part of one single AlN nanowire. The *inset* of bottom-left is the SAED pattern taken from the [100] zone axis [64]

is heated in an ambient of NH_3 under different conditions [64, 75]. Procedure involved heating a 1:1 mixture of Al powder and activated carbon in an alumina or a quartz boat placed in an alumina tube at 1,300°C for 5 h in NH_3 ambient. The basic reaction involved in the formation of AlN nanowires may be written as follows:

$$2\mathrm{Al(s)} + 2\mathrm{NH_3(g)} + \mathrm{O_2(g)} + 2\mathrm{C(s)} \rightleftharpoons 2\mathrm{AlN(s)} + 3\mathrm{H_2(g)} + 2\mathrm{CO(g)}$$

In the contrast experiments done without carbon, no AlN nanowires were found at the applied growth conditions. Therefore, carbon was concluded to play an important role in the reaction, since its presence is essential for the formation of AlN.

Figure 4.14 displays TEM image of the AlN nanowires [64]. From SAED pattern of a typical single AlN nanowire, it can be indexed as a hexagonal phase with lattice constants of $a = 0.3110$ nm and $c = 0.4975$ nm, recorded from [100] zone axis. The measured d-spacings of 0.498 and 0.269 nm are consistent with the values of (001) and (010) planes of hexagonal AlN, respectively. The axis direction of the as-prepared AlN nanowires is along the c-axis of [001]. It was proved that the growth of the as-prepared AlN nanowires obeyed the VS mechanism.

4.3 Doping of AlN Nanostructures

Rare-earth (RE) and transition-metal (TM) dopants often exhibit optical properties that have led to many important photonic and optoelectronic applications [44, 76]. 1D nanostructures, such as nanowires, nanorods, and nanotubes, could be ideal building blocks for nanoelectronics, because they can function both as devices and as the wires that access them [77–79]. Therefore, the assembly of RE and TM dopants and 1D nanostructures will significantly improve the lateral properties and provide a rational pathway for developing nanoscale photonic and optoelectronic devices. Chloride-assisted CVD, rf sputtering, and arc-discharge methods have successfully prepared doped AlN nanostructures.

Ji and coworkers [49, 50] reported Cu- and Fe-doped AlN nanorods arrays by chloride-assisted method. Briefly, $AlCl_3$ (Sigma-Aldrich, 99.99%), $CuCl_2/FeCl_3$, (Sigma-Aldrich), and NH_3 were used as sources. AlN:Cu and AlN:Fe nanorods were grown on Si substrate without catalyst at 1,000°C for about 2 h. A possible route for the formation of the nanorods is that the self-catalytic Al resulted from the decomposition of $AlCl_3$ at the very beginning of the reaction acts as nucleation sites, and then forms AlN with subsequently absorbed N ions resulting from the decomposition of NH_3.

Tb-doped AlN was reported by Liu et al. [44] using reactive rf magnetron sputtering. A 50 mm diameter target of mixture of 97 at.%Al and 3 at.%terbium nitride (TbN) was used. Sputtering was conducted in N_2/Ar mixtures under a pressure of 1.0 Pa (partial pressure of N_2 is about 0.6 Pa) and an rf-discharge power of 240 W. The substrate's temperature was 1,273 K. Figure 4.15 shows a cross-section SEM image of the AlN nanorods on Si(111). The chemical composition of the AlN was analyzed by X-ray energy dispersion spectrum (EDS) as shown in the inset of Fig. 4.15. The composition in atomic percent of 53.3 for Al, 42.9 for N, 3.5 for O, and 0.3 for Tb was obtained by quantitative analysis. It was concluded that Tb was successfully doped into AlN nanorods. The AlN nanorods were Al rich or N deficient.

(Co, Sc, Y)-doped AlN nanorods were synthesized by arc-discharge plasma method [45–47]. High-purity Al and Co/Sc/Y were used as target raw materials. During the growth, Al element and doping element were melted together by arc

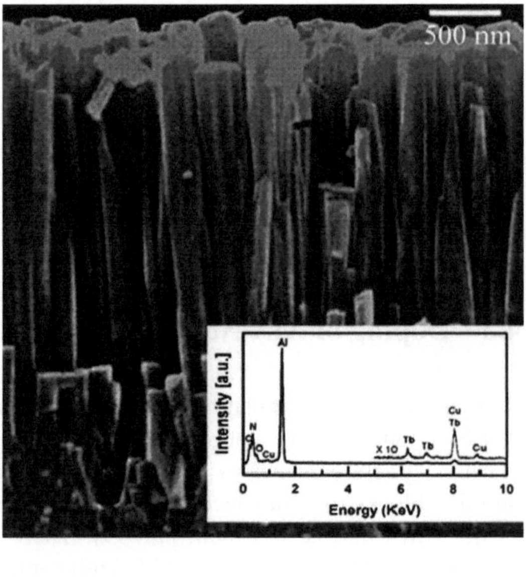

Fig. 4.15 The SEM images of a cross section of TCANs. *Inset* is the X-ray EDS of TCANs [44]

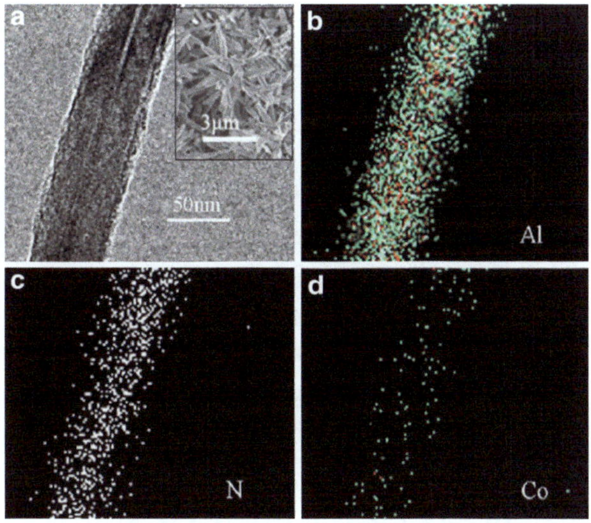

Fig. 4.16 A typical TEM image of a nanorod (**a**) and its corresponding Al (**b**), N (**c**), and Co (**d**) elemental mappings by EDS. The *inset* of (**a**) is SEM image of the flower-shaped AlN nanorod arrays [45]

current. A typical SEM image of the Co-doped AlN nanorod arrays grown on the surface of the target material is shown in the inset of Fig. 4.16a. The as-grown AlN nanostructure has a self-assembled flower shape. A TEM image of an AlN nanorod and its corresponding Al, N, and Co elemental mappings by EDS are displayed in Fig. 4.16b, c and d. It was noted that the atomic ratio of Co in various AlN nanorods varied from 0.9 to 2.1 at. %. The elemental mapping results revealed that the Co element should be doped in the AlN nanorods by substituting the Al atoms. The growth of the flower-shaped Co-doped AlN nanorod arrays should be a VLS process. With the growth of AlN nanorods, Co was doped in the h-AlN lattice by Co substituting Al atoms. It was proved that Co could act as a catalyst in the radial growth, because no similar nanostructure was found in the same experimental condition without Co melt in the raw material.

Tang et al. [51] dispersed Co on Si as catalytic metal. Flowerlike Si-doped AlN nanoneedle array was grown on Si substrate by evaporating $AlCl_3$ and $SiCl_4$ in NH_3 atmosphere. Figure 4.17 shows Si substrate covered with Co catalyst particles [Fig. 4.17a] and SEM images of flowerlike AlN nanoneedle array synthesized at 900°C [Fig. 4.17b]. Co is found to be uniformly dispersed on the Si substrate. AlN microflower consists of numerous AlN nanoneedles grown radically from a seeded Co catalyst particle. These nanoneedles are not absolutely straight, and they are several micrometers long and gradually taper along the growth direction. The catalytic Co particles are believed to be responsible for the formation of flowerlike AlN nanoneedles. The growth of the Si-doped AlN nanoneedles should be a base-model VLS process which is similar to catalytic synthesizing method. It has been revealed that Si element is uniformly distributed in the AlN nanoneedle. The content of Si analyzed from EDS on various nanoneedles is in the range of 2.6 to 4.3 at.%.

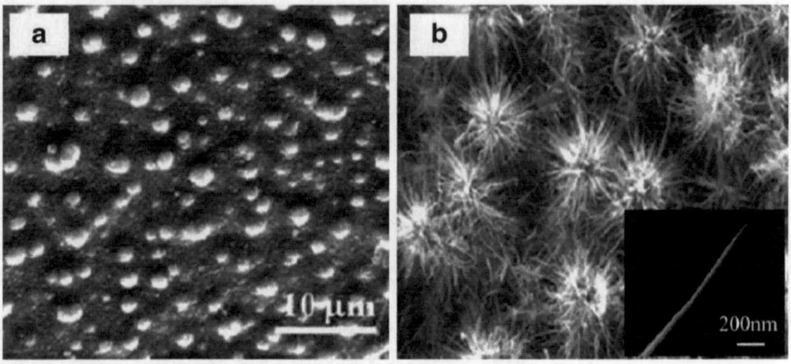

Fig. 4.17 (**a**) Si substrate covered with Co catalyst particles and (**b**) SEM images of flowerlike AlN nanoneedle array synthesized at 900°C [51]

4.4 Physical Properties of AlN Nanostructures

Nanostructured AlN materials are expected to possess novel properties due to the quantum effect. In this section, structural properties in terms of Raman shifting, optical, and ferromagnetic properties of various AlN nanostructures are discussed in detail.

4.4.1 Structural Properties Raman Spectra

Raman spectroscopy is an important tool for identifying the lattice dynamics and the crystal electric field splitting of AlN. Since h-AlN belongs to the space group P63mc, six Raman active modes may be present, i.e., $1A_1(TO) + 1A_1(LO) + 1E_1(TO) + 1E_1(LO) + 2E_2$ [54]. Figure 4.18 shows a typical room temperature (RT) Raman spectrum of AlN nanowires which is taken from reference. Six Raman-active phonon modes, namely, (A_1, E_2), A_1(TO), E_2(high), E_1(TO), A_1(LO), and E_1(LO), at 512.65, 617.70, 659.47, 671.19, 892.96, and 911.20 cm^{-1}, respectively, are observed. Similar results were observed in various AlN nanostructures. [54, 80–83]

Ji et al. [36] studied the stress states in AlN nanorods deposited on Si and its effect on optical properties by means of Raman scattering. Figure 4.19a shows a typical Raman spectrum of the AlN nanorods at RT. Two fingerprint phonons of hexagonal AlN, A_1(TO), and E_2(high) are obviously observed, which are centered at 612 and 656.5 cm^{-1}, respectively. The peak positions and full-width at half-maximum (FWHMs), however, differ from the characteristic frequencies of an unstrained AlN [84]. The Raman shift of E_2 (high) mode indicates the presence of internal stress and the line broadening of the E_2 (high) mode indicates the existence of the defects [85]. E_2 mode of wurtzite AlN would shift to a higher frequency

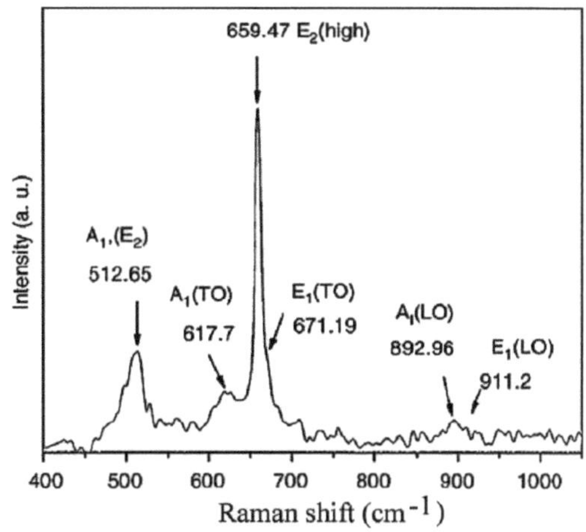

Fig. 4.18 Raman spectrum of h-AlN nanowires [80]

under the biaxial compressive stress within c-axis-oriented AlN. It has been used to estimate the magnitude of the stress between the AlN and substrate. In the linear approximation, the deviation in frequency of a given phonon mode γ under symmetry-conserving stress can be expressed in terms of the biaxial stress σ_{xx} [86]:

$$\Delta\omega_\gamma = K_\gamma\sigma_{xx}, \tag{4.1}$$

where K_γ is the linear stress coefficient and was taken as $3.39\,\text{cm}^{-1}$/GPa. In our case, Raman shift of each active mode was determined using the mixed Gussian and Lorentzian line shape fitting. The blue shift of the E_2 (high) mode corresponding to $655\,\text{cm}^{-1}$ [84] of the fitted value of $656.5\,\text{cm}^{-1}$ demonstrated that AlN nanorods grown on the Si substrate are under compressive stress, which was estimated to be 0.44 GPa. The magnitude of the stress is reasonable in comparison to AlN nanowires on sapphire substrate [86].

The FWHMs of A_1 (TO) and E_2 (high) modes as a function of temperature are shown in the Fig. 4.19c. The FWHM is $\sim 10\,\text{cm}^{-1}$ larger than the bulk AlN. It is well known that the FWHM in Raman spectra is inversely proportional to the phonon lifetime because the reduced dimensions of nanomaterials lead to the confined phonons to scatter severely and to relax at the interface. Furthermore, impurities or defects in semiconductors were found to affect the Raman linewidth, which in turn contribute to the shortening of the phonon lifetime. Therefore, the broadening in Raman spectra could be attributed to both phonon boundary relaxation at the interface in the 1D nanostructures with high aspect ratio, and the additional channel of phonon scattering at impurity or defect centers.

Temperature-dependent Raman scattering of AlN nanorods on Si substrate was further investigated. Figure 4.19b displays the selected Raman spectra of AlN nanorods obtained at different temperatures. The changes in the Raman line position, in the FWHM of the line, and in the intensity are clearly evident. The Raman line position of the A_1 (TO) and E_2 (high) modes as a function of temperature in

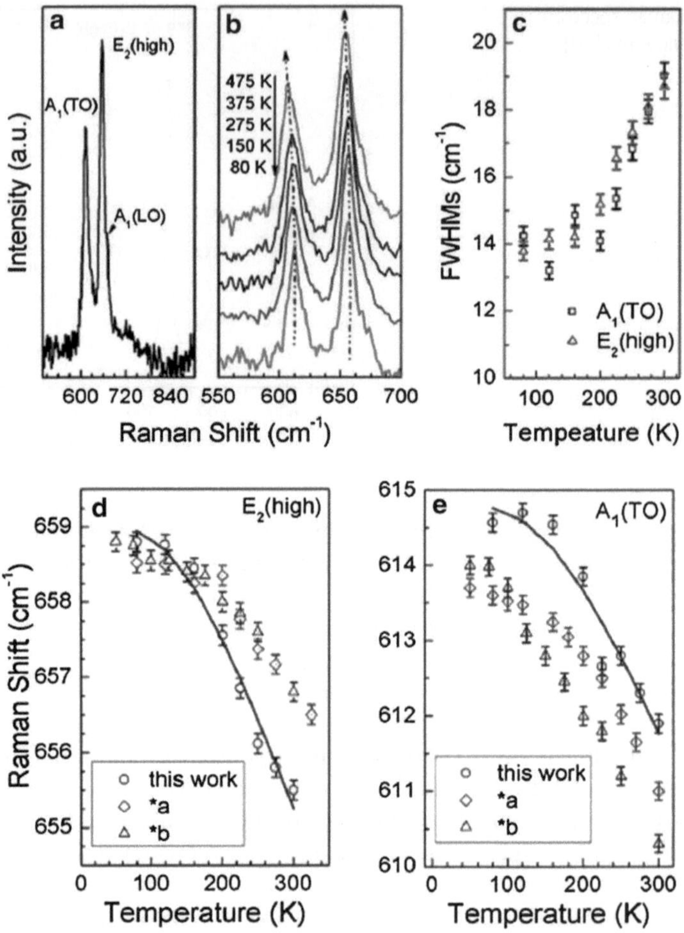

Fig. 4.19 (**a**) Raman spectrum of the AlN nanorods at RT. (**b**) Raman spectra of the AlN nanorods at selected temperatures. (**c**) FMWHs of A_1(TO) and E_2(high) modes in dependence on the temperature. (**e**) and (**d**) Raman shift of A_1(TO) and E_2(high) modes of the AlN nanorods as a function of temperature, data of a* and b* were taken from Refs. [87] and [88], respectively. The solid lines are fitting curves using Equation (4.2) [36]

the range of 80–300 K is shown in Fig. 4.19d, e, respectively. For comparison, the measured temperature-dependent optical frequency shift in the Raman spectrum of the A_1(TO) and the E_2 modes for bulk (stress-free) AlN [87,88] are displayed in the figures.

It is well known that the effects of temperature on the phonon energy measured by Raman scattering are primarily due to the thermal expansion of the lattice; a downshift of phonon frequency with temperature is expected. An improved empirical formula (4.2) has been used by Hayes et al. [88] for bulk AlN, which has a form similar to the band gap renormalization by phonon–electron interaction in the Einstein approximation.

$$\omega(T) = \omega_0 - \frac{A}{\mathrm{e}^{B(hc\omega_0/k_\mathrm{B}T)} - 1}, \tag{4.2}$$

where ω_0 is the Raman photon frequency at 0 K and ω_0, A, and B are fitting parameters. The solid lines in Fig. 4.19d, e indicate that the empirical equation fits well with experiment data for both A_1 (TO) and E_2 (high) modes. The fitting parameters ω_0, A, and B are $\sim 614.9 \pm 0.4, 6.4 \pm 3.8$, and $0.32 \pm 0.14\,\mathrm{cm}^{-1}$ for A1(TO), and $\sim 658.7 \pm 0.4$, 3.9 ± 2.8, and $0.25 \pm 0.15\,\mathrm{cm}^{-1}$ for E_2 (high), respectively. The fitting frequencies at 0 K of both modes are close to those of reported bulk AlN. It is clear that the phonon frequency of the E_2 mode is higher for stress-free AlN than that of AlN nanorods on Si substrate. But for the A_1 (TO) mode, the behavior is reversed. The confinement is anisotropic, which has effect on different phonon modes, and as a result the Raman shift of each mode is different. It is worth to note that the phonon energy difference of the E_2 mode between the stress-free AlN and AlN nanorods on Si appears to increase with increasing temperature as shown in Fig. 4.19e. This suggests that differential thermal expansion between the Si substrate and AlN nanorods is the key contribution to the compressive stress in the nanorods, and the lower the temperature, the less the Si substrate effects on the AlN nanorods phonons.

RE- and TM-doped AlN nanostructures have also been studied. Raman analysis of Cu-doped AlN nanorods has been reported by Ji et al. [49]. The Raman spectra were taken with a Renishaw micro-Raman spectrometer, equipped with a UV 244 nm laser as the excitation source focused on the sample through an optical microscope. The typical UV Raman spectra of AlN and AlN:Cu nanorods are displayed in Fig. 4.20. Two strong phonon modes of hexagonal AlN, E_2(high) and A_1(TO), are centered at 661 and 617 cm^{-1} for AlN nanorods respectively, while a broad peak at 893 cm^{-1} is assigned to A_1(LO). It is worth noting the slight red shift of E_2(high), A_1(TO), and A_1(LO) of the AlN:Cu nanorods relative to those of the AlN nanorods, which may be attributed to the disorder of the crystals due to the incorporation of Cu. Similar observation has been reported in Sc-doped AlN sixfold-symmetrical hierarchical nanostructures as shown in the inset of Fig. 4.20 [46].

4.4.2 Optical Properties of AlN Nanostructures

It is known that nanostructured AlN are promising for light-emitting applications due to their efficient visible luminescence in the 2–4 eV regions [32, 89]. Luminescence properties and energy transfer processes in AlN have been studied thoroughly. The luminescence peaks are mainly due to the presence of the oxygen-related defects in the host lattice [90,91]. The luminescence spectra of AlN ceramic samples in the visible region are characterized with bands around 400, 480, and 600 nm [92].

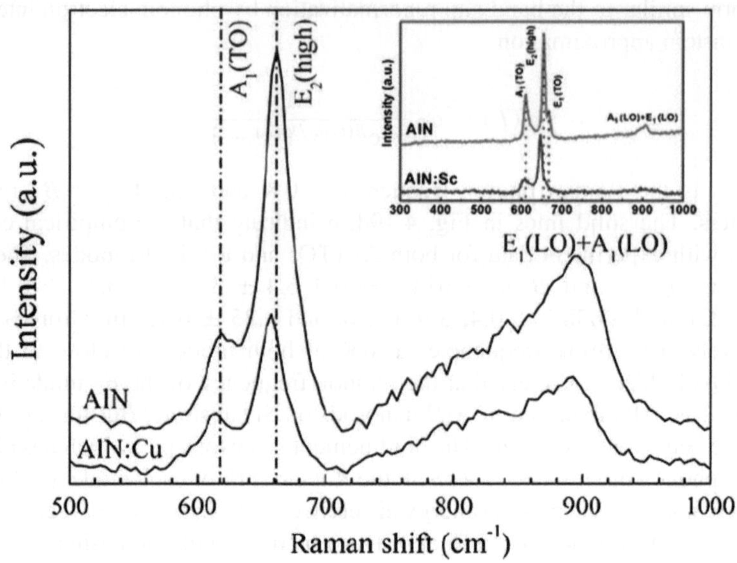

Fig. 4.20 Typical UV Raman spectra of AlN and AlN:Cu nanorods [49]. The *inset* is typical Raman spectra of AlN and AlN:Sc sixfold-symmetrical hierarchical nanostructures [46]

Recently, an AlN-based LED with a wavelength of 210 nm has been successfully realized by carefully controlling the dopant type and content in AlN layers, using a low-pressure metal-organic vapor-phase epitaxy (MOVPE) method [7]. However, it still remains challenging to achieve ultraviolet (UV) emission approaching the band gap in 1D AlN nanostructures due to excessive surface defects. Liu et al. [93] reported the synthesis of AlN nanostructures with low defects, high quality, and good crystallinity. AlN whiskers were synthesized directly by reacting Al vapors with high-purity N_2 gas at 1,700°C in a vertically standing electromagnetic induction furnace. Optical properties of as-synthesized AlN whiskers were characterized in a cathodoluminescence (CL) system. Figure 4.21a shows the CL spectrum measured in the range from 200 to 800 nm for AlN whiskers, under an accelerating voltage of 10 kV and a beam current 1,000 pA. The slit size was fixed at 250 μm × 2,000 μm. A strong ultraviolet emission peak centered at 352 nm (∼3.5 eV) and a broad weak peak in the range of 600–800 nm were observed. The 352 nm peak possibly originates from some high-level defect-related excitation instead of the band gap emission, whereas the weak emission dome may result from some unknown, but probably oxygen-related luminescence center. It was concluded that UV emission (∼3.5 eV) could be tentatively attributed to the transition between a deep impurity center (mainly oxygen) and the valence band. The reason was that although oxygen concentration was reduced to a minimum through cleaning and exceeded the resolution of EDS in the experiment, some residual oxygen in the growth chamber could still penetrate into the AlN lattice due to the higher chemical affinity of Al–O than that of Al–N, thus forming a deep impurity level approaching

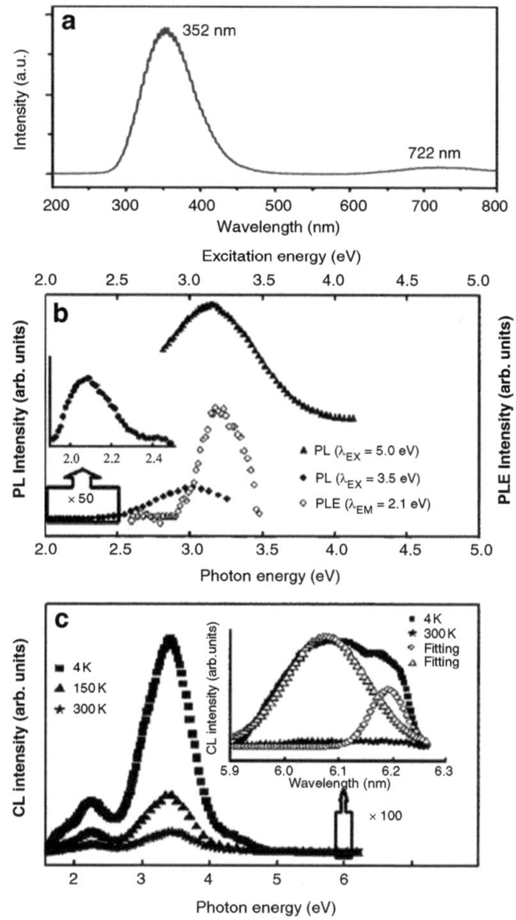

Fig. 4.21 (**a**) Cathodoluminescence (CL) spectrum of AlN whiskers measured in the range from 200 to 800 nm at room temperature [93]. (**b**) PL and PLE spectra of AlNNTs measured at room temperature. PL was measured with (*open triangle*) 5.0 and (*closed circle*) 3.5 eV excitations, respectively. PLE was measured with a detection at (*open circle*) 2.1 eV. (**c**) Temperature-dependent CL spectra from AlNNTs grown on Si substrate measured between 4 and 300 K. The *inset* shows the near band edge transition which can be resolved into two peaks [95]

the conduction band. It should be noted that the wavelength of the light emission is strongly dependent on the oxygen concentration incorporated into the AlN lattice, which has been experimentally demonstrated in bulk AlN crystal [94].

Shi et al. [95] studied the optical properties of aluminum nitride nanotips by CL, PL, thermoluminescence (TL), and UV absorption measurements. Two defect-related transitions at around 2.1 and 3.4 eV and an excitonic feature at 6.2 eV were identified, as shown in Fig. 4.21b, c. Excitation spectral analysis combined with optical absorption measurements reveal that the emission peaks at 2.1 and 3.4 eV are due to O_N–V_{Al} complex in AlN and the same excited to the separated O_N ion-level absorption, respectively. Generally, it is believed that nitrogen deficiency and oxygen point defects may result in the blue or green emissions in the range of 365–520 nm and the peak position is strongly dependent on oxygen concentration. For oxygen point defects, the defect-related peak position usually displays an obvious red shift toward the direction of long wavelength when the oxygen content in

AlN increases gradually. Similar PL characteristic were reported in various AlN nanostructures [66, 95–97].

PL analysis was performed on doped AlN nanostructures as well. Ji et al. [50] reported optical properties of Fe-doped AlN nanorods. Deep ultraviolet PL spectroscopy under a 197 nm laser excitation was employed for the measurement. PL of undoped AlN nanorods was measured for comparison. The PL spectrum of the undoped AlN nanorods comprises two strong impurity-related emissions lines at 3.25 and 4.32 eV, and a weak emission at 5.83 eV, as shown in Fig. 4.22a. The emission lines of 3.25 and 4.32 eV can be attributed to the transitions between free or donor-bound electrons to Al vacancy and vacancy–impurity complex (V_{Al}-complex) with different charge states as discussed above. However, the emission line of 5.83 eV, which is much smaller than the room temperature band

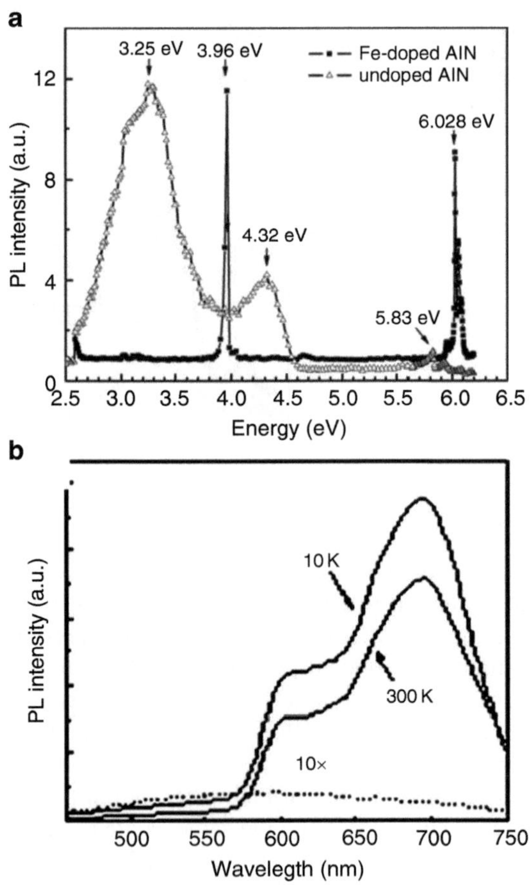

Fig. 4.22 (**a**) PL spectra of the undoped and Fe-doped AlN nanorods [50]. (**b**) PL measured at 10 and 300 K for the Mn-doped AlN nanowires (*solid line*) and the undoped AlN layer (*dashed lines*) [48]

edge emission (for example, 5.96 eV [98]) of AlN, is likely to be related to an unidentified impurity as the peak position does not shift with temperatures from 300 K down to 10 K. It is noted that both impurity emissions are absent from the Fe-doped nanorods, but a 3.96 eV emission becomes apparent in addition to the band edge emission at 6.02 eV. The vanishing of the 3.25 and 4.32 eV emissions may be due to the formation of nonradiative defects complex induced by Fe ions. Heitz et al. [99] reported that the deep $Fe^{3+/2+}$ acceptor level is 3.17 eV above the valence band maximum in GaN. Assuming that band offsets in nitrides following the internal reference rule, the $Fe^{3+/2+}$ acceptor level is expected to be about 3.96 eV above valence band of AlN, where the valence band offset for GaN/AlN is 0.79 eV [100]. Therefore, the emission line of 3.96 eV with narrow linewidth (18.5 meV) might be related to $Fe^{3+/2+}$ acceptor. The narrow UV emission linewidth of the Fe-doped AlN is very similar to that observed in Gd-implanted AlN epilayer [101]. Therefore, strong dopant-related UV emissions can take place not only in AlN thin films, but also in nanostructures. Fe doping seems to play an important role in enhancing the UV emissions in AlN nanorods.

PL measurements on Mn-doped AlN nanowires at both 10 and 300 K were studied by Yang et al. [48]. The as-grown sample exhibited a red luminescence as exemplified by the solid lines in Fig. 4.22b. They consist of two intensive emission peaks around 600 and 695 nm. A small band with the peak at about 670 nm was also observed. The red-orange band at 600 nm is a characteristic of the luminescence from the Mn center in AlN. The other two peaks were also reported in previous work of Mn-doped AlN films with O impurity. The emission could be observed with naked eyes even at room temperature. In contrast, the PL spectra of undoped AlN nanowires were also measured, exhibiting very weak broad bands located between 550 and 600 nm, as shown by dashed lines in Fig. 4.22b with a magnification of 10.

Interesting green emission was reported by Liu et al. [44] from *c*-axis-oriented AlN nanorods doped with terbium (Tb) on silicon (111). The PL spectra were measured at RT using a He–Cd laser as the excitation source with a wavelength of 325 nm. The CL study was done with an electron beam of 5 kV and 200 pA at RT. The PL and CL spectra of Tb-doped AlN nanorods are shown in Fig. 4.23a. The characteristic emission lines of Tb^{3+} ions were observed in both PL and CL spectra. Monochromatic CL images with the high spatial resolution were obtained to investigate whether the luminescent objects were nanorods or not. Figure 4.23b, d shows monochromatic (wavelength = 554 nm) CL images of a cross section, tip surface, and a single nanorod of Tb-doped AlN nanorods, respectively. These CL images directly show that the objects with green emission (554 nm) are the oriented nanorods.

A white light AlN nanowire/p-SiC heterojunction LED [102] has been successfully fabricated by depositing a layer of randomly packed AlN nanowires onto a p-SiC substrate. The sample was prepared inside a horizontal tube furnace in which a mixture of AlN and $MgCl_2$ powders (with weight ratio of 2:1) is the source material. The AlN nanowires were grown on the surface of the SiC substrate. An SEM image of the randomly oriented AlN nanowires was shown in Fig. 4.24a. Closely packed AlN nanowires, with a length of 6 μm and a diameter of 80 nm, form

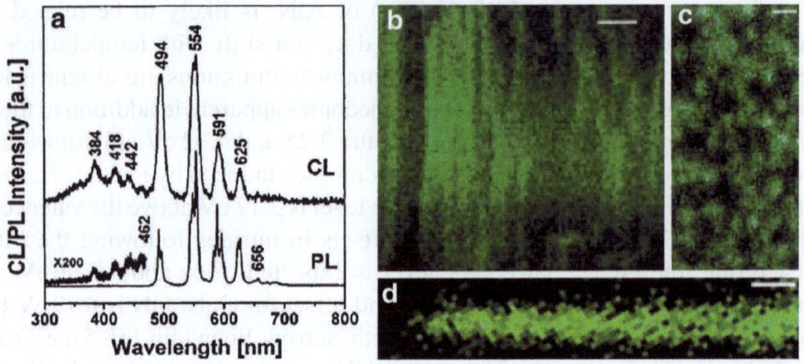

Fig. 4.23 (**a**) The PL and CL spectra of TCANs measured at room temperature. (**b**) Monochromatic (wavelength = 554 nm) CL images of a cross section. (**c**) Monochromatic (wavelength = 554 nm) CL images of a tip surface of TCANs. (**d**) Monochromatic (wavelength = 554 nm) CL image of single Tb-doped AlN nanorod stripped off from the Si substrate [44]

a thin film. Fig. 4.24b shows the cross-sectional SEM image of the heterojunction exhibited good interfacial properties with the SiC substrate primarily due to its lattice match with SiC. It is verified that the nanowire is composed of AlN with a hexagonal structure. The lattice spacing of 0.493 nm in the growth direction corresponds to the (0001) plane of wurtzite AlN. Figure 4.24c, d, e shows EL spectra of the heterojunction LED under bias at different voltages. The spectra were measured by connecting the cathode and anode of a rectangle pulse voltage source (with repetition rate and pulsewidth of 7.5 Hz and 80 ms, respectively) to the ITO and Al/Ti metal contacts, respectively, of the heterojunction. Light was collected from the surface of the ITO by an objective lens. It was noted that at forward bias of 15 V, the corresponding emission spectrum has a FWHM of ~125 nm. The LED exhibited white light emission (Fig. 4.24d). However, no emission was observed from the heterojunction LED under reverse bias. It was suggested that the n-type conduction of the AlN nanowires could be related to defect levels associated with V_N, O_N, or V_{Al}, which was consistent with reported optical properties of nanostructured AlN as discussed above. It was demonstrated that for the LED under forward biased, the corresponding EL emission spectra were observed with multiple peaks at ~400, 420, 468, and 525 nm. By comparing with the PL spectrum as shown in Fig. 4.24e, it was noted that four trap-level states of the AlN nanowires contribute to the effective generation of light. The four trap-level states were related to V_N inside the band gap of the AlN nanowires. It is believed that the high concentration of V_N from the nanowires (i.e., arisen from the large surface-to-volume ratio) leads to the radiative recombination in visible regime.

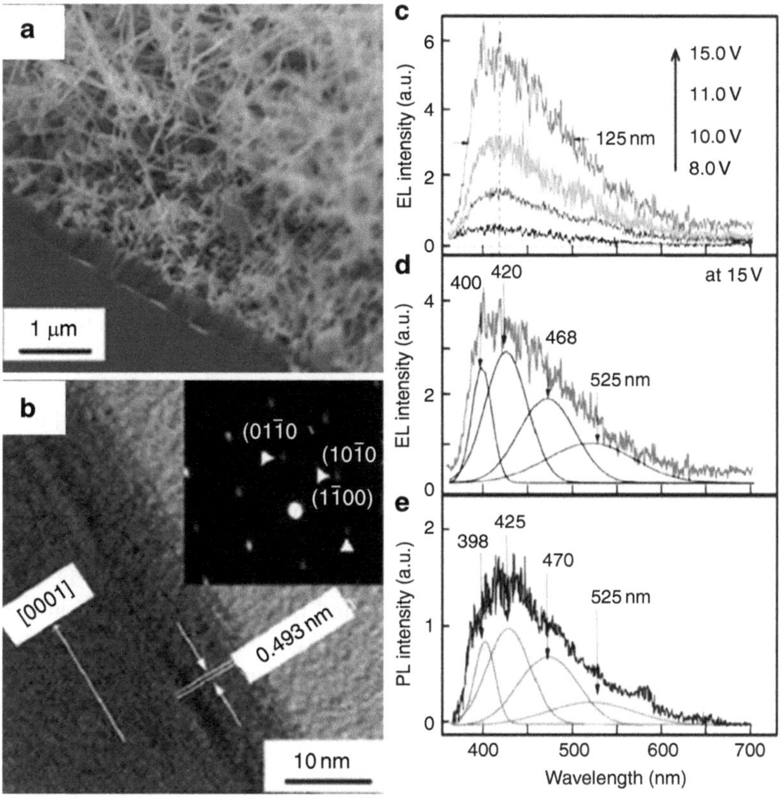

Fig. 4.24 (**a**) Cross-section SEM image of interface between AlN nanowires and SiC substrate. (**b**) An HRTEM image of the AlN nanowire and the SAED pattern. (**c**) Room temperature EL spectra of the AlN nanowires/p-SiC heterojunction. (**d**) Fitting results of the EL spectrum obtained from the LED at forward-biased voltage of 15 V. (**e**) Fitting results of the PL spectrum of the AlN nanowires deposited on p-SiC [102]

4.4.3 Ferromagnetic Properties

The search for magnetic semiconductors with a Curie temperature above RT is currently one of the major challenges in semiconductor spintronics. Spintronic devices are meant to exploit the spin of magnetic materials along with – like in standard electronics – the charge of electrons in semiconductors. It is generally expected that new functionalities for electronic and photonics will arise if the injection, transfer, and detection of carrier spins can be mastered above RT. In this perspective, diluted magnetic semiconductors (DMS) are of particular interest. The elements commonly used as magnetic dopants for the synthesis of DMS belong to the family of transition metals (TMs) (Sc, Ti, V, Cr, Mn, Fe, Co, Ni, and Cu) and rare earths (RE) (Sm, Eu, Gd, Tb, Dy, and Er). The magnetic behavior of these elements is due to the partially filled d- or f-states, respectively, containing unpaired spins. Extensive studies have been carried out on ferromagnetism in DMSs nanomaterials because of their potential applications in spintronic devices such as magneticoptic switches, magnetic sensors, spin valve transistors, and spin LEDs. 1D DMS nanostructures offer the opportunity for studying the dimensionality and size effect on magnetic properties since ferromagnetic DMS nanowires and nanorods were reported to have higher Curie temperature and larger magnetic moment as compared to their bulk and film counterparts [103]. The behavior of carriers in nanostructures may also allow for the enhancement of the ferromagnetic semiconductor behavior, because nanostructures are expected to have longer coherence times than in the bulk, which may provide a pathway for increasing the spin lifetime in ferromagnetic semiconductor devices for practical applications [104].

It has been reported that wide band gap semiconductor AlN exhibits RT ferromagnetism when a few atomic percent of dopant substitute Al [105–109]. Besides AlN bulk and thin films, both magnetic (Co, Mn, and Fe) and nonmagnetic (Cu, Sc, and Y) TMs have been doped in AlN nanostructures. Table 4.2 summarized ferromagnetism properties of TM-and RE-doped AlN nanostructures reported up to now.

Figure 4.25 [50] shows magnetization hysteresis loops of the AlN:Fe nanorods measured at RT with applied field parallel and perpendicular to the sample. The spontaneous saturated magnetization and coercivity of the AlN:Fe are estimated to be 0.64 μ_B/Fe and 116 Oe when the field is applied perpendicular to the sample, which is $\sim$61 and 50% larger than the field applied parallel to the sample, respectively. This indicates that a significant fraction of Fe spins is coupled magnetically in AlN, and the saturated magnetic moment of the AlN:Fe nanorods is significantly larger than Fe-implanted ZnO [110]. The Fe^{3+}–V_N–Fe^{3+} groups are expected in the structure. An electron trapped in the nitrogen vacancy (V_N) constitutes an F center, where the electron occupies an orbital which overlaps with the d shell of both ion neighbors. Since Fe^{3+}, $3d^5$, only has unoccupied minority spin orbitals, the trapped electron will be spin down ($\downarrow$) and the two ion neighbors spin up ($\uparrow$), resulting in the magnetization. It should be noted that although no Fe-related secondary phases can be detected by the XRD and TEM, the possibility of nanosized Fe clusters or Fe oxide phases which may also contribute to the ferromagnetism cannot be ruled out.

Table 4.2 Ferromagnetism properties of TM-doped AlN nanostructures

Dopant	Saturation magnetization (emu/g)	Coercive field (Oe)	Ref.
Y-doped nanorods	0.05	101	[40]
Co-doped nanorods	0.04	400	[38]
Fe-doped nanorods	1.07	116	[43]
Cu-doped nanorods	0.12	100	[42]
Mn-doped nanowires	0.073	44	[41]
Sc-doped nanostructure	0.04	200	[39]

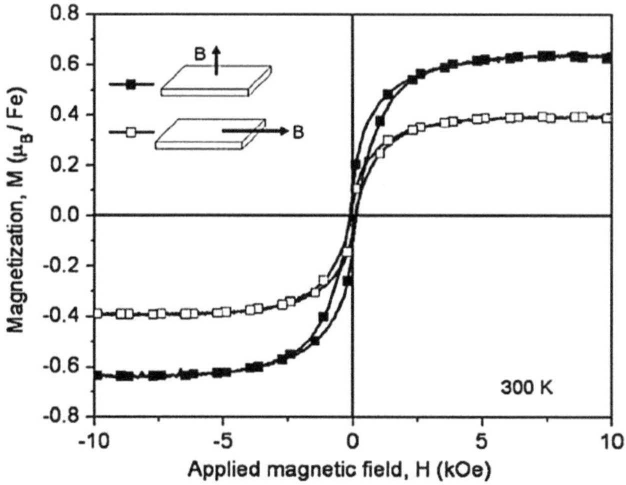

Fig. 4.25 Magnetization hysteresis loops of the AlN:Fe nanorods measured at room temperature with applied field parallel and perpendicular to the sample [50]

Currently, there is no ideal theoretical model which could explain the ferromagnetism of the TM-doped AlN material system. The Zener model of hole-carrier-mediated ferromagnetism has been used to explain the magnetic properties of Mn-doped III–V semiconductors [105, 111], which implicitly assumes that a TM is doped in a semiconductor lattice by substituting a fraction of the host cations. However, AlN is a wide band gap semiconductor with high electric resistance; a mechanism of indirect exchange interaction caused by virtual magnetic acceptor level–valence band transitions [112] might probably be applicable to the case of a low density of carriers based. The percolation network-like model for ferromagnetism in low carrier concentration systems proposed is also another potential mechanism [113, 114].

Although magnetic TM-doped AlN nanorods exhibited ferromagnetic properties as reported, nonmagnetic element doping-induced ferromagnetism deserves further investigation, because it can lead to a better understanding of the origin of ferromagnetism in DMSs and may open up a new class of DMSs without doping magnetic elements. Ferromagnetic AlN nanorods doped with Cu was reported by Ji et al. [49]. The experimental spontaneous magnetization value was 0.38 emu cm^{-3}. Guda et al. [115] studied the copper defects inside these AlN:Cu nanorods by means of X-ray absorption spectroscopy (XAFS) and a full-potential linear-augmented planewave (LAPW) calculation. Experimental XAS spectra above the Cu K-edge for AlN:Cu and reference samples are shown in Fig. 4.26a, b. Energy positions of peaks A, B, C, D, and E are chosen to correspond to the main maxima of bulk copper spectrum. The spectrum of AlN:Cu sample is in agreement with the middle curve on the graphs which corresponds to data for small copper clusters [116]. It was suggested that copper clusters might present in AlN lattice. Average size of cluster was estimated

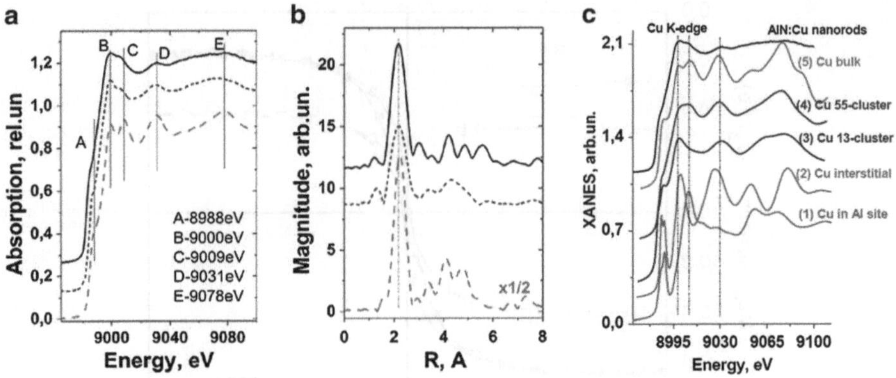

Fig. 4.26 (**a**) Experimental XANES data and (**b**) magnitude of the Fourier transforms of χ(k) data above Cu K-edge. Up to down: data for AlN:Cu nanorods (*solid curve*) copper clusters, grown during radiolysis in water solution of $CuCl_2$ (*short dashes*) and metallic copper (*dashes*). (**c**) Experimental XANES spectrum for AlN:Cu nanorods above copper K-edge (*upper curve*) and calculations for different theoretical models [115]

to be 50–70 atoms [Fig. 4.26c]. Spin-polarized band-structure calculations were made for estimations of magnetic properties for different types of point defects. Calculations predicted that the largest value of magnetic moment 2 μ_B per super cell arise when Cu is placed in Al site. Octahedral and tetrahedral interstitial positions of copper give the values of spin magnetic moment, equal to 0.65 and 0.83 Bohr magnetrons correspondingly. It was proposed that a small fraction of doped Cu substituted Al site in the lattice, and the surface atoms of small copper clusters might contribute the magnetization of inside AlN lattice as well.

4.5 Concluding Remarks and Perspectives

We have presented an up-to-date review of synthesis and physical properties of pristine and doped AlN nanostructures. AlN nanostructures offer some potential in providing electronic, photonic, and spin-based devices, and encouraging progress has been made in the research phase. Despite this progress, there are still a number of important issues that are in need of further investigation before this material can be turned into practical applications. (1) One of the appealing features of nanostructures is that they can be single crystalline, well faceted with low defects and can be grown on lattice mismatch substrates. These features are crucial to prepare high optical quality AlN nanostructures with band edge emission. The growth techniques, which are able to synthesize AlN nanostructures with band edge emission, are urgently needed. The current growth techniques must be improved in order to prepare O-free AlN nanostructures. Hydrogen may be incorporated in the reactive gases for the preparation of O-free AlN nanostructures. (2) Lack of n-type and p-type doping in AlN nanostructures hampers the development of deep UV photonic devices. It is of great importance to demonstrate n- and p-type conduction in AlN nanostructures. Si and Mg are the promising n-type and p-type dopants for AlN nanostructures, respectively. With the rapid progress in synthesis techniques, n- and p-type AlN nanostructures are on the horizon. (3) The large band gap energy of AlN makes it theoretically possible to achieve emission wavelength as short as $\sim$200 nm, but making p–n junction devices that can efficiently operate at such short wavelengths while also achieving a reasonable optical power is a great challenge. Thus, alternative strategy is needed to realize practical deep UV photonic devices. Oto et al. [9] reported a deep UV emission at $\sim$240 nm with output power of 100 mW from AlN-based quantum wells structure pumped by an electron beam. Electron beam pumping of quantum wells is a promising and effective way to generate deep UV light. It is believed that by optimizing the optical properties of AlN nanostructures, a high-power and efficient deep UV light source could also be obtained by electron field emission approach.

Acknowledgements This work was partly supported by The Hong Kong Polytechnic University (Project No. G-U853) and the Research Grants Council of Hong Kong (Project No. PolyU 5013/09P). JXH would like to acknowledge the support of Natural Science Foundation of Guangdong Province, China (Grant No. 9451064101002440) and Doctoral Program Foundation of Institutions of Higher Education of China (No. 20090172120014).

References

1. S. Iijima, Nature **354**, 56 (1991)
2. P.J. Pauzauskie, P. Yang, Today **9**, 36 (2006)
3. J. Appenzeller, J. Knoch, M.T. Bjork, H. Riel, H. Schmid, W. Riess, IEEE Trans. Electron Dev. **55**, 2827 (2008)
4. G. Guisbiers, L. Buchaillot, J. Phys. D: Appl. Phys. **41**, 172001 (2008)
5. A. Husain, J. Hone, H. W.C. Postma, X.M.H. Huang, T. Drake, M. Barbic, A. Scherer, M.L. Roukes, Appl. Phys. Lett. **83**, 1240 (2003); M. Li, R.B. Bhiladvala, T.J. Morrow, J.A. Sioss, K.-K. Lew, J.M. Redwing, C.D. Keating, T.S. Mayer, Nat. Nanotechnol. 3, **88** (2008)
6. F. Patolsky, B.P. Timko, G. Zheng, C.M. Lieber, MRS Bull. 32, **142** (2007)
7. Y. Taniyasu, M. Kasu, T. Makimoto, Nature **441**, 325 (2006)
8. J. Deng, Y. Bilenko, A. Lunev, X. Hu, T.M. Katona, J. Zhang, M.S. Shur, R. Gaska, Jpn. J. Appl. Phys. **46**, L263 (2007)
9. T. Oto, R.G. Banal, K. Kataoka, M. Funato, Y. Kawakami, Nat. Photonics **4**, 767 (2010)
10. M.S. Shur, R. Gaska, IEEE T. Electron. Dev. **57**, 12 (2010)
11. L.W. Yin, Y. Bando, Y.C. Zhu, M.S. Li, Y.B. Li, D. Golberg, Adv. Mater. **17**, 110 (2005)
12. Y. Taniyasu, M. Kasu, T. Makimoto, Appl. Phys. Lett. **84**, 2115 (2004)
13. N. Kuramoto, H. Taniguchi, I. Aso, Am. Ceram. Soc. Bull. **68**, 883 (1989)
14. Y.B. Tang, H.T. Cong, H.M. CHeng, J. Nanosci. Nanotechnol. **2**, 307 (2007)
15. S.C. Jain, M. Willander, J. Narayan, R. Van Overstraeten, J. Appl. Phys. **87**, 965 (2000)
16. I. Ivanov, L. Hultman, K. Järrendahl, P. Mårtensson, J.-E. Sundgren, B. Hjörvarsson, J.E. Greene, J. Appl. Phys. **78**, 5721 (1995)
17. X.D. Wang, Ursula Mazur, K.W. Hipps, J.T. Dickson, Thin solid Films **240**, 45 (1994)

18. K. Kubota, Y. Kobayashi, K. Fujimoto, J. Appl. Phys. **66**, 2984 (1989)
19. S. Strite, H. Morkoc, J. Vac. Sci. Technol. B **10**, 1237 (1992)
20. J.R. Mileham, S.J. Pearton, C.R. Abernathy, J.D. MacKenzie, R.J. Shul, S.P. Kilcoyne, Appl. Phys. Lett. **67**, 1119 (1995)
21. S.N. Mohammad, H. Morkoç, Prog. Quant. Electron. **20**, 361 (1996)
22. O Ambacher, J. Phys. D: Appl. Phys. **31**, 2653 (1998)
23. B.E. Foutz, S.K. O'Leary, M.S. Shur, L.F. Eastman, J. Appl. Phys. **85**, 7727 (1999)
24. L. Liu, J.H. Edgar, Mater. Sci. Eng. R37, **61**, (2002)
25. S. Yoshida, S. Misawa, S. Gonda, J. Appl. Phys. **53**, 6844 (1982)
26. X.S. Miao, Y.C. Chan, E.Y. Pun, Thin Solid Films **315**, 123 (1998)
27. T. Shiosaki, T. Yamamoto, T. Oda, A. Kawabata, Appl. Phys. Lett. **36**, 643 (1980)
28. F. Engelmark, G. Fucntes, I.V. Katardjiev, A. Harsta, U. Smith, S. Berg, J. Vac. Sci. Technol A **18**, 1609 (2000)
29. X.H. Ji, Q.Y. Zhang, S.P. Lau, H.X. Jiang, J.Y. Lin, Appl. Phys. Lett. **94**, 173106 (2009)
30. J. Zheng, Y. Yang, B. Yu, X. Song, X. Li, ACS Nano **2**, 134 (2008)
31. Y.K. Byeun, R. Telle, S.-H. Jung, S.-C. Choi, H.-I. Hwang, Chem. Vap. Depos. **16**, 72 (2010)
32. H.T. Chen, X.L. Wu, X. Xiong, W.C. Zhang, L.L. Xu, J. Zhu, Paul K. Chu, J. Phys. D: Appl. Phys. **41**, 025101 (2008)
33. L. Shen, T. Cheng, L. Wu, X. i Li, Q. Cui, J. Alloys Compd. **465**, 562 (2008)
34. L. Shen, X. Li, Q. Cui, B. Liu, T. Cui, Appl. Phys. A **99**, 111 (2010)
35. X. Song, Z. Guo, J. Zheng, X. Li, Y. Pu, Nanotechnology **19**, 115609 (2008)
36. X.H. Ji, Q.Y. Zhang, Z.Y. Ling, S.P. Lau, Appl. Phys. Lett. **95**, 233105 (2009)
37. H.M. Lv, G.D. Chen, H.G. Ye, G.J. Yan, J. Appl. Phys. **101**, 053526 (2007)
38. Y. Tang, H. Cong, Z. Chen, H. Cheng, Appl. Phys. Lett. **86**, 233104 (2005)
39. S.C. Shi, C.F. Chen, S. Chattopadhyay, Z.H. Lan, K.H. Chen, L.C. Chen, Adv. Funct. Mater. **15**, 781 (2005)
40. H.T. Wang, Z. Xie, Y. Wang, W. Yang, Q. Zeng, F. Xing, L. An, Nanotechnology **20**, 025611 (2009)
41. Q. Wu, F. Zhang, X. Wang, C. Liu, and Z. Hu, Y. Lu, J. Phys. Chem. C **111**, 12639 (2007)
42. C. Liu, Z. Hu, Q. Wu, X. Wang, Y. Chen, W. Lin, H. Sang, S. Deng, N. Xu, Appl. Surf. Sci. **251**, 220 (2005)
43. G.R. Yazdi, P.O.Å. Persson, D. Gogova, R. Fornari, L. Hultman, M. Syvjrvi, R. Yakimova, Nanotechnology **20**, 495304 (2009)
44. Q.L. Liu, T. Tanaka, J.Q. Hu, F.F. Xu, T. Sekiguchi, Appl. Phys. Lett. **83**, 15 (2003)
45. S.L. Yang, R.S. Gao, P.L. Niu, R.H. Yu, Appl. Phys. A **96**, 769 (2009)
46. W.W. Lei, D. Liu, P.W. Zhu, X.H. Chen, Q. Zhao, G.H. Wen, Q.L. Cui, G.T. Zou, Appl. Phys. Lett. **95**, 162501 (2009)
47. W.W. Lei, D. Liu, X. Chen, P. Zhu, Q. Cui, G. Zou, J. Phys. Chem. C **114**, 15574 (2010)
48. Y. Yang, Q. Zhao, X.Z. Zhang, Z.G. Liu, C.X. Zou, B. Shen, D.P. Yu, Appl. Phys. Lett. **90**, 092118 (2007)
49. X.H. Ji, S.P. Lau, S.F. Yu, H.Y. Yang, T.S. Herng, Nanotechnology **18**, 105601 (2007)
50. X.H. Ji, S.P. Lau, S.F. Yu, T.S. Herng, A. Sedhain, J.Y. Lin, H.X. Jiang, K.S. Teng, J.S. Chen, Appl. Phys. Lett. **90**, 193118 (2007)
51. Y.B. Tang, H.T. Cong, Z.M. Wang, H.-M. Cheng, Appl. Phys. Lett. **89**, 253112 (2006)
52. Q. Wu, Z. Hu, X. Wang, Y. Hu, Y. Tian, Y. Chen, Diamond Relat. Mater. **13**, 38 (2004)
53. J.A. Haber, P.C. Gibbons, W.E. Buhro, J. Am. Chem. Soc. **119**, 5455 (1977); J.A. Haber, P.C. Gibbons, W.E. Buhro, Chem. Mater. **10**, 4062 (1998)
54. C. Liu, Z. Hu, Q. Wu, X.Z. Wang, Y. Chen, H. Sang, J.M. Zhu, S.Z. Deng, N.S. Xu, J. Am. Chem. Soc. **127**, 1318 (2005)
55. N. Takahashi, Y. Matsumoto, T. Nakamura, J. Phys. Chem. Solids **67**, 665 (2006)
56. M. Yoshioka, N. Takahashi, T. Nakamura, Mater. Chem. Phys. **86**, 74 (2004)
57. L. Yu, N. Liu, C. He, Q. Wu, Z. Hu, J. Phys. Chem. C **113**, 14245 (2009)
58. J.H. Duan, S.G. Yang, H.W. Liu, J.F. Gong, H.B. Huang, X.N. Zhao, R. Zhang, Y.W. Du, J. Phys. Chem. B **109**, 3701 (2005)

59. R.K. Paula, K.-H. Leeb, B.-T. Leea, H.-Y. Song, Mater. Chem. Phys. **112**, 562 (2008)
60. M. Lei, H. Yang, P.G. Li, W.H. Tang, J. Alloys Compd. **459**, 338 (2008)
61. S.-C. Shi, S. Chattopadhyay, C.-F. Chen, K.-H. Chen, L.-C. Chen, Chem. Phys. Lett. **418**, 152 (2006)
62. Z.H. Yuan, S.Q. Sun, Y.Q. Duan, D.J. Wang, Nanoscale Res. Lett. **4**, 1126 (2009)
63. J. Liu, X. Zhang, Y. Zhang, R. He, J. Zhu, J. Mater.Res. **16**, 3133 (2001); Y. Zhang, J. Liu, R. He, Q. Zhang, X. Zhang, J. Zhu, J. Mater.Chem. **13**, 3899 (2001)
64. T. Yang, Y. Zhu, Q. Ruan, X. Du, Y. Zeng, F. Xu, J. Nanosci. Nanotechnol. **10**, 421 (2010)
65. Y. Zhang, J. Liu, R. He, Q. Zhang, X. Zhang, J. Zhu, Chem. Mater. **13**, 3899 (2001)
66. J.H. He, R. Yang, Y.L. Chueh, L.J. Chou, L.J. Chen, Z.L. Wang, Adv. Mater. **18**, 650 (2006)
67. F. Zhang, Q. Wu, X. Wang, N. Liu, J. Yang, Y. Hu, L. Yu, X. Wang, Z. Hu, J. Zhu, J. Phys. Chem. C **113**, 10 (2009)
68. X. Song, Z. Guo, J. Zheng, X. Li, Y. Pu, J. Phys. Chem. C **111**, 12639 (2007)
69. V.N. Tondare, C. Balasubramanian, S.V. Shende, D.S. Joag, V.P. Godbole, S.V. Bhoraskar, M. Bhadbhade, Appl. Phys. Lett. **80**, 4813 (2002)
70. W. Lei, D. Liu, J. Zhang, P. Zhu, Q. Cui, G. Zou, Cryst. Growth Des. **9**, 3 (2009)
71. W. Lei, D. Liu, P. Zhu, Q. Wang, G. Liang, J. Hao, X. Chen, Q. Cui, G. Zou, J. Phys. Chem. C **112**, 13353 (2008)
72. W. Lei, D. Liu, L. Shen, J. Zhang, P. Zhu, Q. Cui, G. Zou, J. Cryst. Growth **306**, 413 (2007)
73. W. Lei, D. Liu, P.n. Zhu, X. Chen, J. Hao, Q. Wang, Q. Cui, G. Zou, Cryst. Eng. Comm. **12**, 511 (2010)
74. H. Shen, X.F. Li, J. Zhang, Y.M. Ma, F. Wang, G. Peng, Q.L. Cui, G.T. Zou, Appl. Phys. A **84**, 73 (2006)
75. C.N.R. Rao, Gautam Gundiah, F.L. Deepak, A. Govindaraja, A.K. Cheetham, J. Mater. Chem. **14**, 440 (2004)
76. A.J. Steckl, J.M. Zavada, MRS Bull. **24**, 16 (1999)
77. A.M. Morales, C.M. Lieber, Science **279**, 208 (1998)
78. X.F. Duan, Y. Huang, Y. Cui, J.F. Wang, C.M. Lieber, Nature **409**, 66 (2001)
79. Y. Cui, C.M. Lieber, Science **291**, 851 (2001)
80. Y. Tian, Y. Jia, Y. Bao, Y. Chen, Diamond Relat. Mater. **16**, 302 (2007)
81. Q. Wu, Z. Hu, X.Z. Wang, Y. Chen, Y.N. Lu, J. Phys. Chem. B **107**, 9726 (2003)
82. Q. Zhao, H. Zhang, X. Xu, Z. Wang, J. Xu, D. Yu, Appl. Phys. Lett. **86**, 193101 (2005)
83. L.-W. Yin, Y. Bando, Y.-C. Zhu, M.-S. Li, Y.-B. Li, D. Golberg, Adv. Mater. **17**, 110 (2005)
84. C. Bungaro, K. Rapcewicz, J. Bernholc, Phys. Rev. Lett. **61**, 6720 (2000)
85. M. Kuball, F. Demangeot, J. Frandon, M.A. Renucci, N. Grandjean, O. Briot, MRS Internet J Nitride Semicond Res, **4S1** (1999)
86. D.G. Zhao, S.J. Xu, M.H. Xie, S.Y. Tong, H. Yang, Appl. Phys. Lett. **83**, 677 (2003)
87. M. Kazan, Ch. Zgheib, E. Moussaed, P. Masri, Diamond Relat. Mater. **15**, 1169 (2006)
88. J.M. Hayes, M. Kuball, Y. Shi, J.H. Edgar, Jpn. J. Appl. Phys. **39**, L710 (2000)
89. J. Yang, T.-W. Liu, C.-W. Hsu, L.-C. Chen, K.-H. Chen, C.-C Chen, Nanotechnology **17**, S321 (2006)
90. G.A. Slack, J. Phys. Chem. Solids **34**, 321. (1973)
91. R.A. Youngman, J.H. Harris, J. Am. Ceram. Soc. **73**, 3238 (1990)
92. L. Trinkler, B. Berzina, A. Auzina, M. Benabdesselam, P. Iacconi, Phys. Status Solidi C **4**, 1032 (2007)
93. B. Liu, Y. Bando, A. Wu, X. Jiang, B. Dierre, T. Sekiguchi, C. Tang, M. Mitome, D. Golberg, Nanotechnology **21**, 075708 (2010)
94. J.H. Harris, R.A. Youngman, R.G. Teller, J. Mater. Res. **15**, 1763 (1990)
95. S.C. Shi, C.F. Chen, S. Chattopadhyay, K.H. Chen, B.W. Ke, L.C. Chen, L. Trinkler, B. Berzina, Appl. Phys. Lett. **89**, 163127 (2006)
96. Y.B. Tang, H.T. Cong, F. Li, H.M. Cheng, Diamond Relat. Mater. **16**, 537 (2007)
97. T. Marttila, R.M. Nieminen, Phys. Rev. B **55**, 9571 (1997)
98. J. Li, K.B. Nam, M.L. Nakarmi, J.Y. Lin, H.X. Jiang, Appl. Phys. Lett. **81**, 3365 (2002)

99. R. Heitz, P. Maxim, L. Eckey, P. Thurian, A. Hoffmann, I. Broser, K. Pressel, B.K. Meyer, Phys. Rev. B **55**, 4382 (1997)
100. N. Nepal, M.L. Nakarmi, H.U. Ang, J.Y. Lin, H.X. Jiang, Appl. Phys. Lett., **89**, 192111 (2006)
101. J.M. Zavada, N. Nepal, J.Y. Liu, H.X. Jiang, E. Brown, U. Hömmerich, J. Hite, G.T. Thaler, C.R. Abernathy, S.J. Pearton, R. Gwilliam, Appl. Phys. Lett. **89**, 152107 (2006)
102. H.Y. Yang, S.F. Yu, Y.Y. Hui, S.P. Lau, Appl. Phys. Lett. **97**, 191105 (2010)
103. Y. Jun, Y. Jung, J. Cheon, J. Am. Chem. Soc. **124**, 615 (2002)
104. M.H. Kane, M. Strassburg, A. Asghar, Q. Song, S. Gupta, J. Senawiratne, C. Hums, U. Haboeck, A. Hoffmann, D. Azamat, W. Gehlhoff, N. Dietz, Z.J. Zhang, C.J. Summers, I.T. Ferguson, Proc. SPIE **5732**, 389 (2005)
105. T. Dietl, H. Ohno, F. Matsukura, J. Cibert, D. Ferrand, Science 287, 1019 (2000)
106. K. Sato, H. Katayama-Yoshida, Physica E **10**, 251 (2001)
107. V.I. Litvinow, V.K. Dugaev, Phys. Rev. Lett. **86**, 5593 (2001)
108. Q. Wang, Q. Sun, P. Jena, Phys. Rev. B **70**, 052408 (2004)
109. A.Y. Polyakov, N.B. Smirnov, A.V. Govorkov, R.M. Frazier, J.Y. Liefer, G.T. Thaler, C.R. Abernathy, S.J. Pearton, J.M. Zavada, Appl. Phys. Lett. **85**, 4067 (2004)
110. K. Potzger, Shengqiang Zhou, H. Reuther, A. Mücklich, F. Eichhorn, N. Schell, W. Skorupa, M. Helm, J. Fassbender, T. Herrmannsdörfer, T.P. Papageorgiou, Appl. Phys. Lett. **88**, 052508 (2006)
111. H.-J. Choi, H.-K. Seong, J. Chang, K.-I. Lee, Y.-J. Park, J.-J. Kim, S.-K. Lee, R. He, T. Kuykendall, P. Yang, Adv. Mater. **17**, 1351 (2005)
112. A.Y. Polyakov, N.B. Smirnov, A.V. Govorkov, R.M. Frazier, J.Y. Liefer, G.T. Thaler, C.R. Abernathy, S.J. Pearton, J.M. Zavada, Appl. Phys. Lett. **85**, 4067 (2004)
113. M. Bercui, R.N. Bhatt, Phys. Rev. Lett. **87**, 107203 (2001)
114. A. Kaminski, S.D. Sarma, Phys. Rev. Lett. **88**, 247202 (2002)
115. A.A. Guda, S.P. Lau, M.A. Soldatov, N. Yu Smolentsev, V.L. Mazalova, X.H. Ji, A.V. Soldatov, J. Phys.: Conference Series **190**, 012136 (2009)
116. S. Jayanetti, R.A. Mayanovic, A.J. Anderson, W.A. Bassett, I.-Ming Chou, J. Chem. Phys. **115**, 954 (2001)

Chapter 5
Semiconductor Nanowire Heterostructures: Controlled Growth and Optoelectronic Applications

Chuanwei Cheng and Hong Jin Fan

Abstract Nanowire heterostructures with a modulated composition, structure, and interface represent an ideal system for the investigation of their synergetic physical properties arising from dissimilar material combination and for the construction of a wide range of novel nanoscale optoelectronic and photonics devices. In this chapter, the research progress on the controlled synthesis of a wide variety of nanowire heterostructures including segmented and superlattices, core/shell and core/multishell radial heterostructures, as well as branched heterostructures are reviewed. The fabrication methods cover vapor-phase, solution-phase, and template-based methods. The optoelectronic applications of semiconductor nanowire heterostructures in photoluminescence, photovoltaics, and photoelectrochemical water splitting as well as in photodetectors are also discussed.

5.1 Introduction

Since the discovery of carbon nanotubes by Iijima in 1991 [1], one-dimensional (1D) and quasi-one-dimensional nanostructures, including nanowires (NWs), nanotubes (NTs), nanorods (NRs), and nanobelts (NBs), have sparked a worldwide interest due to their unique optical, electrical, and magnetic properties as well as novel application potentials in nanoelectronics, catalysis, field emission, drug delivery, chemical and biological sensing, medical diagnostics and treatment, energy harvesting and conversion, and so forth [2–20]. Semiconductor NWs, as the key

C. Cheng (✉)
Department of Physics, Tongji University, 1239 Siping Road, 200092, Shanghai City, China
e-mail: cwcheng@tongji.edu.cn

H.J. Fan
Division of Physics and Applied Physics, School of Physical and Mathematical Sciences, Nanyang Technological University, 21 Nanyang Link, 637371, Singapore
e-mail: FANHJ@ntu.edu.sg

G.-C. Yi (ed.), *Semiconductor Nanostructures for Optoelectronic Devices*, NanoScience and Technology, DOI 10.1007/978-3-642-22480-5_5,

representative of 1D nanomaterials, are especially significant for the investigation of their fundamental physical properties and potential technological applications. First, the radial dimension of these structures are at or below the characteristic length scale of various fundamental solid-state phenomena, for example, the exciton Bohr radius, wavelength of light, phonon mean free path, critical size of magnetic domains, and exciton diffusion length [21]. As a result, many physical properties of semiconductors are significantly altered within the confinement in the radial direction of the NW. Second, the longitudinal unconstrained direction of a NW allows the transport of quantum particles such as electrons, phonons, and photons to be dimensional, and in extreme conferment case, ballistic. Driven by the above motivations, the research on NWs has expanded from conventional nonoxide Si, Ge, and III–V semiconductors to various metal oxides, as well as even to ternary compound semiconductors.

In the past two decades, various approaches have been developed to fabricate 1D nanostructures, ranging from chemical vapor deposition (CVD), metal-organic chemical vapor deposition (MOCVD), molecular beam epitaxy (MBE), pulsed laser deposition (PLD), thermal evaporation, and laser ablation-catalytic growth to solution-phase growth. The advance in research of single-component NWs set the stage for the present thrust of research on heterostructured NWs. One-dimensional heterostructures composed of two or more dissimilar materials are of particular interest due to their synergetic physical properties that arise but differ from the individual constituents, which can yield novel and/or enhanced functions, and thus are more useful for constructing new-concept nanoscale functional devices [22–30]. For example, a different semiconductor shell could be used to generate internal fields perpendicular to interface to provide confinement potentials in the core. By combining a radial heterojunction with a layer dopant atom, a two-dimensional cylindrical quantum well (QW) could be populated with high-mobility charge carriers. In addition to longitudinal superlattice or radial core-shell heterostructured NWs, hierarchical branch heteronanostructures with further increased specific surface areas in contrast to unidirectional NWs are also of wide interest. The advantage of nanostructural complexity includes, but not limited to, the improvement of light harvest and thus effectively reduces physical thickness of the photovoltaic (PV) and photoelectrochemical (PEC) devices.

This chapter provides a review of the state-of-the-art research advance in the controlled synthesis of semiconductor NW heterostructures and physical properties as well as the optoelectronic device applications. We first introduce various synthetic methodologies developed for the synthesis of 1D semiconductor heterostructures, including segmented NW heterostructures, core/shell and core/multishell NW heterostructures, and branched NW heterostructures (as illustrated in Fig. 5.1). In the second part, the physical properties and optoelectronics devices built in 1D semiconductor heterostructures are presented, which include optical properties such as photoluminescence (PL) and lasing, PV and PEC devices, as well as photodetectors. The chapter ends with a conclusion together with some perspectives and outlook to the future direction in the 1D semiconductor heterostructures research area.

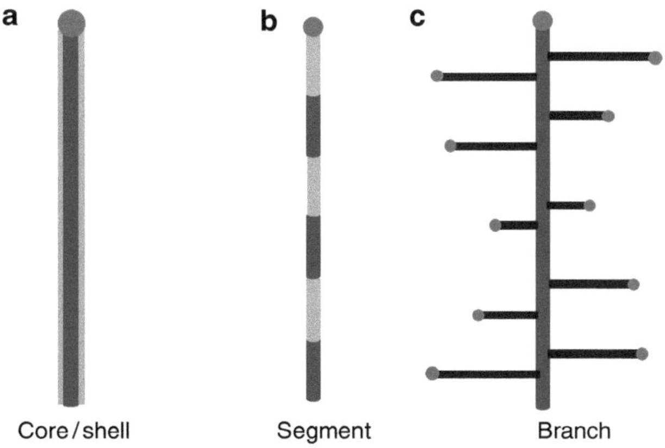

Fig. 5.1 Three different types of nanoscale 1D heterostructures: (**a**): Core/shell; (**b**): Segment; (**c**): Branch

5.2 Synthesis of Semiconductor NW Heterostructures

5.2.1 Segmented NW Heterostructures

Segmented heterostructures are typically called axial 1D heterostructures, which consist of more than two different constitutes. Based on their structures, it can be classified into two main forms, i.e., end-to-end and superlatticed NWs. There have been a few demonstrations of different strategies to fabricate segmented semiconductor NW heterostructures in the recent literature; we attempt to summarize some representative examples in four categories: catalyst-assisted vapor–liquid–solid (VLS) growth, catalyst-free vapor-phase methods, solution-based methods, template-induced growth methods, and lithography-based methods.

5.2.1.1 Catalyst-Assisted VLS Growth

The VLS growth is one of the most common routes to fabricate various semiconductor NWs. In 1964, Wagner made the pioneering studies on the silicon whiskers with modulated doping concentration in order to elucidate the VLS growth mechanism [31]. Later, Haraguchi et al. made substantial progress on the growth of GaAs whiskers in the early 1990s by using MOCVD method, including the fabrication of p–n junctions within whiskers [32,33]. After that, a significant step in the controlled synthesis of axial NW heterostructures was demonstrated by three leading groups (Lieber in Harvard, Samuelson in Lund, and Yang in Berkeley) with the growth of NW superlattices in a number of systems.

For example, Lieber and coworkers synthesized NW superlattices with laser ablation-catalytic growth route [34]. They obtained GaP–GaAs NW superlattices via alternating laser ablation of solid GaP and GaAs targets at temperature of

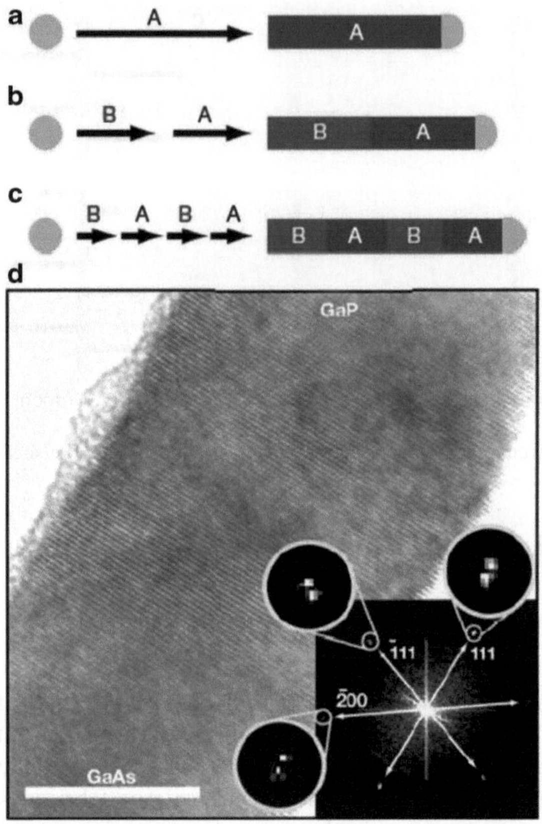

Fig. 5.2 Schematics of the synthesis of NW superlattices: (**a**) A nanocluster catalyst (shown gold) nucleates and directs semiconductor NW of the first growth step, a different material can be grown from the end of the NW. (**c**) Repetition of steps a and b leads to a compositional superlattice within a single NW. (**d**) HRTEM image of a GaAs/GaP junction grown from a 20-nm gold nanocluster catalyst. *Scale bar*, 10 nm. *Inset*, two-dimensional Fourier transforms of the entire image show a splitting of the reciprocal lattice peaks along the [111]. Reprinted with permission from [34]. Copyright@2002 Nature Publishing Group

700–850°C at 100 Torr in a continuous argon flow. Au particles on silicon served as a nucleation site, and segments were formed by switching between the targets. The lengths of the segments were controlled by the numbers of pulses delivered to each other. The synthesis process for creating segmented junctions within the NW is schematically described in Fig. 5.2a–c. High-resolution TEM (HRTEM) image of a typical GaAs/GaP junction region (Fig. 5.2d) exhibits a crystalline NW core without obvious defects. By VLS growth on 20 nm Au catalyst particles, Lieber group also synthesized p–i–n three-layered Si NWs, in which the p-Si and n-Si were obtained by growth in the presence of diborane and phosphine, respectively [35].

Tutuc et al. used similar method to fabricate p–n junctions within a Ge NW, where a boron source was used as dopant to form p-type segment [36]. Yang and coworkers combined CVD using silane tetrachloride ($SiCl_4$) and laser ablation with a solid germanium target to synthesize Si/SiGe segmented NW heterostructures[37]. The NW growth was at 850–950°C and 1 atm total pressure. During the growth, the silane tetrachloride ($SiCl_4$) vapor is constantly flowing though the chamber, while the ablation of Ge is controlled by programmed laser pulse to form SiGe at various segments (laser on) of the wire the other parts of NW consisting of Si only (laser off). Figure 5.3a shows an SEM image of the synthesized NW array. The diameters of these NWs range from 50 to 300 nm. Figure 5.3b shows a scanning transmission electron microscopy (STEM) image of two NWs in bright-field mode. Along the wire axis, dark stripes appear periodically, which originate from the periodic deposition of the SiGe alloy and Si segments. The chemical composition of the darker area is examined using energy-dispersive X-ray spectroscopy (EDS) (Fig. 5.3c), which shows a strong Si peak and apparent Ge doping (12 wt.% Ge). The periodic modulation of Ge doping is further confirmed by scanning a focused electron beam along the NW growth axis and tracking the change of X-ray signal from Si and Ge atoms in the wires (Fig. 5.3d). Both Si and Ge X-ray signals show periodic modulation.

Samuelson and coworkers produced InAs/InP NW superlattices with very abrupt interfaces using ultrahigh vacuum MBE [38]. The NW growth catalyzed by Au particles was performed at 420°C and $\sim$3 Torr using trimethylindium (TMIn), tertiarybutylphosphine, and tertiarybutylarsine as precursors. They demonstrated sequential growth NW heterojunctions by changing the source materials. By using similar methods, they also reported the synthesis of GaAsP/GaP [39,40], GaP/GaAs [41], and InAs/InAsP [42] segmented NWs. Zakharov and coworkers synthesized Si/Ge NW superlattices by MBE on a Au-coated Si(111) substrate [43], where Ge segments grew during interruption of the Si source.

5.2.1.2 Catalyst-Free Vapor-Phase Methods

Yi and coworkers developed a catalyst-free MOCVD method to prepare ZnO and various ZnO-based NR heterostructures. For example, they fabricated $ZnO/Zn_{0.8}Mg_{0.2}O$ NR quantum structures using catalyst-free MOVPE [44]. While MOCVD offers facile control in the well widths and number of well layers, other simple thermal evaporation methods also can produce segmented NWs. Zhan and coworkers adopted a simple thermal evaporation of a mixture of In and SiO powders for the fabrication of indium–silicon (In–Si) end-to-end NW heterostructures sheathed with amorphous silica [45]. The Si NWs sheathed with silica grow by the so-called oxide-assisted growth, while silica NTs grow in the opposite direction at the junction between the Si NWs and In droplets. Vapor-phase In is drawn into the silica nanotubes and condenses to form the final In–Si segmented junctions sheathed with amorphous silica. Similarly, Hu et al. synthesized Ga/ZnS segmented NWs encapsulated within a silica tube using thermal evaporation of a mixture of precursor powders of ZnS, Ga_2O_3, and Si heated at 1,400–1,500°C [46].

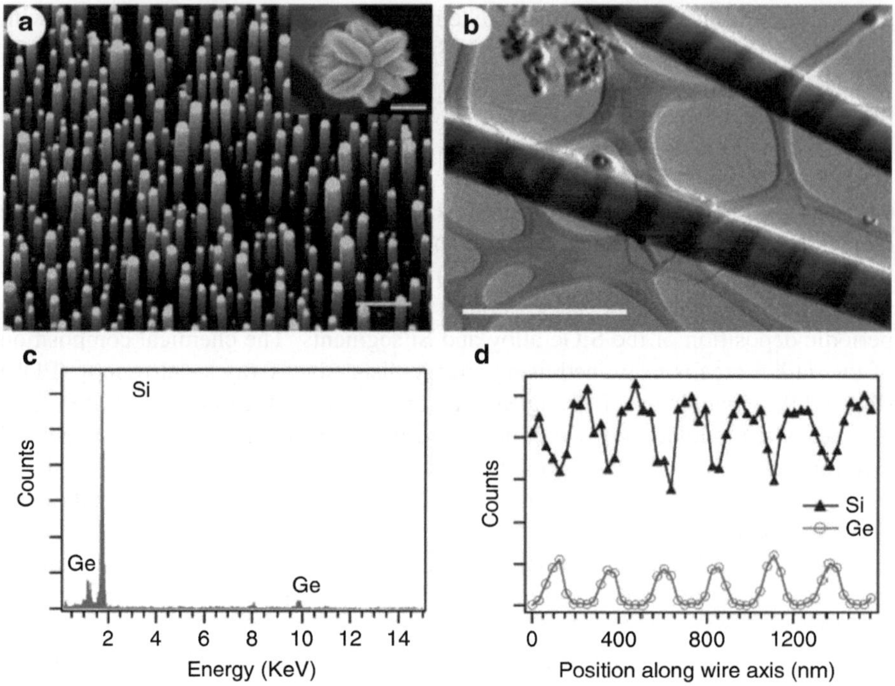

Fig. 5.3 (**a**) SEM image of Si/Ge superlattice NW array on Si(111) substrate. The *scale bar* is 1 μm. The *inset* shows the tip of one NW. The *scale bar* is 100 nm. (**b**) STEM image of two NWs in bright-field mode. The *scale bar* is 500 nm. (**c**) EDS spectrum of the Ge-rich region on Si/SiGe superlattice NWs. (**d**) Line profile of EDS signal from Si and Ge components along the NW growth axis. Reprinted with permission from [37]. Copyright@2002 American Chemical Society

5.2.1.3 Solution-Phase Methods

Solution methods can be both bottom-up and top-down. While the major literature about solution-phase growth of heterogeneous nanostructures is on metal NWs, semiconductors have also been attempted. Analogous to the epitaxial growth in a vapor phase, liquid epitaxy can also result in heterogeneous NRs of lattice-matching materials. For example, the Alivisatos group obtained CdS/CdSe and CdSe/CdTe segmented NRs in the solution phase using CdO mixed with alkylphosphonic acids as cation precursors and Se, S, or Te dissolved in a tri(*n*-alkylphosphine) as anion precursors [47]. As for the top-down strategy, Peng and coworkers demonstrated an interesting chemical etching method for fabricating p–n junction in Si NWs [48]. They used p–n junction Si wafer prepared by deposition of n-type silicon film on p-type Si wafers, followed by selective etching the wafer in HF solution with the addition of $AgNO_3$ at 50°C. Due to the so-called metal-assisted Si etching phenomenon, p–n Si NW arrays were formed. The NW length can be controlled by etching time. Similar to this approach, Geyer et al. fabricated Si/Ge superlattice

NWs by a combination of the growth of Si/Ge superlattices on Si wafer by MBE, surface patterning by anodic alumina oxide (AAO) templates, and metal-assisted wet chemical etching [49].

5.2.1.4 Template-Induced Growth Methods

Micro- and nanoporous templates are routinely used in the fabrication of multi-segmented NWs, in which the pores can be selectively filled with a variety of materials by a number of techniques. The size uniformity and high packing density of the pores of the templates facilitate the synthesis of monodisperse 1D nanomaterials in high yield. In the early 1990s, Martin and coworkers pioneered the use of hard templates for the synthesis of metal and polymer tubulars, rods, and wires by electrochemical deposition in the porous membranes followed by dissolution of the membrane [50, 51]. Since then, these early studies on single-component materials have been extended to multiple components nanostructures. Various heterojunctions have been formed by sequential deposition of metal, semiconductor, and carbon in the pores of membranes. For example, Pena and coworkers synthesized Au/CdSe/Au and Ni/CdSe/Ni segmented NWs by sequential deposition of materials in polycarbonate membranes using commercially available plating solutions for the Au and Ni segments and acid solution of $CdSO_4 + SeO_2$ for deposition of the semiconducting CdSe segment [52]. The fabrication process was illustrated schematically in Fig. 5.4a–f. Shown in Fig. 5.4g, h are SEM images of a 350-nm-diameter Au–CdSe–Au NWs and a 70-nm-diameter Ni-CdSe-Ni NW. These images show that the NWs can be removed intact from the template. A TEM image of a 70-nm-diameter CdSe wire shows that it is composed of grains approximately 10 nm in size. Zhu and coworkers fabricated metal–semiconductor end-to-end heterojunctions such as Ag/Si[53] and Pt_6Si/Si [54] by first depositing Ag or Pt segments in AAO templates from common plating solutions and then depositing Si segments on the tips of metal by CVD through the reaction of $SiCl_4$ and H_2.

5.2.1.5 Lithography-Based Methods

Lithography, including photolithography, electron beam lithography, and electrochemical dip-pen lithography (E-DPN), represents another kind of method to synthesize 1D segmented heterostructures. For example, Wu et al. synthesized striped metal–semiconductor NW heterostructures containing NiSi and Si segments [55]. They synthesized the Si NWs first by VLS methods and then used photolithography to form striped NiSi segments. Maynor and coworkers fabricated GaN/Ga_2O_3 NW heterojunctions by E-DPN [56]. The method involves performing local electrochemical reactions on a GaN NW underneath an atomic force microscopy. The NWs were typically $\sim$100 nm in diameter and several micrometers in length. Figure 5.5 shows an AFM image of fabricated GaN/Ga_2O_3 heterojunctions created

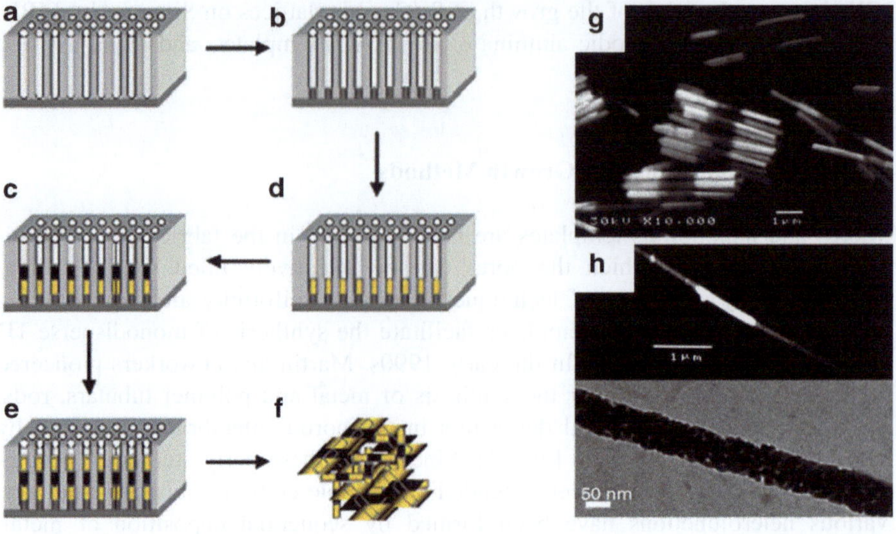

Fig. 5.4 Template-assisted growth of Au–CdSe–Au segmented NWs. (**a–f**) Schematics of the fabrication process: An AAO or polycarbonate template is coated on one side with Ag to provide the electrode. The template is then filled sequentially with approximately 1 μm of Ag, Au, CdSe layer, and another Au layer. Finally, the Ag backing and the membrane are dissolved, leaving free-standing NWs. (**g**) SEM images of 350-nm-diameter Au–CdSe–Au NWs. (**h**) A 70-nm-diameter Ni–CdSe–Ni NW. The *lower panel* in (**h**) shows a TEM image of a 70-nm-diameter CdSe segment. Reprinted with permission from [52]. Copyright@2002 American Chemical Society

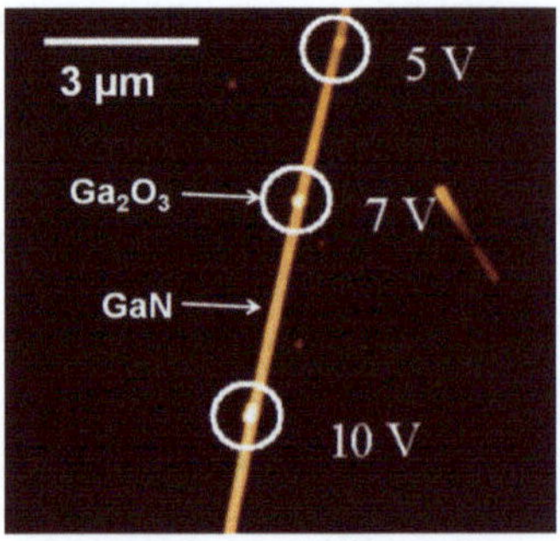

Fig. 5.5 AFM image of a GaN/Ga_2O_3 NW heterojunction fabricated using E-DPN. Reprinted with permission from [56]. Copyright@2004 American Chemical Society

by applying a voltage of 5, 7, and 10 V to the surface. The applied voltage and the amount of current passed controls the amount of GaN oxidized. The uniqueness of lithography method is the capability of fabrication on a single NW with arbitrary segment lengths and locations.

5.2.2 Coaxial and Core/Multishell Semiconductor NW Heterostructures

A solid NW, when wrapped with one or more outer layers of different materials, turns into radial NW heterostructures. Like the conventional planar multilayer thin-film devices, rational design and synthesis of functional radial semiconductor NW heterostructure are very important for constructing new nano-optophotonic devices such as p–n junction photovoltaics and LED and even vertical field-effect transistors. Again, the fabrication approaches mentioned in Sect. 5.2.1 are also workable for radial NW heterostructures, as summarized in the following.

5.2.2.1 Multistep Vapor-Phase Growth by Alternating Precursors

The most straightforward, although not necessarily the simplest way, is to start with a pregrown NW and conduct uniform CVD deposition of targeted materials. The radial growth of a second material can be realized by altering the growth kinetics through changing the pressure, flow rate, temperature, and reactant precursors, which favors a homogeneous vapor-phase deposition on the NW surface of the first material, rather than VLS growth (unless the metal catalyst used for the VLS growth was intentionally removed before the secondary deposition). MOCVD is the most-used method for forming high-quality core/shell and multishell semiconductor heterostructures. Lieber and coworkers synthesized i-Si/p-Si NWs by conducting a radial growth mode during CVD. Following the VLS i-Si core synthesis, a core/shell formed with borane added as a p-type material [57]. They also used the similar procedure to fabricate various core/shell and core/multishell heterostructures, including Ge/Si [58,59] and n-GaN/$In_xGa_{1-x}N$/GaN/p-GaN [60]. Similarly, Samuelson and coworkers fabricated GaAs/$Ga_xIn_{1-x}P$ core/shell NWs by growth of a GaAs core at 450°C followed by the formation of a $Ga_xIn_{1-x}P$ shell at 600°C [61]. Fukui et al. synthesized various core/shell heterostructures by catalyst-free selective-area MOVPE, where the precursors were introduced in an alternating fashion [62–65]. Yi et al. used MOVPE and alternating reaction conditions to fabricate ZnO/GaN [66], as well as ZnO/$Mg_{0.2}Zn_{0.8}O$ core/shell structures.

In addition to CVD, a good alternative to create coaxial and core/multishell heterostructures is PLD. In the PLD process, the thickness of the shell could be controlled by the target-to-substrate distance, deposition duration, pulse repetition frequency, and laser energy density. Recently, we combined the VLS and PLD method to synthesize ZnO/ZnCdO core/shell and core/multishell QWNW heterostructures [69]. Figure 5.6a, b shows the SEM images of both bare ZnO NWs and ZnCdO/ZnOMQW structures. The bare ZnO NWs have an average diameter of 65 nm and a length of $\sim$1 μm, and a vertical alignment due to the epitaxial growth. After the PLD of MQWs, there was no significant change in morphology except that the NW diameters were enlarged, which is ascribed to the continuous growth of the ZnCdO (well) and ZnO (barrier) layers. The single-crystalline

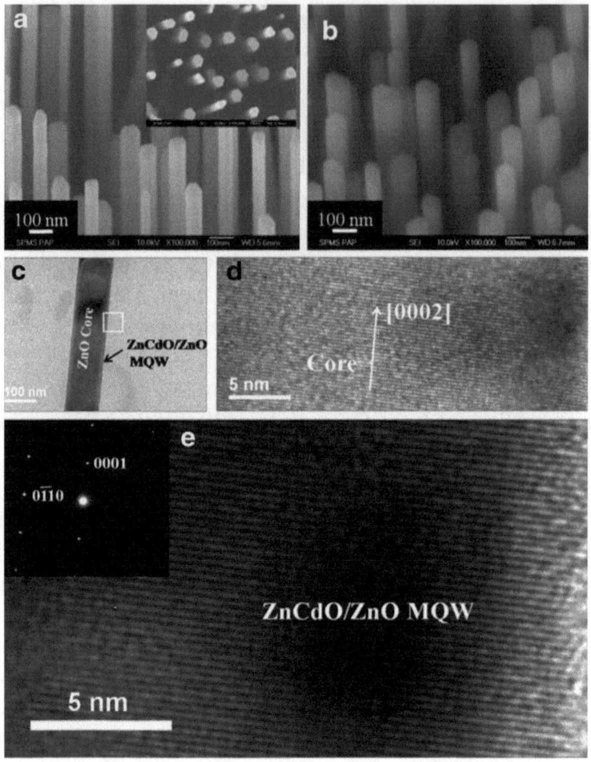

Fig. 5.6 20° tilted SEM images of (**a**) bare ZnO NW arrays. *Inset*: top view. (**b**) ZnCdO/ZnO MQW NW arrays. (**c**) TEM image of the ZnCdO/ZnO NW MQW. (**d**) HRTEM image taken from (**c**) the ZnO NW core, (**e**) the MQWs shell. *Inset*: Corresponding selected area electron diffraction pattern (SAED) recorded from the shell region [69]

ZnO NW cores enable an epitaxial and dislocation-free growth of highly uniform (ZnCdO/ZnO) QWs. TEM images (Fig. 5.6c) show that the core-shell structures are smooth both in their surface and interface. The total thickness of the multishell layers is about 22 nm. HRTEM image and electron diffraction pattern (Fig. 5.6d e) reveal that the total NW is single crystalline without interfacial defects. Because of the small percentage (3.8%) of Cd, the lattice structure is nearly unchanged; thus, the diffraction pattern of ZnCdO/ZnO MQWs NW is similar to the pristine ZnO NW. Other demonstrations of PLD of outer shells were made by Wang et al. for ZnSe/ZnO core-shell NW arrays [70].

Atomic layer deposition (ALD) is a cyclic self-limiting deposition method, which is capable of conformal and uniform coating of thin films at atomic level. Due to the nature of layer-by-layer deposition and the shell thickness is directly proportional to the number of precursor/purge cycles, ALD allows a very precise control in the shell thickness. ALD has been used largely for gate-oxides in thin-film or NW-based field-effect transistors, as well as deep-trench capacitors. Recently, ALD is being

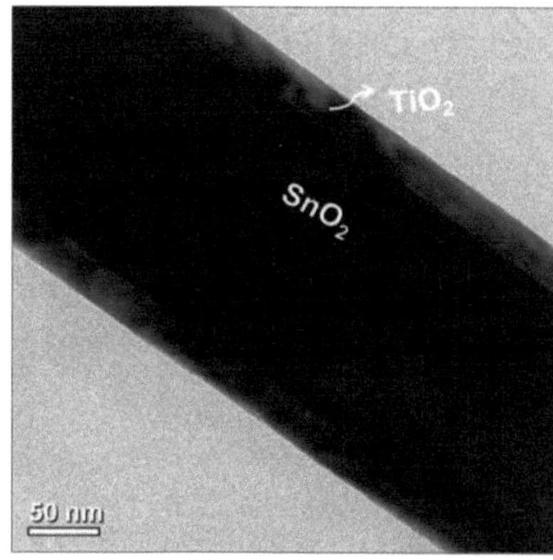

Fig. 5.7 A TEM image of SnO_2/TiO_2 core-shell structures created by ALD

employed as a new tool to deposit a variety of materials including oxides, nitrides, and metals on NWs surfaces to form core/shell heterostructures. For example, we employed ALD to produce uniform Al_2O_3 and TiO_2 shells onto VLS-grown ZnO cores [71, 72]. Chang et al. prepared Si/ZnO NW heterojunctions by ALD ZnO thin films on Si NWs [73]. We also fabricated SnO_2/TiO_2 core/shell heterojunctions by depositing TiO_2 shell onto VLS-grown SnO_2 NW using $TiCl_4$ and H_2O as the sources of titanium and oxygen, respectively. Figure 5.7 shows a typical TEM image of SnO_2/TiO_2 core-shell structures created by coating TiO_2 layer with ALD onto VLS-grown SnO_2 NWs. Unlike CVD, the shell by ALD is normally amorphous (e.g., for Al_2O_3, SiO_2, TiO_2, and Fe_2O_3) or polycrystalline (e.g., ZnO) and thus there is no interface epitaxy; an additional annealing step is usually needed to convert the amorphous shell into crystalline for optoelectronic applications.

5.2.2.2 One-Step Vapor-Phase Growth by Introducing all Precursors Together

Magic can frequently occur. In simple one-step evaporation growth by thermal evaporation of a mixture of different precursors, core-shelled NWs are also obtained from a large variety of independent experiments. Li et al. used a single-step thermal evaporation using CdSe powders and a thin oxide layer covered Si wafer as substrate, and observed Si/CdSe coaxial nanocables [74]. The growth of the nanocable was proposed to involve two main stages: In the first stage, silicon NWs were formed on the silicon substrate via an oxide-assisted mechanism [75], where silicon oxide plays important roles including serving as reactant in the disproportionation reaction and terminating specific Si surfaces and thus inducing the anisotropic NW growth. The CdSe starts to sublime at 1,100°C and is carried by the Ar to a lower temperature zone, where it condenses. In the second stage, the as-formed silicon NWs serve as the template for the deposition of CdSe, resulting in the Si-core/CdSe-sheath nanocable heterostructures as shown in Fig. 5.8. Light/dark contrast is observed in the corresponding core/shell region of each

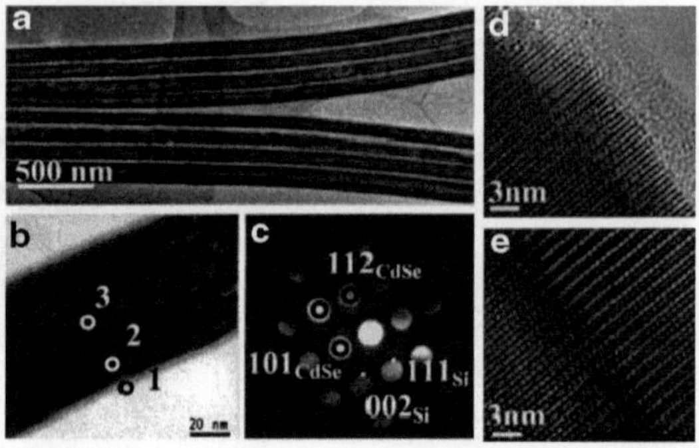

Fig. 5.8 (**a**) Low-magnification TEM image of Si/CdSe nanocables. (**b**) TEM image showing a single nanocable; (**c**) μ− diffraction taken from the nanocable in (**b**), suggesting crystalline Si and CdSe. (**d**) HRTEM image of the interface between areas 1 and 2 as marked in (**b**). (**e**) HRTEM image showing the interface between areas 2 and 3 as marked in (**b**). Reprinted with permission from [74]. Copyright@2008 American Chemical Society

individual nanocables (Fig. 5.8a). Microdiffraction (μ-diffraction) taken from the same nanocable shown in Fig. 5.8b reveals two sets of diffraction spots (Fig. 5.8c), which can be indexed to hexagonal CdSe and silicon, respectively. High-resolution images shown in Fig. 5.8d, e reveal the crystalline layer of both Si-core and CdSe shell, as well as a ∼5-nm-thick outmost amorphous layer. By adopting the similar procedure, they also fabricated Si/ZnSe bicoaxial NW heterostructures by simple evaporation of ZnSe powders in the presence of H_2 gas [76]. Mathur and coworkers reported a single-step synthesis of 1D Ge/SiC_xN_y coreshell nanocables by CVD of the molecular precursor [Ge{N(SiMe$_3$)$_2$}$_2$] [77]. Single-crystalline Ge NWs (diameter ∼60 nm) embedded in uniform SiC_xN_y shells were obtained in high yield on various substrates such as Si, Fe, Al_2O_3, and MgO. Uemura and coworkers prepared InS/SiO_2 core/shell NWs 30–200 nm in diameter and tens of micrometers by a single-step physical vapor deposition method [78]. The InS and silicon powders were used as precursors and heated to 900–1,000°C for 1 h and the product was collected from the wall of the alumina crucible.

5.2.2.3 Solution-Phase Methods

Solution-based methods are favorable due to its simplicity and low cost and thus widely used for synthesis of semiconductor NW core/shell heterostructures. There are two main types of solution synthesis route: one is by mixing all the precursors at once in one step and the other is a two-step one: the core NW was chemically synthesized first and then transferred to a new solution for the growth of the shell. A few examples are given below.

Wang and coworkers synthesized SiC–C coaxial nanocables with orientation accumulation of β-SiC tapered crystallite NWs coated in an amorphous carbon sheath via one-step and low-temperature (250°C) solvothermal process using $SiCl_4$, C_6Cl_6, and Na as starting materials [79]. The length of these nanocables is up to 1 μm and the diameter is around 50 nm. The same group also reported a facile and environmental friendly one-step hydrothermal route for fabrication of Te/C nanocables with semiconducting Te as core and carbon as shell [80]. Investigation of the formation process of the nanocable revealed that t-Te NWs were obtained first through a solid–solution–solid transformation process and then dextran gradually carbonized on the surfaces of the Te NWs to form the Te/C nanocables.

Goebl and coworkers demonstrated an effective solution-phase method for synthesis of II–VI core/shell NW heterostructures via solution–liquid–solid (SLS) growth mechanism [81]. Three core/shell systems, CdS/CdSe, CdSe/CdS, and CdSe/ZnTe NW heterostructures, have been synthesized by slowly introducing shell precursors into a solution of premade core NWs dispersed in a noncoordinating solvent at moderate temperature (215–250°C). Representative TEM micrographs of CdS/CdSe NWs are shown in Fig. 5.9. A survey of such images reveals that ~95% of the wires appear coated. In the thin-shell image in Fig. 5.9c, the shell is epitaxial with no misfit dislocations. Thus, instead of generating such defects during shell deposition, lateral strain relaxation of the growing islands must relieve most of the misfit strain present. In contrast, the thicker shell in Fig. 5.9d is polycrystalline, exhibiting lattice fringes oriented in various directions. Thus, there exists a transition from epitaxial to nonepitaxial shell growth as the shell thickness increases.

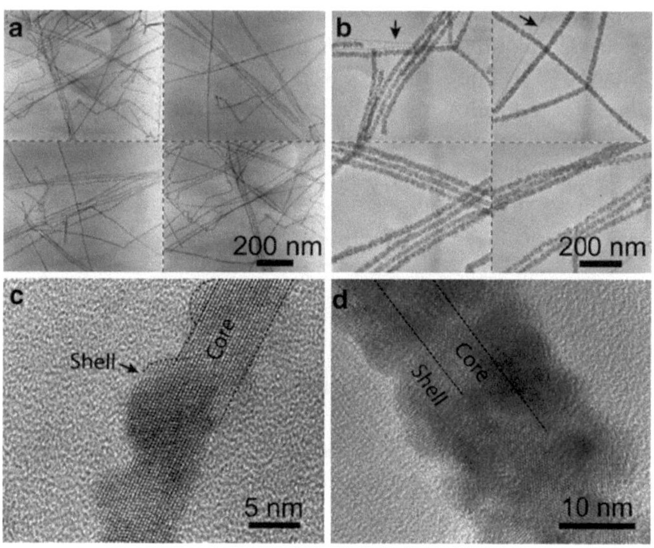

Fig. 5.9 TEM images of CdS/CdSe core/shell NWs with (**a**) and (**c**) thin (~1 nm) and (**b**) and (**d**) thick shells (~10 nm). Reprinted with permission from [81]. Copyright@2008 American Chemical Society

5.2.3 *Branched Semiconductor NW Heterostructures*

5.2.3.1 Metal-Catalyzed Growth with Sequential Catalyst

As the growth of unidirectional NWs which needs a metal nanocluster as catalyst, the growth of secondary branches or even hyperbranches can be induced by introducing new metal catalyst onto the NW surfaces. Therefore, the methods used in early reports of branched heterogeneous NWs involve sequential depositions of metal catalyst, followed by VLS or SLS growth. Obviously, this process involves three steps. The first procedure is the standard VLS or SLS growth of semiconductor NW as the backbone. The second step is deposition of metal catalyst onto the NW. The third step is the growth of the branched NW onto the previously prepared NW backbone. Lieber and colleagues synthesized branched and hyperbranched Si and GaN NWs via a multistep Au nanocluster-catalyzed VLS approach [82]. Samuelson and coworkers obtained GaAsP/GaP branched nanotrees vertically aligned on the GaP (111)B substrate in which the Au aerosol particles were deposited onto the GaP stems for the subsequent growth of GaAsP branched (see Fig. 5.10a) [83, 84]. The GaAsP branches with ~40 nm in diameter and up to several hundreds of nanometers long formed on GaP stems. The EDX on various parts of the nanotrees (Fig. 5.10a) shows that the As component is present only in the branches grown during the second stage. The density of the branches can be varied by controlling the amount of the Au particles deposited on the first NW. Wan et al. synthesized $In_2O_3/Sn : In_2O_3$ [85] (as shown in Fig. 5.10b) and $SnO_2/Sb : SnO_2$ [86] branched NWs also by a two-step Au-catalyzed VLS process. Zhou and coworkers reported the synthesis of sixfold-symmetry CdS/ZnS branched NW heterostructures with a simple two-step evaporation method [87]. The ZnS NWs were synthesized in first step and were then used as templates for the following growth of sixfold symmetrical CdS NBs or NWs at varied temperature in the second step. Besides the VLS growth, Buhro has also demonstrated the sequential SLS growth of homo- and heterostructured NWs [88]. This approach allows different types of metal catalyst, or different sizes of metal catalyst, to be used for alternate generations of branches, providing an additional control.

5.2.3.2 Solution Growth Combined with Other Methods

As the solution growth of NWs or NRs can take place on nonplanar substrate, it can be modified to produce hierarchical structured NWs by using preformed NWs as the growth substrate. Recently, various types of branched metal oxide NW heterostructures were realized in our group by a combination of VLS growth for the backbone NWs and the subsequent solution growth of branches. In the following, we give a material-by-material summary of our recent work on branched oxide NWs.

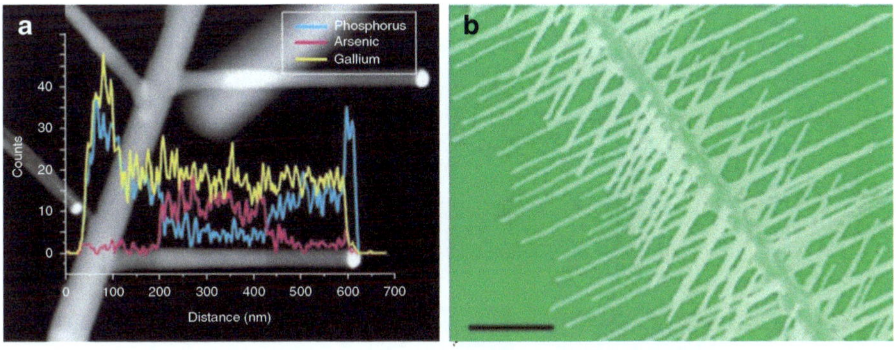

Fig. 5.10 (**a**) STEM image of branched GaP/GaAsP NWs with EDX elemental line scanning of As, Ga, and P along a branch. Reprinted with permission from [83]. Copyright@2004 Nature Publishing Group. (**b**) SEM image of branched In_2O_3 NWs grown on ITO NWc backbones. Reprinted with permission from [85]. Copyright@2006 American Chemical Society

SnO_2/ZnO

Figure 5.11a–c shows schematically the two-step growth process for growing SnO_2/ZnO hybrid NWs [89]. First, the VLS-grown SnO_2 NWs were attached by dip coating with a layer of ZnO seed nanoparticles as specific nucleate sites. In the early stage, ZnO NRs nucleate and crystallize along their [0001] direction on the ZnO seed sites. With time increases, the initial NRs absorb Zn^{2+} and OH^- from the solution and crystallize following the well-documented chemical reactions in the presence of zinc salts and HMT. A typical image of such structure is shown in Fig. 5.11d. The NR branches stand perpendicular to the side surfaces of the SnO_2 NWs as multiple rows in a parallel manner, forming a fourfold symmetry structures. The main interface relation for SnO_2/ZnO hierarchical nanostructures is $(0002)ZnO||(\bar{1}01)SnO_2$ revealed by HRTEM and SAED. Moreover, the number density and morphology of the secondary ZnO can be tailored by changing the precursor concentration, reaction time, and by adding surfactants.

Dendritic TiO_2/SnO_2

Dendritic SnO_2/TiO_2 nanoheterostructures with a twofold point symmetry was fabricated directly by solution heteroepitaxial growth of TiO_2 NR branched onto VLS-growth SnO_2 NWs [90]. The hydrothermal process was conducted at 180°C with tetrabutyl titanate, HCl, and toluene as precursors. A typical SEM image of the as-obtained SnO_2/TiO_2 structures is shown in Fig. 5.12a. In general, the TiO_2 NR branches assembling in a bundle structure stand with a fixed ~70° angle to the two side surfaces of the SnO_2 NWs as multiple rows in a parallel manner, forming a twofold point symmetry. The length of the TiO_2 NRs ranges from 300 to 400 nm and the diameter of individual TiO_2 NR is about 10 nm. The SAED

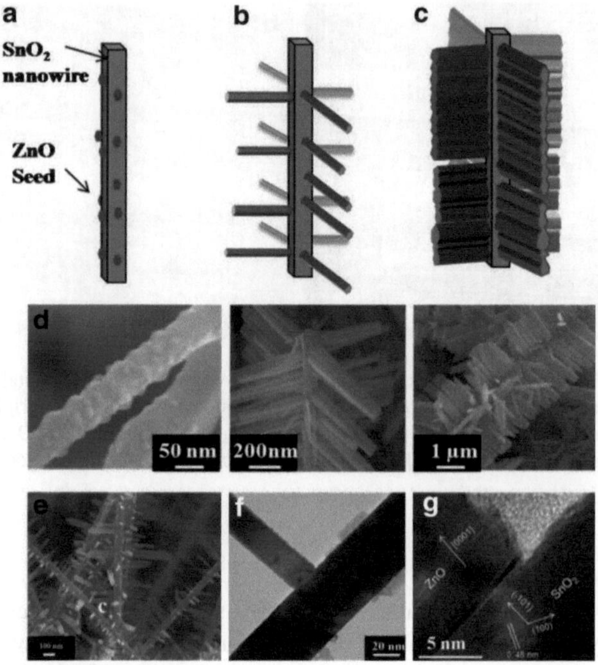

Fig. 5.11 Growth of SnO_2/ZnO hierarchical nanostructures. (**a–c**) Schematics of the fabrication process. (**a**) Before the growth. The SnO_2 NW surfaces are coated with ZnO seed nanoparticles. Subsequent solution epitaxial growth of ZnO NRs on the four side faces of SnO_2 NWs. (**b**) With a low $Zn(NO_3)_2$ precursor concentration the branches are individual ZnO NR arrays, whereas (**c**) with a high precursor concentration the NRs tend to merge. (**d**) Corresponding SEM images of the product. (**e**) A typical SEM image of hierarchical SnO_2/ZnO nanostructures with ZnO NRs branch on SnO_2 NWs backbone. (**f**) TEM image taken near the junction. (**g**) HRTEM lattice image of the junction [89]

pattern (Fig. 5.12b) consists of two sets of clearly correlated patterns, from which one can see that both the SnO_2 and TiO_2 are in the same zone axis and have the same orientation, revealing the perfect interfacial lattice relationship between the stem and branch. Based on the HRTEM images taken from the SnO_2 backbone (Fig. 5.12c) and the interface (Fig. 5.12d), the lattice spacings are measured to be 2.62 and 2.38 Å, which are consistent with the d-spacings of $(10\bar{1})$ planes of tetragonal SnO_2 and $(10\bar{1})$ of TiO_2 respectively. From the HRTEM and SAED analysis, we can conclude that the SnO_2 NW stems grow in the [101] direction bounded by $\{10\bar{1}\}$ and {010} surface facets. The longitudinal direction of the TiO_2 branch is then determined to be along $[\bar{1}01]$. The stem-branch angle determined from these reciprocal lattice peaks (Fig. 5.11b), 70.6°, is consistent with the 68° angle between the two growth directions for the stem and branch. The adopted epitaxial orientation relationship between the stem and branch is as follows: $TiO_2[010]//SnO_2[010]$ and $TiO_2(10\bar{1})//SnO_2(10\bar{1})$.

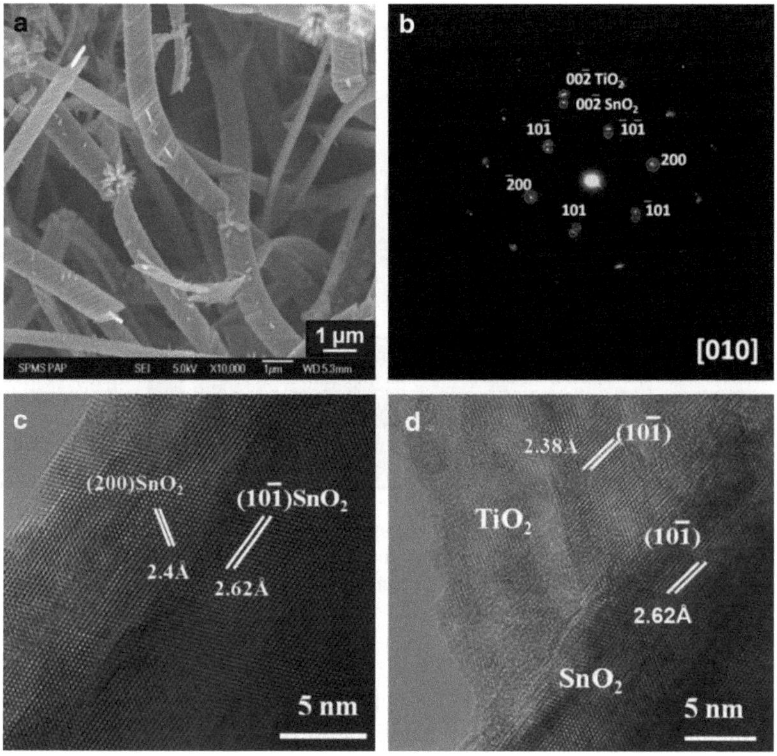

Fig. 5.12 (**a**) Typical SEM image of the dendritic SnO_2/TiO_2 nanostructures. (**b**) SAED pattern recorded from the SnO_2/TiO_2 truck-branch interface with the [010] zone axis. The diffraction spots from rutile SnO_2 and TiO_2 are labeled in *blue* and *red*, respectively. (**c**) and (**d**) HRTEM images taken from the SnO_2 NW trunk and the truck–branch interface, respectively, with the determined lattice spacings [90]

SnO_2/Fe_2O_3

By combining a vapor transport deposition and a facile hydrothermal method, we also successfully synthesized a novel sixfold-symmetry branched $\alpha - Fe_2O_3/SnO_2$ nanoheterostructures composed of SnO_2 NW stems and $\alpha - Fe_2O_3$ NR branches [91]. First, sixfold symmetry $FeOOH/SnO_2$ heterostructures were obtained by solution expitaxy, followed by thermal treatment at 450°C for 2 h to convert FeOOH into $\alpha - Fe_2O_3$, during which the overall morphology was conserved. The lengths of FeOOH NRs can be simply tuned by adjusting the concentration of the reactant and/or the growth time. A typical SEM image of the $\alpha - Fe_2O_3/SnO_2$ composite structures is shown in Fig. 5.13a. It demonstrates that multiple rows of secondary NRs were formed on the major NWs in a parallel fashion. The majority of these hierarchical structures exhibit a sixfold symmetry; i.e., the NR branches grow along six directions on the tetragonal SnO_2 NWs with the angle of approximately 60°

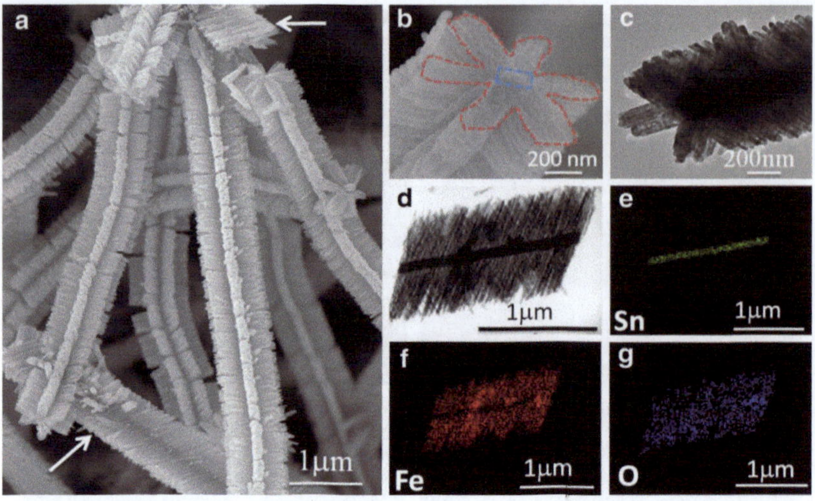

Fig. 5.13 (**a**) Typical SEM image of the sixfold-symmetry branched $\alpha - Fe_2O_3/SnO_2$ nanostructure. (**b**, **c**) High-magnification SEM and TEM images of single branched NW. (**d**–**g**) TEM image of single branched NW and the corresponding elemental mapping [91]

between adjacent branches (Fig. 5.13b). The magnified SEM and TEM images (Fig. 5.13b–d) also reveal that the branches are assembled in bundles, with an average diameter of about 50 nm and length of 500 nm, of well-aligned porous NRs. The TEM elemental mapping of a "twofold-symmetry" branched NW is conducted to clearly identify the spatial distributions of Fe, Sn, and O in the backbone and branches (as shown in Fig. 5.13e–g).

It should be noted that in previous reports the sixfold-symmetry branched nanoheterostructures are induced by the anchoring of branches onto six identical surfaces of the stems [92–94]. In our case, surprisingly, the secondary FeOOH NRs are epitaxially grown on the four side surfaces of the SnO_2 NWs with a sixfold-symmetry, which is a result of the minimization of lattice mismatches on distinct SnO_2 surfaces. When tested as lithium-ion battery anode materials, such branched $\alpha-Fe_2O_3/SnO_2$ composites show superior performance to both SnO_2 and $\alpha-Fe_2O_3$ individual components in terms of low initial irreversible loss and high reversible capacity [91].

Si/ZnO Nanotrees

In addition to the VLS-based vapor-phase growth technique, conventional lithography can also be applied to prepare the backbone NWs, and finally ordered tree-like heteronanostructures [95]. In the example of Si/ZnO, three steps were involved. First, the patterned Si nanopillar arrays were prepared by standard top-down photolithography followed by dry etching. This step defines the height and interdistance

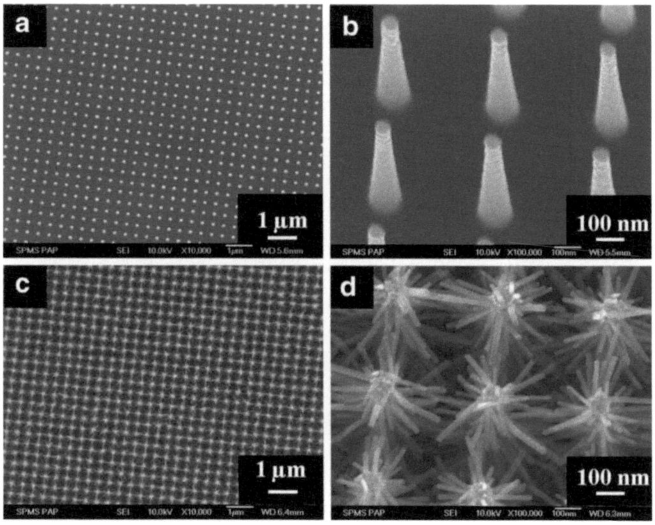

Fig. 5.14 SEM images of (**a**, **b**) the Si nanopillar arrays and (**c**, **d**) Si/ZnO nanotrees; (**a**, **c**) top view and (**b**, **d**) 20° titled view

of the trunks. Subsequently, ZnO branches were grown on the Si nanopillars through a routine hydrothermal process. The lengths of the NRs are adjustable by controlling the growth time (1–5 h). The ZnO seed layer needed for the hydrothermal growth was coated using ALD, which is particularly advantageous in the case of high aspect ratio Si pillars. Figure 5.14a–d shows the top-view and tilted-view SEM images of the as-prepared Si nanopillar arrays on Si substrate after etching. The pillars are oriented perpendicular to the substrate surface with good uniformity. The pillar shows a tower-like morphology with a typical height of 300 nm, the average diameter of Si pillar is about 150 nm and gradually decreases from the root to the tip. The nearest-neighbor spacing of the pillars is 400 nm, as predefined by shadow mask. After applying the solution growth of ZnO NRs, the initially smooth Si pillars branch out, forming tree-like nanostructures. The typical SEM images of such structures are shown in Fig. 5.14a–d. The ZnO NR branches mainly stand perpendicular to the side surfaces of the Si pillars. From the magnified SEM images (Fig. 5.14c, d), the diameter and length of secondary ZnO NRs are determined to be about 30 nm and 150–200 nm, respectively.

5.2.3.3 Phase Transition-Induced Branching

Another interesting approach toward semiconductor branched heterostructures is through phase transition during the growth without the addition of intentional branching catalyst [96, 97]. Generally, these materials are limited to those that can crystallize in either wurtzite (WZ) or zinc-blende close-packed crystal structure or

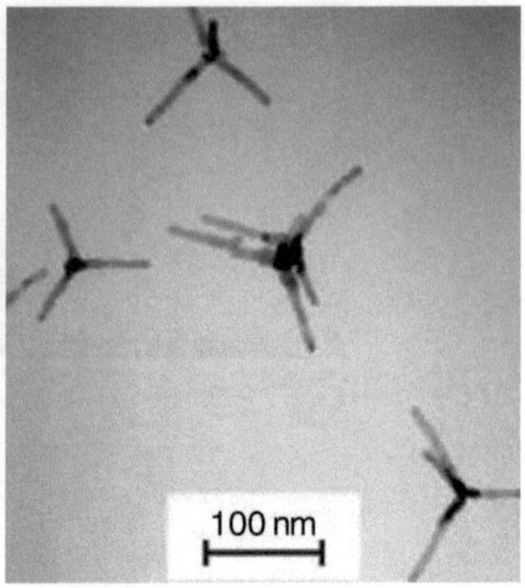

Fig. 5.15 A TEM of the CdTe tetrapods. Reprinted with permission from [97]. Copyright@2003 Nature Publishing Group

even mixed polytype under specific conditions. For example, CdTe [97] has cubic-close-packed zinc-blende polymorph with very similar thermodynamic stability to its hexagonal-close-packed wurtzite form. They also have a relatively small stacking fault energy to switch between the zinc-blende <111> or wurtzite <0001> epitaxial interface. One example of such crystallographic phase change-induced CdTe nanotetrapod is given in Fig. 5.15. A zinc-blende structure is more favored at smaller nanoscale size, whereas wurtzite is the more stable in bulk phase. Therefore, in a homogeneous solution, the initial CdTe first nucleates with a tetrahedral zinc-blende core; when the core reaches a certain critical size, branched NWs with two, three, or up to four equivalent <111> directions form, resulting in a tetrapod geometry.

The variation of crystal phase could also be mediated by chemical solvents during solution growth. Chu et al. reported the synthesis of different branched CdS products by adjusting the ratio of two solvents, ethylenediamine and ethylene glycol, under the solvothermal condition [98]. There the ethylenediamine is propitious to the formation of wurtzite CdS with high aspect ratio, and ethylene glycol is favorable for the formation of zinc-blende CdS with low aspect ratio.

5.3 Applications of Semiconductor NW Heterostructures

5.3.1 Optical Properties

NW heterostructures provide a possibility for the modulation of the optical properties of semiconductors. In the past few years, optical properties of various semiconductor NW heterostructures have been studied by PL spectroscopy. Recently, we have reported the interesting PL property of ZnO/ZnCdO MWQ radial NW heterostructures [99]. Figure 5.16a shows the 10 K PL spectra of the pristine ZnO

NWs and ZnO/ZnCdO MQW NW heterostructures. The dominant peaks at 3.354 and 3.358 eV are ascribed to the neutral donor-bound exciton (D^0X) in ZnO NWs. The A-free exciton emission (FX_A) at 3.368 eV is also observed. At a lower energy side of D^0X, the peak at 3.306 eV is ascribed to the longitudinal optical (LO) phonon of FX_A. The other lower energy peaks resolved at 3.236, 3.166, 3.095, 3.023, and 2.950 eV are from the donor–acceptor pair (DAP) as well as its four orders LO-phonon replica with a constant energy separation of 70-73 meV. For the ZnCdO/ZnO MQW NWs, a new peak centered at 3.2 eV dominates the PL spectrum, as shown in Fig. 5.16a, which is believed to be the ZnCdO QW-related emission due to an evident quantum confinement effect. Further investigation on the effect of exciton localization in the MQWs is conducted by temperature-dependent PL spectra. Figure 5.16b shows the normalized PL spectra measured from 10 to 300 K. The intensity of the ZnCdO QW-related emission and near-band-gap emission of ZnO decreases with increasing temperature, which is partly due to the increased nonradiative recombination. Interestingly, the QW-related PL emission shows a different energy shift behavior with increasing temperature compared with those of FX_A, D^0X, and FX_A-1LO: the former exhibits an S-shaped trend. The temperature-dependent PL peak for the FX_A emission of ZnO is consistent with the estimated energy decrease of about 93 meV at 300 K. However, the ZnCdO QW-related PL emission does not follow the conventional trend due to the carrier localization effect. Instead, it exhibits an anomalous S-shaped temperature dependence of the peak energy (E_p). As the temperature increases from 10 to 80 K, E_P red shifts by 12 meV and then blue shifts by 22 meV in the temperature range of 80–180 K. When the temperature is further increased above 180 K, the peak red shifts again. This interesting S-Shape behavior can be ascribed to the carrier localization effect caused by the inhomogeneous alloy composition and/or roughness interface, in addition to the confinement of carriers within the QW [99, 100]

The stimulated emission and lasing of NW heterostructures have also attracted much attention. We reported the random lasing action in SnO_2/ZnO branched heterostructures with SnO_2 NW as stem and ZnO NRs as branches. It is observed that when excitation power exceeds a threshold of $\sim 0.18\,\text{MW/cm}^2$, sharp peaks in a linewidth as narrow as 0.4 nm emerged from the single-broad spontaneous emission spectrum. The properties of emission spectra are qualitatively consistent with the random lasing behavior with a coherent feedback [101].

Qian and colleagues reported the controlled synthesis and lasing property of GaN/InGaN MQW core/shell NW heterostructures [102]. Optical studies on individual NWs demonstrate lasing action with InGaN QW composition-dependent emissions ranging from 365 to 494 nm.

5.3.2 Photovoltaics and Photoelectrochemical Water Splitting

Nanostructured materials offer opportunities for solar energy conversion that promise lower fabrication cost and higher conversion efficiency [103]. Especially,

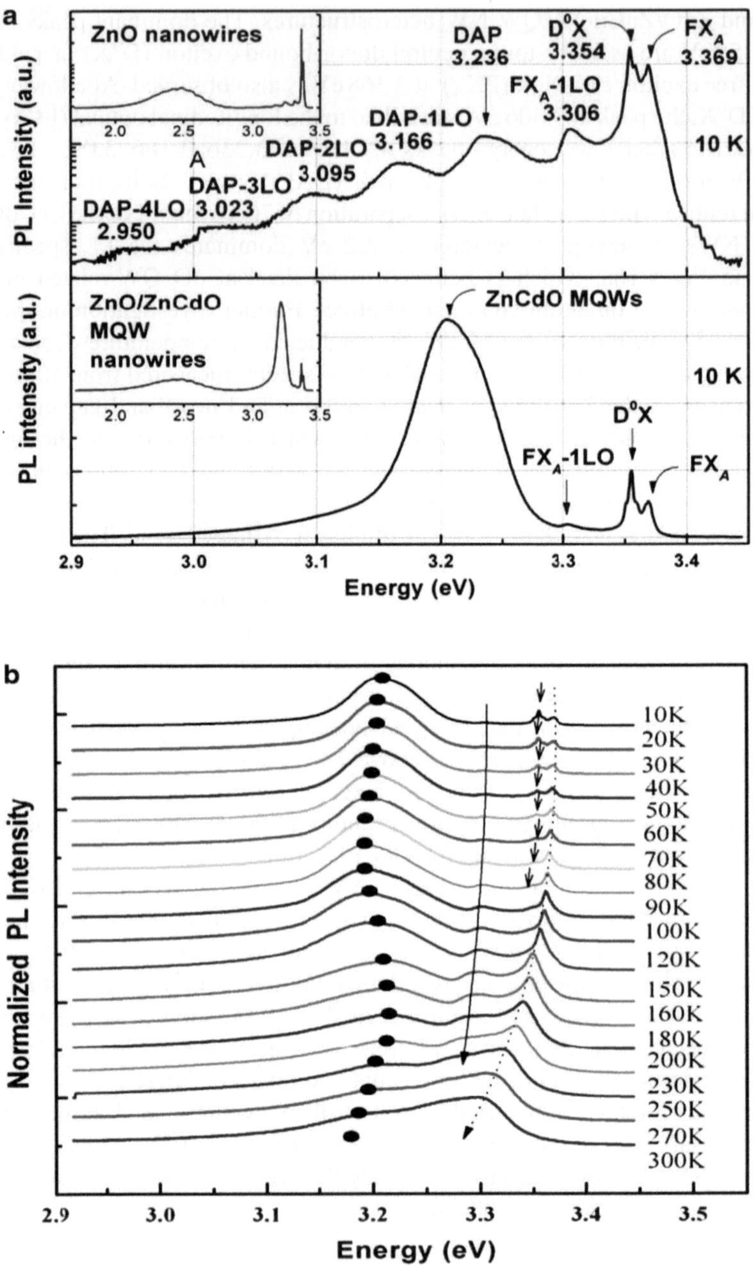

Fig. 5.16 (**a**)10 K PL spectra of bare ZnO NW arrays and ZnO/ZnCdO MQW NW arrays. (**b**) Temperature-dependent PL spectra of the NW MQWs in the range from 10 to 300 K. *Solid circles* show the evolution of the QWrelated emission. The *dashed line* indicates the peak evolution of the FX_A emission and the *solid line* shows the trend of FX_A-1LO. The *arrows* denote the D^0X peak [69]

1D semiconductor heterostructures show great promise in solar energy conversion which is advantageous over conventional planar thin-film PV or PEC cells. For example, in a radial NW heterostructure, the light absorption and carrier separation occur in orthogonal directions, so solar light absorption occurs along the longer dimensions of NWs (several microns or more), while the carrier separation occurs by diffusion across the short radial distance (tens of nanometers). As a result, overall device performance does not significantly suffer if lower quality materials are used. By introducing branches in the NWs, it can further increase the light scattering and therefore improve the light harvesting ability. To date, many research efforts have been paid to the PV or PEC application of semiconductor heterostructure NWs, ranging from single axial or radial NW PV device to vertical NW arrays and 3D hierarchical NWs. For example, Lieber's group fabricated single NW-based PV device using core/shell p–i–n silicon NW [104]. Under 1 sun illumination, the single NW device yields a maximum power output of up to 200 pW and energy conversion efficiency of $\sim$ 3.4%. They also explored other PV devices, such as single and tandem axial p–i–n NW PV devices and core/shell group III nitride NW PV devices. Yang et al. have demonstrated a room temperature aqueous etching method followed by low-temperature thin-film deposition and a rapid thermal annealing crystallization step to make wafer-scale radial p–n junction solar cells [105]. Figure 5.17a shows a schematic of the Si n/p core shell solar cells. Electrical output characteristic studies (Fig. 5.17c, d) demonstrate that the overall efficiency is only 0.5%, due to the low open circuit voltage V_{oc} (0.29 V) and fill factor FF (0.35). In their following work, they fabricated radial Si NW arrays with controllable NW

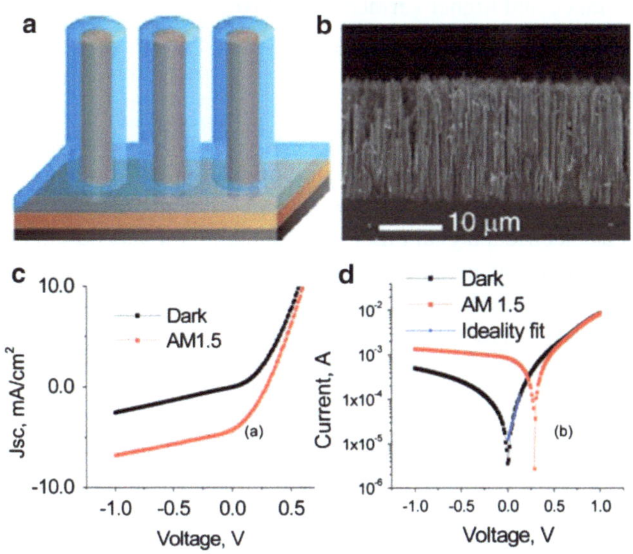

Fig. 5.17 (**a**) Schematics of solar cell design with n/p core/shell Si NW arrays. (**b**) TEM image of a single n/p core/shell Si NW. (**c, d**) p–n junction Si NW solar cell electrical performance. Reprinted with permission from [105]. Copyright@2008 American Chemical Society

density and diameter by using deep reactive ion etching to form NW arrays and diffusion to form p–n junction [106]. These ordered Si NW array solar cells showed very promising efficiency up to 5–6% by using 20 and 8 μm silicon absorber layers with different roughness factors. Besides, Fan et al. reported the direct growth of highly regular CdS/CdTe heterostructured nanopillar arrays inside AAO templated by CVD method and further demonstrated its photovoltaic application [107]. An efficiency η of 6% is obtained with an open circuit voltage (V_{oc}) 0.62 V, short circuit current density $J_{sc} \sim 21\text{mA/cm}^2$, and FF 0.43 under AM1.5G illumination.

Similar to PV device application, the photo-to-chemical energy conversion research such as PEC water spitting by utilizing 1D heterostructures has also attracted increasing attention. The Wang's group reported a $TiSi_2/TiO_2$ core/shell heterostructure that was fabricated by a combination of CVD and ALD methods as well as their application in water splitting [108]. Figure 5.18a shows the schematic of the $TiSi_2/TiO_2$ heterostructures. The core-shell structure is clearly revealed by TEM image (Fig. 5.18b). The network structures of $TiSi_2$ NWs serve as a scaffold with high surface area to improve the photon absorption of TiO_2. The highly conductive $TiSi_2$ was explored to facilitate the charge transport. The combined advantages led to a high performance in PEC measurements, and a peak incident photon conversion efficiency (IPCE) of 16.7% was achieved under monochromic UV illuminations (as shown in Fig. 5.18d). The same group extended the idea to fabricate $TiSi_2/Fe_2O_3$ nanonet heterostructures and obtained photocurrents of 1.6 and 2.7 mA/cm^2 at 1.23 and 1.53 V vs. RHE, respectively [109]. Yang et al. fabricated highly dense Si/TiO_2 core/shell NW arrays and found that the Si/TiO_2 NW arrays have 2.5 times higher photocurrent density than the planar Si/TiO_2 due to lower reflectance and higher surface area [110].

5.3.3 Photodetectors

Another type of important optoelectronic application of semiconductor NW heterostructures is photodetectors (PDs), which can efficiently detect optical inputs and process them as electrical signal output. Specifically, the physics and technology of NW PDs offer numerous insights and opportunities for nanoscale optoelectronics, photovoltaics, plasmonics, and emerging negative index metamaterial devices. Semiconductor NW heteronanostructures for constructing photodetectors offer several advantages over their bulk or thin-film counterparts, such as the following:

1. Large surface-to-volume ratios and Debye length comparable to their small size as well as internal photoconductive gain, thus leading to superior light sensitivity
2. Possibility to integrate functionality with single NW devices, such as NW avalanche photodiodes
3. Heterogeneous integration NWs of different material for enhanced or specific spectra sensitivity

Herein, we present some recent representative examples of the photodetection application of both homo- and heterogeneous NWs. In 2006, the Lieber's group first

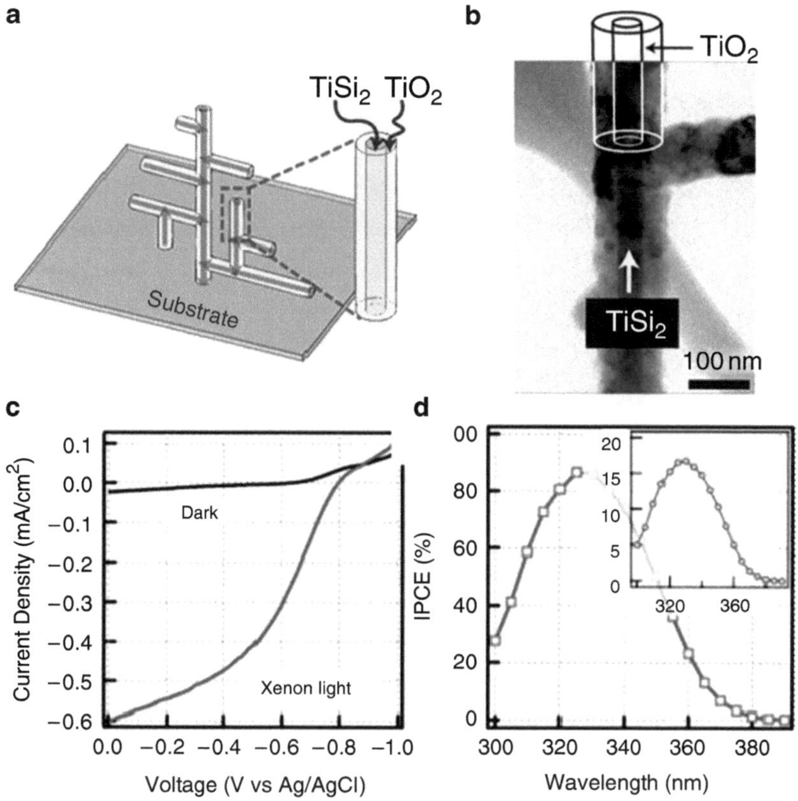

Fig. 5.18 (**a**) A schematic sketch of $TiSi_2/TiO_2$ core/shell heterostructures. (**b**) A TEM image revealed the core/shell structures. (**c, d**) PEC properties of $TiSi_2/TiO_2$ heteronanostructures: (**c**) I–V plots under different illumination conditions. (**d**) IPCE plots; *inset*: overall efficiency under monochromic illuminations. Reprinted with permission from [108]. Copyright@2009 American Chemical Society

reported the fabrication of axial modulation-doped p-type/intrinsic/n-type (p–i–n) silicon NWs and their application in nanoscale avalanche photodetectors [35]. Spatially resolved photocurrent measurements showed that the largest photocurrent is generated at the intrinsic region located between the electrode contacts, with multiplication factors in excess of ca. 30, and demonstrated that single p–i–n NWs function as avalanche photodiodes. The observation of largest photocurrent in the intrinsic Si region is consistent with the large potential drop and corresponding strong electric field in i-region, which can efficiently separate photogenerated carriers to create a photocurrent. Bugallo et al. demonstrated a visible-blind UV photodetector based on p–i–n GaN NW ensembles on Si(111) substrate [111]. The detector peak responsivity is $0.47\,AW^{-1}$ at -1 V exceeding that of thin-film GaN p–i–n photodetectors. The UV-to-visible rejection ratio is 2×10^2. The good performance is due to the efficient electron–hole pair separation by the built-in

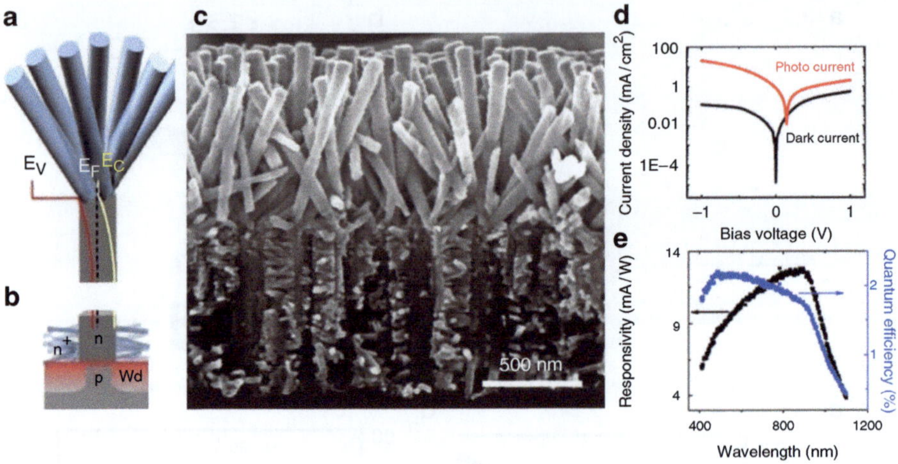

Fig. 5.19 ZnO/Si branched NW heterostructures for photodetection. (**a**, **b**) Schematics of the heterostructures: (**a**) top junction and energy band diagram; (**b**) bottom junctions and depletion layer in the silicon substrate. (**c**) A typical SEM image of ZnO/Si branched NW heterostructures; (**d**) I–V characteristics measured in the dark and under xenon lamp illumination; (**e**) spectral photoresponsivity and external quantum efficiency of ZnO/Si NW heterostructure photodetectors. Reprinted with permission from [112]. Copyright@2010 American Chemical Society

field at the junction region and carrier collections by the top ITO p-contact and bottom Si n-contact. Recently, Sun and coworkers reported a low-cost solution fabrication of wafer-scale 3D branched ZnO/Si NW heterostructures and their application in photodetection [112]. Figure 5.19c shows the SEM image of branched ZnO/Si NWs. The current–voltage (I–V) characteristics measured in the dark and under Xenon lamp illumination at room temperature indicated that both dark and photocurrent voltage characteristics were rectifying and the best ON/OFF ratio achieved at -1 V was about 250 (Fig. 5.19d). The maximum responsivity measured for this ZnO/Si heterostructure was as high as 12.8 mA/W at around 900 nm, and the maximum quantum efficiency was 2.20% (Fig. 5.19e). Such branched structures promises enhanced light-trapping due to the aperiodically arranged NW array and high refractive index material filling, as well as advantages in broadband photon detection.

5.4 Conclusions and Perspective

In summary, we have reviewed a wide range of methods for the fabrication of semiconductor NW heterostructures, ranging from vapor phase, solution, and template-based growth to combined methods such as multistep fabrication. The physical properties and applications of individual or ensembles of NW heterostructures are highlighted. Applications in photovoltaic and PEC water splitting devices as well as photodetectors based on NW heterojunctions are described. The potential optoelectronic applications of heterostructures NWs are not only limited to above, but also demonstrated in plasmonic lasers when coupled with metal surface plasmons. Applications in energy storage such as lithium-ion battery and supercapacitors, where high surface areas are desirable, are also a major venue for heterostructured NWs, but they are not covered in this chapter.

Overall, compared with the fast progress in the research of single-component semiconductor NW systems, the synthesis and application of heterostructured NWs with well-defined interfaces have lingered far behind. The hindrance, among others, is the lack of a good growth control coupled with high crystal quality (low defects, doping, and desired crystallographic phases).

Further development in this research field requires improvement in synthetic methods and novel fabrication processes to provide better control of the dimension, composition, structure, and interface, as well as the yield, of 1D semiconductor heterostructures. Currently, the dominating techniques toward high-quality semiconductor heterostructures are vapor-phase epitaxy growth such as MOCVD and MBE, which require expensive equipment and toxic source materials. Therefore, developing simple and environment-friendly benign growth methods such as solution-based epitaxy for the fabrication of 1D semiconductor heterostructures are of great interest. In addition, it is equally important, although challenging, to develop better characterization methods and tools to study the fundamental physical properties of individual or ensembles of heterostructured NWs. All in all, the ultimate goal is to integrate NW heterostructures into functional electronic and optoelectronic devices toward practice technological applications, which remains the focus of current worldwide research.

References

1. S. Iijima, Nature **354**, 56 (1991)
2. J.T. Hu, T.W. Odom, C.M. Lieber, Accounts Chem. Res. **32**, 435 (1999)
3. Y. Xia, P. Yang, Y. Sun, Y. Wu, B. Mayers, B. Gates, Y. Yin, F. Kim, H. Yan, Adv. Mater. **15**, 353 (2003)
4. C.N.R. Rao, F.L. Deepark, G. Gundiah, A. Govindaraj, Prog. Solid State Chem **31**, 5 (2003)
5. X.F. Duan Y. Huang, Y. Cui, J.F. Wang, C.M. Lieber, Nature **409**, 66 (2001)
6. M.H. Huang, S. Mao H. Feick, H.Q. Yan Y.Y. Wu, H. Kind, E. Weber, R. Russo, P.D. Yang, Science **292**, 1897 (2001)
7. X.F Duan, Y Huang, R Argarawal CM. Lieber, Nature **421** 241 (2002)
8. N. Wang, Y. Cai, R.Q. Zhang, Mat. Sci. Eng. R **60**, 1 (2008)
9. H.J. Fan, P. Werner, M. Zacharias, Small **2**, 700 (2006)
10. W. Lu, C.M. Lieber J. Phys. D: Appl. Phys. **39**, R387 (2006)
11. M. Law, J. Goldberger, P. Yang, Annu. Rev. Mater. Res. **34**, 83 (2004)
12. H.J. Fan, Y. Yang, M. Zacharias, J. Mater. Chem. **19**, 885 (2009)
13. Y. Huang, C.M. Lieber, Pure Appl. Chem. **76**, 2051 (2004)
14. S.J. Hurst, E.K. Payne, L. Qin, C.A. Mirkin, Angew. Chem. Int. Ed. **45**, 2672 (2006)

15. J.G. Lu, P. Zhang, Z. Fan, Mat. Sci. Eng. R**52**, 49 (2006)
16. C.W. Cheng, G.Y. Xu, H.Q. Zhang, Y. Luo, J.Nanosci.Nanotechnol. **7**, 4439 (2007)
17. C.W. Cheng, G.Y. Xu, H.Q. Zhang, Y. Luo, P.G. Zhang, K. Shen, Mater. Res. Bull. **43**, 3506 (2008)
18. J. Kong, N.R. Franklin, C.W. Zhou, M.G. Chapline, S. Peng, K. Cho, H.J. Dai, Science, **287**, 622 (2000)
19. Z.L. Wang, Mat. Sci. Eng. R. **64**,33 (2009)
20. A. Hochbaum, P. Yang, Chem. Rev. **110**, 527 (2010)
21. A.P. Alivisatos, Science **271**, 933 (1996)
22. A.J. Miezawska, T. Jalilian, G.U. Smanasekera, F.P. Zamborini, Small **5**, 722 (2007)
23. Y. Li, F. Qian, J. Xiang, C.M. Lieber, Mater. Today **9**, 18 (2006)
24. G. Shen D. Chen, Sci. Technol. Adv. Mat **1**, 213 (2009)
25. L.J. Lauhon, M.S. Gudikssen, C.M. Lieber, Philos. Trans. R. Soc. A**362**, 1247 (2004)
26. J. Hu, Y. Bandob, D. Golberg, J. Mater. Chem. **19**, 330 (2009)
27. C.M. Lieber, Nano Lett. **2**, 81 (2002)
28. Y. Li, J. Xiang, F. Qian, S. Gradeèak, Y. Wu, H. Yan, D.A. Blom, C.M. Lieber, Nano Lett. **6**, 1468 (2006)
29. M.J. Bierman, J. Song, Energ. Environ. Sci. **2** 1050 (2009)
30. X. Liu, Y. Lin, S. Zhou, S. Sheehan, D. Wang, Energy **3** 285 (2010)
31. R.S. Wagner, W.C. Ellis, Appl. Phys. Lett. **4**, 89 (1964)
32. K. Haraguchi, K. Hiruma, K. Hosomi, M. Shirai, T. Katsuyama, J. Vac. Sci. Technol. B **15**, 1685 (1997)
33. M. Shirai, K. Haraguchi, K. Hiruma, T. Katsuyama, Gold Bull. **32**, 80 (1999)
34. M.S. Fuhrer, L.J. Lauhon, J. Wang, D.C. Smith, C.M. Lieber, Nature **415**, 617 (2002)
35. C. Yang, C.J. Barrelet, F. Capasso, C.M. Lieber, Nano Lett.**6**, 2929 (2006)
36. E. Tutuc, J. Appenzeller, M.C. Reuter, S. Guha, Nano Lett. **6**, 2070 (2006)
37. Y. Wu, R. Fan, P. Yang, Nano Lett. **2**, 83 (2002)
38. M.T. Björk, C. Thelander, A.E. Hansen, L.E. Jensen, M.W. Larsson, L.R. Wallenberg, L. Samuelson, Nano Lett. **4**, 1621 (2004)
39. C.P.T. Svensson, W. Seifert, M.W. Larsson, L.R. Wallenberg, J. Stangl, G. Bauer, L. Samuelson, Nanotechnology **16**, 936 (2005)
40. T. Mårtensson, C.P.T. Svensson, B.A. Wacaser, M.W. Larsson, W. Seifert, K. Deppert, A. Gustafsson, L.R. Wallenberg, L. Samuelson, Nano Lett. **4**, 1987 (2004)
41. B.J. Ohlsson, M.T. Björk, A.I. Persson, C. Thelander, L.R. Wallenberg, M.H. Mangusson, K. Deppert, L. Samuelson, Phys. Rev. E **13**, 1126 (2002)
42. E. Lind, A.I. Persson, L. Samuelson, L.-E. Wernersson, Nano Lett **6**, 1842 (2006)
43. N.D. Zakharov, P. Werner, G. Gerth, L. Schubert, L. Sokolov, U. Gösele, J. Cryst. Growth **290**, 6 (2006)
44. W.I. Park, G.C. Yi, M. Kim, S.J. Pennycook, Adv. Mater. **15**, 526 (2003)
45. J. Zhan, Y. Bando, J. Hu, Z. Liu, L. Yin, D. Goberg, Angew. Chem. Int. Edit. **44**, 2140 (2005)
46. J. Hu, Y. Bando, J. Zhan, D. Golberg, Adv. Mater. **17**, 1964 (2005)
47. D.J. Milliron, S.M. Hughes, Y. Cui, L. Manna, J. Li, L.W. Wang, A.P. Alivisatos, Nature **430**, 190 (2004)
48. K. Peng, Z. Huang, J. Zhu, Adv. Mater. **16**, 73 (2004)
49. N. Geyer, Z.P. Huang, B. Fuhrmann, S. Grimm, M. Reiche, T. Nguyen-Duc, J. Boor, H.S. Leipner, P. Werner, U. Gösele, Nano Lett. **9**, 3106 (2009)
50. C.J. Brumlik, C.R. Martin J. Am. Chem. Soc**113** 3174 (1991)
51. C.R. Martin, Chem. Mater.**8** 1739 (1996)
52. D.J. Pena, K.N. Mbindyo, A.J. Carado, T.E. Mallouk, C.D. Keating, B. Razavi, T.S. Mayer, J. Phys. Chem. B **106** 7458 (2002)
53. J. Luo, J. Zhu, Nanotechnology **17** S262 (2006)
54. J. Luo, Y. Zhang, J. Zhu Adv. Mater **14**, 1413 (2002)
55. Y. Wu, J. Xiang, C. Yang, W. Lu, C.M. Lieber, Nature **61**, 430 (2004)
56. B.W. Maynor, J. Li, C. Lu, J. Liu, J. Am. Chem. Soc. **126**, 6409 (2004)

57. L.J. Lauhon, M.S. Gudiksen, D. Wang, C.M. Lieber, Nature **420**, 57 (2002)
58. J. Xiang, W. Lu, Y. Hu, Y. Wu, H. Yan, C.M. Lieber, Nature **44**, 489 (2006)
59. W. Lu, J. Xiang, B.P. Timko, Y. Wu, C.M. Lieber, P. Natl. Acad. Sci. U.S.A. **102**, 10046 2005)
60. F. Qian, S. Gradeak, Y. Li, C.Y. Wen, C.M. Lieber, Nano Lett. **5**, 2287 (2005)
61. N. Sköld, L.S. Karlsson, M.W. Larsson, M.E. Pistol, W. Seifert, J. Trägårdh, L. Samuelson, Nano Lett. **5**, 1943 (2005)
62. P. Mohan, J. Motohisa, T. Fukui, Appl. Phys. Lett. **88**, 133105 (2006)
63. P. Mohan, J. Motohisa, F. Fukui, Appl. Phys. Lett. **88**, 013110 (2006)
64. J. Motohisa, J. Takeda, M. Inari, J. Noborisaka, T. Fukui, Phys. Rev. E. **23**, 298 (2004)
65. J. Noborisaka, J. Motohisa, S. Hara, T. Fukui, Appl. Phys. Lett. **87**, 093109 (2005)
66. S.J. An, W. Park, G.C. Yi, Y.J. Kim, H.B. Kang, M. Kim, Appl. Phys. Lett. **84**, 3612 (2004)
67. E.S. Jang, J.Y. Bae, J. Yoo, W. Park, D.W. Kim, G.C. Yi. Appl. Phys. Lett. **88**, 023102 (2006)
68. W. Park, J. Yoo, D.W. Kim, G.C. Yi, M. Kim, J. Phys. Chem. B **110**, 1156 (2006)
69. C.W. Cheng, B. Liu, E.J. Sie, W.W. Zhou, J.X. Zhang, H. Gong, C.H.A. Huan, T.C. Sum, H.D. Sun, H.J. Fan. J. Phys. Chem. C **114**, 3863 (2010)
70. K. Wang, J. Chen, W.L. Zhou, Y. Zhang, Y. Yan, J. Pern, A. Mascarenhas, Adv. Mater. **20**, 3248 (2008)
71. H.J. Fan, M. Knez, R. Scholz, K. Nielsch, E. Pippel, D. Hesse, M. Zacharias, U. Gösele, Nat. Mater. **5**, 627 (2006)
72. Y. Yang, R. Scholz, H.J. Fan, D. Hesse, U. Gösele, M. Zacharias, ACS Nano **3**,555 (2009)
73. Y.M. Chang, S.R. Jian, H.Y. Lee, C.M. Lin, J.Y. Juang, Nanotechnology **21**,385705 (2010)
74. Q. Li, C. Wang, J. Am. Chem. Soc. **125**, 9892 (2003)
75. N. Wang, Y.F. Zhang, Y.H. Tang, C.S. Lee, S.T. Lee. Appl. Phys. Lett. **73**, 3902 (1998)
76. C. Wang, J. Wang, Q. Li, G.C. Yi, Adv. Funct. Mater. **15**, 1471 (2005)
77. S. Mathur, H. Shen, N. Donia, T. Rügamer, V. Sivakov, U. Werner. J. Am. Chem. Soc. **129**, 9746 (2007)
78. Y.B. Li, Y. Bando, D. Golberg, Y. Uemura, Appl. Phys. Lett.**42**, 6723 (2003)
79. X. Wang, H. Zhai, C. Cao, H. Cai, Y. Wang, H.W. Chan, J. Mater. Chem. **19**, 2958 (2009)
80. W. Wang, L. Sun, Z. Fang, L. Chen, Z. Zhang, Cryst. Growth Des.**9**, 2119 (2009)
81. J.A. Goebl, R.W. Black, J. Puthussery, J. Giblin, T.H. Kosel, M. Kuno, J. Am. Chem. Soc. **130**, 14822 (2008)
82. D. Wang, F. Qian, C. Yang, Z. Zhong, C.M. Lieber, Nano Lett. **4**, 871 (2004)
83. K.A. Dick, K. Deppert, M.W. Larsson, T. Mårtensson, W. Seifert, L.R. Wallenberg, L. Samuelson, Nat. Mater.**3**, 380 (2004)
84. W. Seifert, M. Borgström, K. Deppert, K.A. Dick, J. Johansson, M.W. Larsson, T. Mårtensson, N. Sköld, C.P.T. Svensson, B.A. Wacaser, L.R. Wallenberg, L. Samuelson, J. Cryst. Growth 272, 211 (2004)
85. Q. Wan, E.N. Dattoli, W.Y. Fung, W. Guo, Y.B. Chen, X.Q. Pan, W. Lu, Nano Lett.**6**, 2909 (2006)
86. Q. Wan, J. Huang, Z. Xie, T.H. Wang, E.N. Dattoli, W. Lu, Appl. Phys. Lett. **92**, 102101 (2008)
87. W.C. Zhou, A.L. Pan, Y. Li, G.Z. Dai, Q. Wan, Q.L. Zhang, B.S. Zou, J. Phys. Chem. C **112**, 9253 (2008)
88. A.G. Dong, R. Tang W.E. Buhro, J. Am. Chem. Soc **129** 12254 (2007)
89. C.W. Cheng, B. Liu, H.Y. Yang, W.W. Zhou, L. Sun, R. Chen, S.F. Yu, J.X. Zhang, H. Gong, H.D. Sun, H.J. Fan, ACS Nano **3**, 3069 (2009)
90. C.W. Cheng, Y.Y. Tai, H.H. Hng, H.J. Fan, J. Mater. Res. (in press)
91. W.W. Zhou, C.W. Cheng, J.P. Liu, Y.Y. Tay, J. Jiang, X.T. Jia, X.J. Zhang, H. Gong, H.H. Hng, T. Yu, H.J. Fan, Adv. Funct. Mater. **21**, 2439 (2011)
92. M.T. Niu, F. Huang, L.F. Cui, P. Huang, Y.L. Yu, Y.S. Wang, ACS Nano **4**, 681 (2010)
93. F. Zhang, L.D. Sun, C.J. Jia, Z.G. Yan, L.P. You, C.H. Yan, J. Am. Chem. Soc. **127** 13492 (2005)
94. S.J. May, J.G. Zheng, B.W. Wessels, L.J. Lauhon, Adv. Mater. **17** 598 (2005)

95. CW. Cheng, B. Yan, S.M. Wong, T. Yu, Z.X. Shen, H.Y. Yu, H.J. Fan *ACS*. Appl. Mater. Inter, **7**, 1824 (2010)
96. L Manna, E.C. Scher, A.P. Alivisatos J. Am. Chem. Soc. **122**, 12700 (2000)
97. L. Manna, DJ. Milliron, *A*. Meisel, E.C. Scher, A.P. Alivisatos, Nat. Mater. **2**, *3*82 (2003)
98. H. Chu, X. Li, G. Chen, W. Zhou, Y. Zhang, Z. Jin, J. Xu, Y. Li, Cryst. Growth Des. **5**, 1801 (2005)
99. Y.H Cho, G.H. Gainer A.J. Fischer J.J. Song, S. Keller U.K. Mishra, S.P. DenBaars, Appl. Phys. Lett **73**, 1370 (1998)
100. A. Bell, S. Srinivasan, C. Plumlee, H. Omiya, F.A. Phonce, J.S. Tanaka A. Fujioka, Y. Nakagawa J. Appl. Phys. **95**, 4670 (2004)
101. H.Y. Yang, S.P. Lau, S.F. Yu, A.P. Abiyasa, M. Tanemura, T. Okita, H. Hatano, Appl. Phys. Lett. **89**, 011103 (2006)
102. F. Qian, Y. Li, S. Gradeak, H.G. Park, Y.J. Dong, Y. Ding, Z.L. Wang, C.M. Lieber, Nat. Mater. **7**, 701 (2008)
103. I. Gonzalez-Valls, M. Lira-Cantu, Energ. Environ. Sci **2**, 19 (2009)
104. B. Tian, X. Zheng, T.J. Kempa, Y. Fang, N. Yu, G. Yu, J. Huang, C.M. Lieber, Nature **449**, 885 (2007)
105. E.C. Garett, P.D. Yang, J. Am. Chem. Soc. **130**, 9224 (2008)
106. E.C. Garett, P.D. Yang, Nano Lett. **10**, 1082 (2010)
107. Z.Y. Fan, H. Razavi, J.-W. Do, A. Moriwaki, O. Ergen, Y.-L. Chueh, P. Leu, J. Ho, T. Takahashi, L.A. Reichertz, S. Neale, K. Yu, M. Wu, J. Ager, A. Javey, Nat. Mater. **8**, 648 (2009)
108. Y.J. Lin, S. Zhou, X.H. Liu, S. Sheehan, D.W. Wang, J. Am. Chem. Soc. **131**, 2772 (2009)
109. Y.J. Lin, S. Zhou, S. Sheehan, D.W. Wang, J. Am. Chem. Soc. **133**, 2398 (2011)
110. Y.J. Hwang, A. Boukai, P.D. Yang, Nano Lett. **9**, 410 (2009)
111. A.L. Bugallo, M. Tchernycheva, G. Jacopin, L. Rigutti, F.H. Julien, S.T. Chou, Y.T. Lin, P.H. Tseng, L.W. Tu, Nanotechnology **21**,315201 (2010)
112. K. Sun, Y. Jing, N. Park, C. Li, Y. Bando, D. Wang, J. Am. Chem. Soc. **132**, 15465 (2010)

Chapter 6
Hybrid Semiconductor Nanostructures with Graphene Layers

Won Il Park, Jung Min Lee, Dong Hyun Lee, and Gyu-Chul Yi

Abstract We review current research activities on the hybrid system that combines nanostructured semiconductor materials with two-dimensional (2D) graphene materials as emerging building blocks for future electronics and photonics. Within this category are three distinct subgroups: (1) 0D–2D nanoparticle–graphene hybrids, (2) 1D–2D nanorod–graphene hybrids, and (3) 2D–2D hybrid lamellar composites. Particular attention is paid to the 1D–2D hybrids, with an emphasis on superior properties and potential applications in nanoelectronics and photonics.

6.1 Introduction

The word "hybrid" is often used to refer to a mixture of numerous separate things. In particular, materials scientists and chemists have often used the words "hybrid" and "hybrid materials" to denote materials that can be obtained by mixing different types of materials. However, distinguished from the conventionally known "composites" that are mere mixtures, it is necessary to define "hybrid materials" when the individual components are interactively linked to reveal distinct and/or extraordinary properties. Based on the level of hybridization (i.e., chemical-bond mode), Makishima categorized hybrid materials into four cases:

W.I. Park (✉) · J.M. Lee · D.H. Lee
Department of Material Science and Engineering, Hanyang University, 222 Wangsimni-ro, Seongdong-gu, Seoul 133-791, Korea
e-mail: wipark@hanyang.ac.kr

G.-C. Yi (✉)
National Creative Research Initiative Center for Semiconductor Nanostructures, Department of Physics and Astronomy, Seoul National University, 599 Gwanak-ro, Gwanak-gu, Seoul 151–747, Korea
e-mail: gyuchul.yi@gmail.com

G.-C. Yi (ed.), *Semiconductor Nanostructures for Optoelectronic Devices*, NanoScience and Technology, DOI 10.1007/978-3-642-22480-5_6,

1. composites as a mixture of materials consisting of matrix- and micron-level dispersions;
2. nanocomposites as a submicron-level mixture of similar kinds of materials;
3. hybrids as a submicron-level mixture of different kinds of materials; and
4. nanohybrids at the atomic or molecular level [1]. Alternatively, Nanko proposed systematic criteria for hybrid materials based on the purpose of hybridization [2].

The categories are

1. structurally hybridized materials,
2. materials hybridized with a chemical bond, and
3. functionally hybridized materials.

Different categories of hybrid materials can be designed using different concepts of hybridization.

A variety of composite materials have long been studied in the fields of structural materials, electrochemistry, and semiconductor technology. More recently, with recent developments in nanotechnology, increasing research interests have been placed on hybrid nanostructures [3]. There are several key factors that attributed to the rapid increase (awareness) in this field. First, rapid technological development in the last few years enabled precise design and control of structures as well as their characterization at the nanometer level or below. Second, nanometer-scale hybridization may enable unique and fascinating properties and high performances superior to those of their counterpart individual components and/or bulk materials. Third, compared with the discovery of a totally new material, hybrid use could present substantial opportunities for achieving new types of materials with versatile and tailor-made properties.

"Hybrid semiconductor nanostructures (hSNs)" include various types of hybrid material systems in which at least one component is based on nanostructured semiconductor materials. One type of a nanostructured semiconductor material is the one-dimensional (1D) form. These 1D materials can be used for efficient transport of electrons and optical excitations and are thus expected to be critical to the function and integration of high-performance electronics and photonics [4]. On the other hand, graphene, a two-dimensional (2D) layer of carbon atoms, is emerging as a promising family of materials due to its unique properties, including good electrical and thermal conduction, mechanical strength and elasticity, and optical transparency [5]. Recent achievements in large-area synthesis using several methods and success in transferring graphene films onto arbitrary substrates demonstrate significant advances toward their implementation in next-generation electronics and optoelectronics. However, compared with other materials, the study on hybridization of materials and processing by using graphene as one component is still in an early stage, although it can promote the construction of novel three-dimensional (3D) architectures and multifunctionalities. Thus, this chapter focuses on "hSNs with graphene layers." Particular attention is paid to the systematic design and synthetic strategies that have been exploited to create new types of materials with the best components or with superior properties and new functions.

6.2 Graphene: 2D Materials for Transparent Conducting Layers

Graphene is a one-atom-thick, 2D planar sheet of sp^2-bonded carbon atoms arranged in a honeycomb crystal lattice. As the basic structural element of some carbon allotropes including graphite, charcoal, carbon nanotubes (CNTs) and fullerenes, graphene can be transformed into either graphite, consisting of vertically stacked graphene sheets, or a single-walled carbon nanotube (SWNT) through formation of a seamless cylinder [6]. Although it contradicts the theory of Landau, Peierles, and Mermin, who argued that 2D crystals were thermodynamically unstable, a single sheet of graphene was successfully experimentally isolated by Andre Geim and Konstantin Novoselov in 2004 [7]. Immediately after the discovery of graphene, peculiar and outstanding properties of graphene, including a half-integer quantum hall effect and a remarkably high electron mobility (in the excess of 15,000 $cm^2V^{-1}\,s^{-1}$ even at room temperature), were demonstrated independently by Andre Geim et al [8] and Philip Kim et al [9]. This discovery has triggered enormous interest and activity on graphene-based fundamental and applied research. The Nobel Prize in Physics for 2010 was jointly awarded to Andre Geim and Konstantin Novoselov "for groundbreaking experiments regarding the 2D material graphene" [4]. In the following paragraph, properties and synthesis methods of graphene are introduced. Then, the application of graphene as a transparent conducting layer is discussed (Fig. 6.1).

6.2.1 Physical Properties of Graphene

The structure of graphene strongly influences the electronic and electrochemical properties. The graphene honeycomb lattice is connected by two equivalent lattices A and B (shown in Fig. 6.2a) of carbon atoms bonded together with sp^2 bonds.

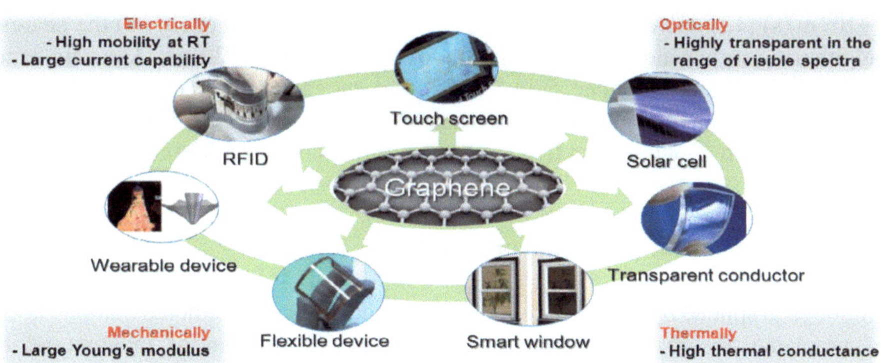

Fig. 6.1 Potential applications of graphene utilizing its outstanding electrical, optical, thermal, and mechanical properties

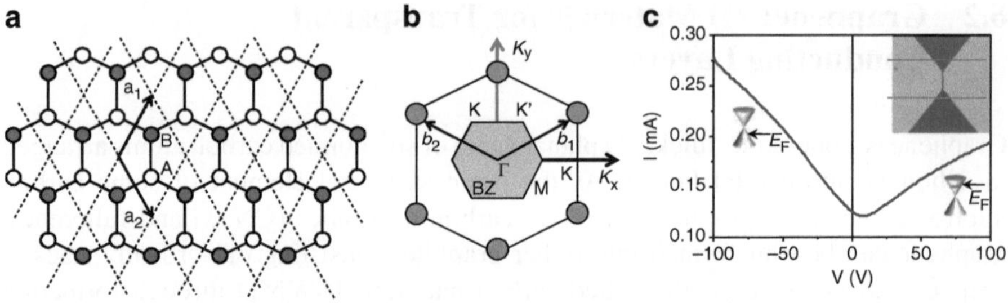

Fig. 6.2 (**a**) Graphene lattice with unit vectors $\vec{a}_1$ and $\vec{a}_2$. (**b**) Reciprocal lattice of graphene, and corresponding first Brillouin zone (*the shaded hexagon*) with reciprocal lattice vectors $\vec{b}_1$ and $\vec{b}_2$. (**c**) Ambipolar electric field effect in a CVD-grown graphene

Each carbon atom in the lattice forms a delocalized network of electrons caused by the π orbital. Figure 6.2b shows the first Brillouin zone of graphene with the high-symmetry points K, K', the two inequivalent points in the Brillouin zone (called Dirac points) [10, 11].

The electronic properties of graphene are based on fundamental studies on its structure. The dispersion of carbon atoms caused by the π orbital can be described using the tight-binding model based only on the first nearest-neighbor interactions [10, 11]. According to this theory, the electronic states near Dirac points are composed of states belonging to the different sublattices, and their relative contributions have to be considered using two-component wave functions. Thus, the effective Hamiltonian near K/K' can be expressed using the Dirac equation with zero mass. The equation describing the E–k relation is

$$E \pm (k) = \pm\gamma|\vec{k}| = \frac{h}{2\pi} v_{\mathrm{F}}|\vec{k}|,$$

where $\vec{k}$, h is the momentum measured relative to the Dirac point and Planck's constant, and $\gamma = (h/2\pi)v_F$, v_F is the Fermi velocity given by $v_{\mathrm{F}} = \sqrt{3}a\gamma 0/2$ ($\sim 10^6$ m/s) [12]. Here, γ_0, the transfer integer, is the hopping energy to the nearest neighboring carbon atoms with a magnitude of approximately 2.8 eV. As a zero band gap semiconductor, graphene showed an ambipolar electric field effect, and charge carriers can be tuned continuously between electrons and holes in concentrations as high as 10^{13} cm^{-2}, with room temperature mobilities up to 15,000 cm^2 V^{-1} s^{-1} [8].

6.2.1.1 The Optical Properties of Graphene

Although graphene is only one atom thick, it is theoretically predicted that one layer of graphene absorbs a significant ($\pi\alpha = 2.3\%$) fraction of incident white light, where α, the fine-structure constant, is a fundamental physical coupling constant which characterizes the strength of an electromagnetic interaction [6].

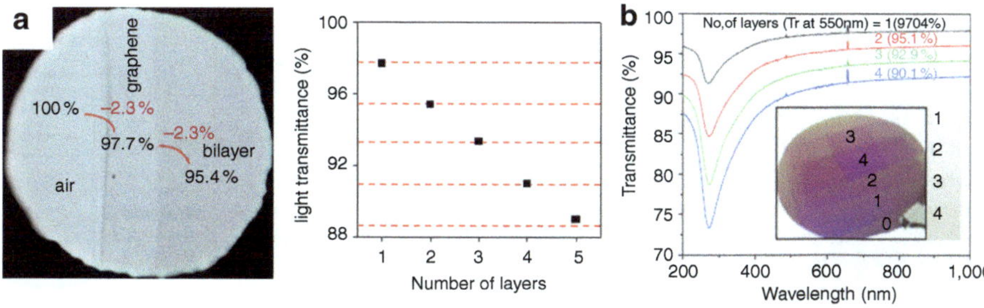

Fig. 6.3 (**a**) *Left*: photograph of an aperture partially covered by single- and bilayer graphene. *Right*: transmittance of white light as a function of the number of graphene layers. (**b**) Transmittance spectra of layer-by-layer transferred CVD–graphene films as a function of the number of graphene layers *Inset*: optical image of graphene (adapted from [14])

This was further experimentally confirmed from graphene membranes prepared via micromechanical cleavage of graphite that was then mounted onto a metal scaffold with small apertures (Fig. 6.3a) [13]. Measurements from chemical vapor deposition (CVD)-grown graphene films after transfer to glass substrates (Fig. 6.3b) [14, 15] also exhibited similar results, although some deviation may exist due to defects or wrinkles [6]. Despite an unexpectedly high opacity for a one-atom-thick planar sheet, single- or multiple-layer graphene sheets still exhibit high transparency for visible light, with optical transmittance exceeding 90%, only if the number of graphene layers is less than four. This relatively high optical transmittance, together with a high electrical conductivity [14, 15], makes graphene a candidate for transparent conducting electrodes, which is explicitly discussed in the next section.

The mechanical properties of graphene including the Young's modulus and fracture strength, have been issued, in order to investigate the mechanical properties of a true 2D material. Using an atomic force microscope (AFM), the Young's modulus of monolayer graphene was studied with force–displacement [17] and/or force–volume [18] measurements on a strip trench and/or circular membranes, respectively. Furthermore, the elastic properties and intrinsic breaking strengths of free-standing graphene monolayers were measured using AFM-based nanoindentation. In those studies, the Young's modulus and fracture strength of defect-free graphene were 1.0 TPa and 130 GPa, respectively.

6.2.2 *Synthesis and Application of Graphene*

6.2.2.1 The Synthesis of Graphene

In order to implement graphene films in practical applications, it is essential to develop new technology to produce large-area, continuous films with a controlled

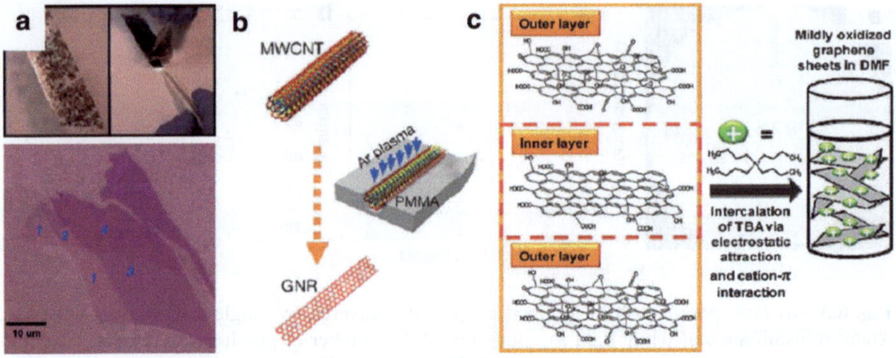

Fig. 6.4 Top-down approaches to synthesize graphene films. (**a**) Photo (*top*) and optical (*bottom*) images of few-layer graphene flakes made by a mechanical exfoliation process (adapted from [20]). (**b**) Etching multiwalled CNTs in order to prepare graphene nanoribbons adapted from [7]. (**c**) Intercalation of tetrabutylammonium ions in large graphite oxide sediments and unreacted graphite particles to obtain mildly oxidized single-layer graphene sheets in dimethylformamide (DMF) adapted from [21]

number of layers. Following successful isolation of graphene flakes via mechanical exfoliation of graphite [7, 19, 20], several routes including chemical [20–22], Si sublimation from a single crystal 4-H SiC (0001) surface [23, 24], and CVD on catalytic metals [25, 26] were introduced to achieve large-area, free-standing graphene films. In general, these synthesizing methods can be categorized into two approaches: top-down and bottom-up.

On the basis that graphene can form better known graphite by weak bonds between its individual layers or exist in its 1D form of a CNT through edge bonding, the top-down approach essentially utilizes the breakdown of a higher structure into its counterpart subsystem (it refers often to the synthesizing process from graphite to graphene, but that from CNT to graphene nanoribbons would also be included in this category). One representative top-down method is the (1) micromechanical exfoliation of graphite, in which sticky tape is first used repeatedly to split the highly ordered pyrolytic graphite (HOPG) crystals into atomically thin graphene layers (Fig. 6.4a) [7, 19, 20]. Other methods involving (2) unzipping of multiwalled CNTs via etching for the fabrication of nanoribbons [23, 28] and (3) chemical oxidation of graphite to graphite oxide to obtain mildly oxidized graphene [21, 29] are schematically described in Fig. 6.4b, c, respectively. Exfoliation of graphite in a solvent combines nitric acid oxidation and small ion intercalation within graphene sheets.

In the bottom-up approach, graphene film is usually made from hydrocarbon gas vapor. Indeed, the bottom-up approaches include CVD growth of graphene films, which is attractive because it is easy to scale for the large-area synthesis of continuous films. The growth of graphene monolayers on single-crystalline transition metals such as Co [30], Pt [31], Ir [32], Ru [33], and Ni [26, 34] has been well studied. In particular, CVD growth of graphene on a Ni substrate has the advantage of producing continuous large-area films. As for the growth mechanism, it has been proposed that the graphene growth process involves the dissolution of carbon in Ni at high temperature and segregation from the Ni surface during substrate cooling, and a fast cooling rate and/or thin Ni film (less than 200 nm) has been suggested to be critical to achieve single- or few-layer graphene by suppressing the formation of multiple layers [26,35]. Although continuous graphene films can be obtained on polycrystalline Ni substrates with a preannealing step (at 900–1,000°C), they do not form a uniform monolayer as they have multidomain structures with a wide variation in thickness over the film area. Alternatively, Li et al suggested Cu as an excellent candidate for synthesis of near-single-layer graphene films with uniform thickness as it has lower solubility than does C [27, 36]. While graphene growth on Ni is ascribed to C segregation or a precipitation process, based on a carbon isotope study, it has been suggested that the growth on Cu is rather due to a surface adsorption growth mechanism in which a "self-limiting process" yields high-quality, single-layer graphene (Fig. 6.5a, b) [36]. This process was further developed for extremely large-area (30 in.) graphene films by Bae et al (Fig. 6.5c) [14].

Graphene as a transparent conductor has been recently demonstrated as an integral part of many electronic and optoelectronic devices, including touch screens, flexible displays, lightemitting diodes (LEDs), and photovoltaic cells.

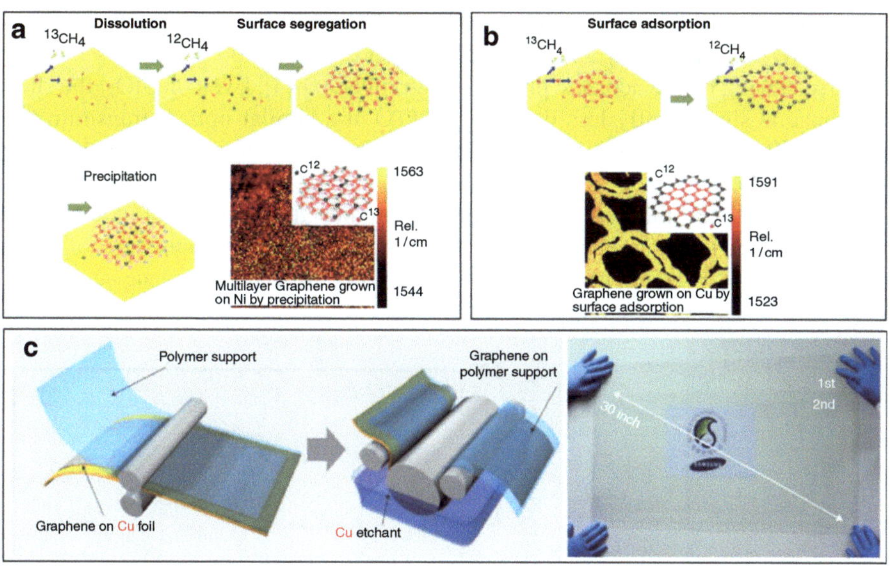

Fig. 6.5 Bottom-up approaches to synthesize graphene (CVD process). Growth mechanisms of a multilayer graphene film on Ni (**a**) and single-layer graphene film on Cu (**b**) (adapted from [36]). (**c**) Schematic of the roll-to-roll process of graphene films grown on a copper foil, and ultra large-area graphene film transferred onto a 35-inch polyethylene-terephthalate (PET) substrate (adapted from [14])

The most important criteria for choosing a transparent conducting film are optical transparency in the visible spectra and high electrical conductivity (i.e., low sheet resistance); however, a favorable chemical stability and low fabrication cost are other important factors when considering a practical application. Indium tin oxide (ITO) has long been the market leader of transparent conductors due to its good optical transparency greater than 90% even at sheet resistances less than 50 Ω/square [37]. However, indium is expensive and scarce; therefore, worldwide demand for a replacement for ITO is increasing. In addition, ITO is rigid and fragile, and it may fracture or lose its conductivity upon bending [38]. Accordingly, it is essential to find alternative transparent conducting materials that exhibit both market stability and mechanical flexibility with an endurable electrical conductivity for future device applications.

Over the past decade, various materials have been introduced as alternative transparent conductors such as thin metal foils [39], metal grids [40], conductive polymers [41], and CNT network films [42]. Among these, CNT network films have shown much promise as a replacement material. However, due to high internanotube contact resistance, the measured sheet resistance of CNT film cannot reach those of ITO at a similar optical transmittance. Graphene, which has high electrical conductivity, high carrier mobility, and excellent optical transmittance in the visible light range, is considered to be an ideal frontrunner to replace ITO. Assuming a carrier density of $10^{12}\,cm^{-2}$, even single-layer graphene (with an optical transmittance of 97.7%) can have a very low sheet resistance value of ~100 Ω/square with carrier mobilities of a 40,000 cm/Vs extrinsic limit (scattering by the SiO_2 surface) [15] and of ~ 20 Ω/square with a 200,000 cm/Vs intrinsic limit (scattering by the acoustic phonons of graphene) [43]. Even on SiO_2, the sheet resistance is significantly less than that of ITO at a similar optical transmittance, and the difference between them increases as the optical transmittance surpasses 85% (Fig. 6.6).

Although these theoretical calculations predict the substantial potential of graphene-based transparent conducting electrodes, the large-area synthesis of single crystal monolayer graphene films is essential to achieve performance superior to

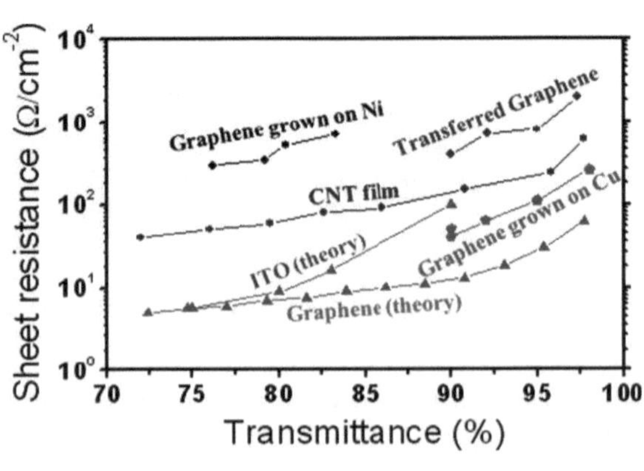

Fig. 6.6 Comparison of sheet resistance versus transmittance plots

that of ITO. As shown in Fig. 6.6, several graphene-based materials have been investigated for use as transparent conducting materials. In particular, by using Cu foil and a chemical doping process, Bae et al have synthesized large-area graphene with good characteristics using Cu foil and a chemical doping process. The measured sheet resistance of p-doped, multilayer graphene film with ∼90% optical transmittance was as low as ∼30 Ω/square, which is substantially superior to those of ITO-based transparent electrodes and CNT network films [14].

6.3 Hybrid Semiconductor Nanostructures with Graphene: 0D–2D, 1D–2D, and 2D–2D Hybrids

To satisfy needs from industry and academia, the characters and applications of nanostructured materials have been vigorously researched. However, the applications of preexisting materials are limited by their intrinsic properties. Thus, to create high performance or high functionality that cannot be achieved with preexisting materials, a totally new material is needed. Alternatively, by hybridizing different types of materials at the nanometer scale, the possibility increases for obtaining extraordinary improvements in material characteristics.

In general, graphene-based hSNs can be categorized by the dimensionality of the individual components: 0D–2D, 1D–2D, and 2D–2D hybrids (Fig. 6.7). Among these hybrid systems, we first briefly introduce the simplest 2D–2D hybrids, e.g., hybrid lamellar composites that consist of alternating 2D layers of organic materials and a graphene layer [44, 45], and then move onto more complicated structures of 0D–2D nanoparticle–graphene hybrids [46, 47] and 1D–2D nanorod–graphene hybrids [48–55].

6.3.1 Hybrid Lamellar Composites: 2D–2D Hybrids

As the most simple structure among hSNs with graphene, 2D–2D hybrid systems include the "lamellar composite" structure (Fig. 6.7c). As an alternative to conven-

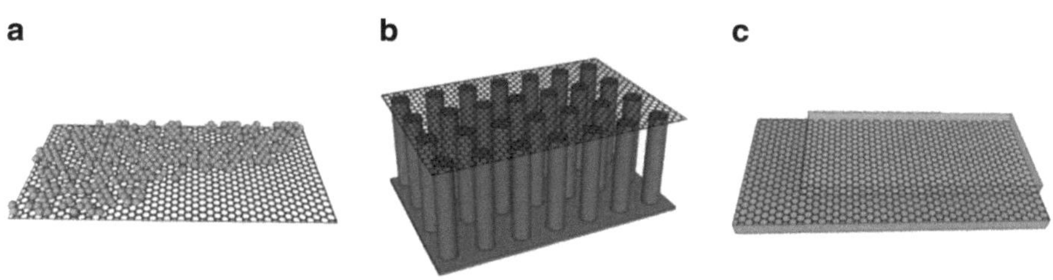

Fig. 6.7 Schematics illustrating three types of graphene-based hybrid nanostructures: (**a**) 0D–2D hybrid, (**b**) 1D–2D hybrid, and (**c**) 2D–2D hybrid

tional conducting metal oxide film, ultrathin graphene film is generally embedded in vertically stacked organic semiconductor layers.

With increasing demand for "low-power consumption, ultrathin, and flexible devices," organic semiconductor optoelectronics are emerging as competitors to existing inorganic semiconductor optoelectronics. Despite the fact that transparent conducting oxides (TCOs), such as ITO, are ubiquitously employed as window electrode layers in those areas [56], the exclusive use of these oxides becomes increasingly problematic due to the dwindling supply, mechanical brittleness, and chemical instability in the presence of acid or base.

Encouraged by the excellent properties of graphene [8, 9], a number of research groups have been using ultrathin graphene films as window electrodes for organic electronics [44, 57]. For example, Wang et al synthesized graphene films through acid oxidation of flake graphite, followed by dip coating and thermal reduction (Fig. 6.8a). The graphene films were then used for transparent window electrodes for solid-state dye-sensitized solar cells [44]. Alternatively, Eda et al proposed Cl-doped, reduced graphene oxide (RGO) films [59, 60] as transparent electrodes in organic photovoltaics (OPV) [58]. Graphene can also be used as an anode or acathode for organic lightemitting diodes (OLEDs) Matyba etal. utilized chemically derived graphene for the transparent cathode in an all-plastic sandwich-structure device, called a light-emitting electrochemical cell (LEC) (Fig. 6.8b) [45]. Using a screen-printable conducting polymer as a partially transparent anode and a

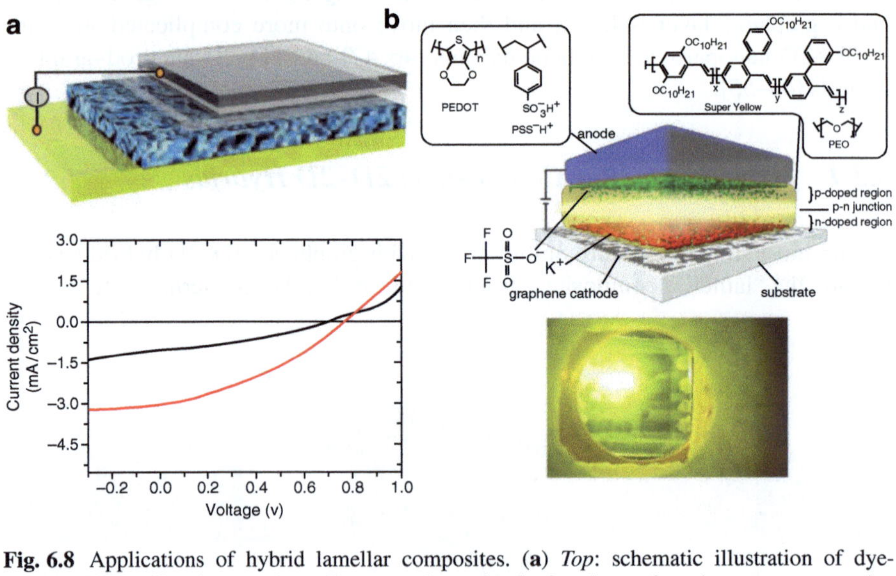

Fig. 6.8 Applications of hybrid lamellar composites. (**a**) *Top*: schematic illustration of dye-sensitized PV cells using graphene film as an electrode; the four layers from bottom to top are Au, dye-sensitized heterojunction, compact TiO_2, and graphene film. *Bottom*: current–voltage ($I-V$) curves of PV cells with graphene (*black*) and with fluorine-doped tin oxide (*red*), illuminated under AM solar light (1 sun) (adapted from [44]). (**b**) *Top*: schematic illustration of all-plastic, solution-processed LECs. *Bottom*: photograph of an LEC during light emission (adapted from [45])

micrometer-thick active layer solution deposited from a blend of a light-emitting polymer and polymer electrolyte, they demonstrated an LED based solely on solution-processable carbon-based materials. This result demonstrates that low-voltage, inexpensive, and efficient LEDs can be created without using metals.

However, the initial performances obtained with RGO films were relatively poor, with a conversion efficiency of 0.4 and 0.1% according to Wang et al and Eda et al, respectively. The poor performances of these devices may not be intrinsic to graphene but rather may be attributed to the poor electronic property of RGO due to (1) incomplete restoration of the π-conjugation, (2) surface roughness of the stacked graphene flakes, and (3) significant flake to flake resistance in the films. To this end, it is highly desirable to produce large-area, continuous films with a controlled number of layers. With recent advances in the CVD growth of high-quality graphene monolayers [14, 27], the performances of graphene-incorporated flexible OPV and OLED devices are approaching those of its counterparts with metal oxide electrodes.

6.3.2 Nanoparticle–Graphene Hybrids: 0D–2D Hybrids

Graphene is an excellent substrate to host active nanomaterials due to its high conductivity, large surface area, flexibility, and chemical stability [8, 9, 26]. Inspired by this idea, Wang et al. developed hybrid materials based on Mn_3O_4 nanoparticles and RGO sheets for a high capacity anode material for lithium ion batteries (Fig 6.9a, b) [46]. Although the use of Mn_3O_4 as an anode material [61] is limited due to its electrically insulating properties, Mn_3O_4–graphene hybrid nanostructures have provided an unprecedented high capacity ($\sim$900 mAh/g based on the total mass of Mn_3O_4), good rate capability, and cycling stability. In this hybrid structure, the Mn_3O_4 particles were electrically wired to an underlying graphene layer, thereby increasing the capacity of Mn_3O_4 as comparable to its theoretical value of $\sim$ 936 mAh/g.

In addition to the electrochemical performance, the hybrid approach would be very useful for imposing other optical and/or electrical functions. For example, hybrid nanostructures consisting of graphene, poly-*N*-vinylcarbazole (PVK)–ZnO quantum dot (QD) composite, and poly(3,4-ethylenedioxythiophene):poly(styrene-sulfonate) (PEDOT:PSS) layers were employed to achieve enhanced photocurrent in ultraviolet photodetectors [47]. While ZnO QDs embedded in the PVK layer were used to generate electron and hole pairs via ultraviolet (UV) light illumination, adjacent layered graphene film would provide an efficient pathway for photocarriers (i.e., photoexcited electrical carriers) (Fig 6.9c, d).

6.3.3 Nanorod–Graphene Hybrids:1D–2D Hybrids

Vertical arrays of semiconductor nanowires [62] and nanorods [66] can exhibit unique electrical, optical, and mechanical characteristics that may be useful in future

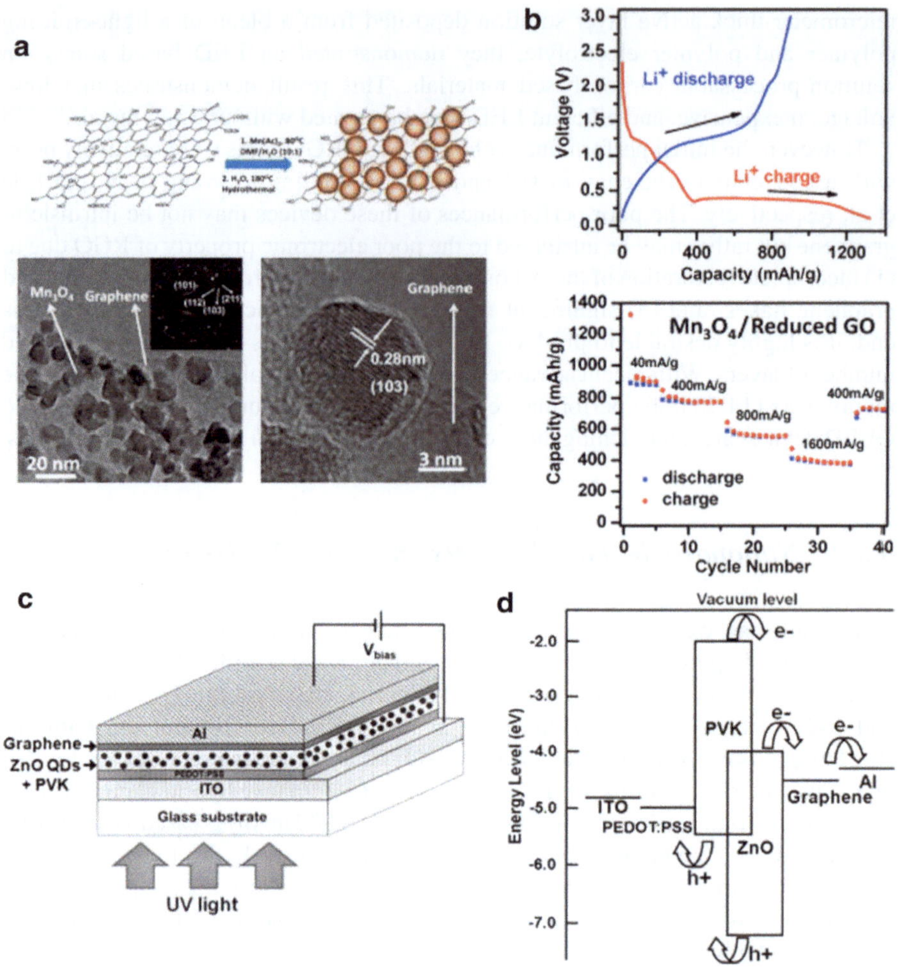

Fig. 6.9 Applications of nanoparticle–graphene hybrids. (**a**) Schematic and TEM images of Mn_3O_4 nanoparticles grown on RGO. (**b**) *Top*: charge (*red*) and discharge (*blue*) curves of Mn_3O_4/RGO for the first cycle at a current density of 40 mA/g. *Bottom*: capacity retention of Mn_3O_4/RGO at various current densities (adapted [46]) (**c**) Schematic diagram of the structure of the UV photodetector and (**d**) energy bands corresponding to the carrier transport mechanisms under UV illumination (adapted from [47])

electronic and optoelectronic devices. Such arrays, when integrated into pieces of plastic, for example, can effectively relieve the strain at the heterointerface and also accommodate flexural deformation.

Despite the superior properties of these nanostructures, site-selective and aligned nanorod arrays and their 3D interconnections are still highly desirable for future electronics and optoelectronics. In this regard, free-standing graphene provides

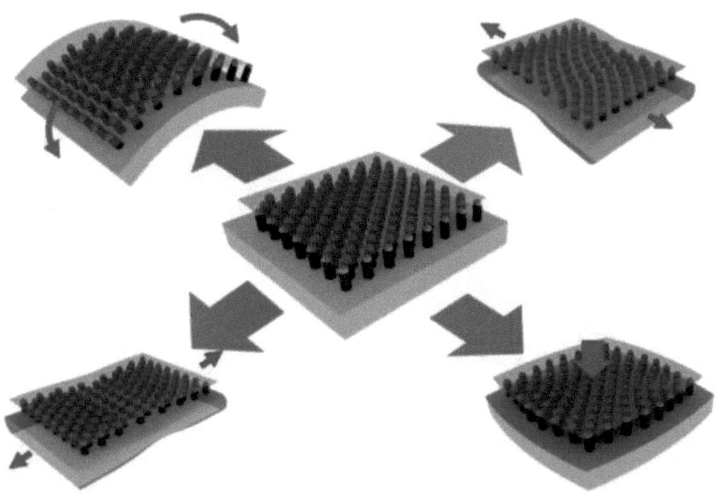

Fig. 6.10 Schematic illustrating the mechanical deformations of 1D–2D hybrid bending, tensile, and compressive forces

a new way to fabricate contact electrodes that can bridge the gaps between the nanorods if the materials are to be used for real-world applications [52, 55]. Specifically, the concept of hybridization of a nanorod as a 1D structure and a fewlayer graphene sheet as a 2D structure could present a new method for constructing 3D architectures and corresponding functionalities.

Advantages of these 3D nanoarchitectures (NAs) consisting of such two building blocks include the followings: first, the combination of a periodic array of 1D nanocrystals (with appropriate diameter and separation) and a very thin 2D graphene layer opens up new opportunity to build a type of light sources, e.g., LEDs or laser diodes (LDs), by exploiting the photonic crystal effect [53–55]; second, the hybrid architectures can maintain sufficient spacings between the monolithic 1D nanorods to allow for easy and fast access of analytes to the active part, thereby enabling highly sensitive sensor devices with relatively fast response and recovery times; third, soft (flexible and stretchable) electronics and optoelectronics based on these hybrids have higher resistance to mechanical deformation caused by bending, tensile and compressive forces (see Fig. 6.10) [52,53].

6.4 1D–2D Nanorod–Graphene Hybrids for Electronics and Optoelectronics

In Sect. 6.3, possible graphene hybrid nanostructures were categorized into three cases, and related techniques and applications were briefly discussed based on the dimensionality of each individual component. In this section, we focus our

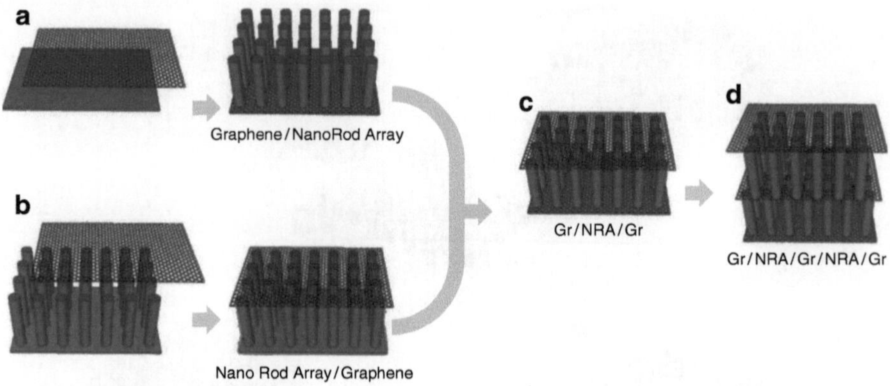

Fig. 6.11 Schematic illustration of constructions of the 1D–2D hybrids. (**a**) Type I hybrid constructed via growth/fabrication of a 1D nanostructure array on a graphene surface. (**b**) Type II hybrids constructed via transfer of free-standing graphene onto a preexisting 1D nanostructure array. (**c**, **d**) Multistage hNAs constructed via combination of type I and type II hybrids

discussion more on 1D–2D nanorod–graphene hybrids with an emphasis on superior properties and new functions that can be obtained with systematically designed hybrid geometries and synthetic strategies [48–55]. The "hybrid geometry" in this context refers to the configuration, ordering, and orientation of the individual components; representative hybrid geometries of the 1D–2D hybrids and their characteristics are schematically described in Fig. 6.11.

Constructions of these hybrid structures have been demonstrated mainly using two approaches: (1) growth/fabrication of a 1D nanostructure array on a graphene surface (Fig. 6.11a) [48–51, 53, 54] and (2) transfer of free-standing graphene onto a preexisting 1D nanostructure array (Fig. 6.11b) [52, 55]. Combining both approaches, more complicated hybrids, i.e., multistage hybrid nanoarchitectures (hNAs), could be attained (Fig. 6.11c, d).

6.4.1 *Vertical 1D Nanostructures on 2D Graphene*

The first type of 1D–2D hybrids can be achieved by growing 1D nanostructures on 2D graphene layers in a preferred orientation (Fig. 6.11a). In particular, considering that a vertical 1D nanostructure array allows implementation into a 3D integrated device platform, vertical growth of a 1D nanostructure array on graphene with site selectivity is highly desirable for electronic and photonic applications.

Lee et al. have achieved carbon hybrid films made of vertically aligned CNTs grown on RGO films using the fabrication process illustrated in Fig. 6.12a [48]. First, mechanically flexible and optically transparent graphene oxide platelets were spin-coated onto a thermally oxidized silicon wafer from aqueous colloidal suspensions and were then thermally reduced to achieve a good electrical conductivity.

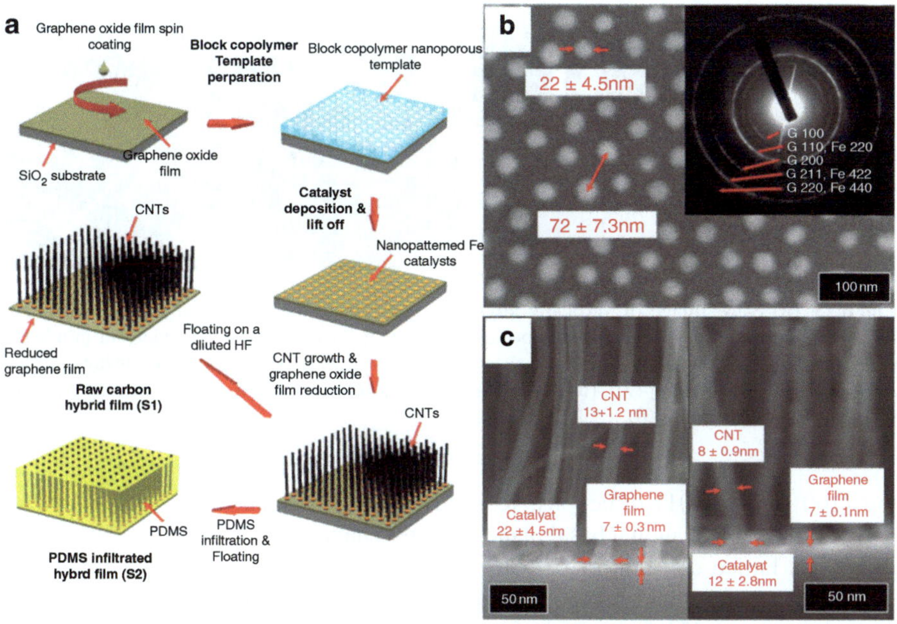

Fig. 6.12 (**a**) Schematic illustration of the fabrication process for carbon hybrid film. (**b**) Scanning electron microscopy (SEM) image of iron catalyst particles on graphene film. *Inset*: electron diffraction pattern. (**c**) Cross-sectional SEM images of carbon hybrid films (adapted from [48])

Second, the RGO films were decorated with patterned catalyst particles using the block-copolymer lithography method. Third, highly aligned vertical CNTs were grown via plasma-enhanced CVD (PE-CVD) at exceptionally high growth rates and temperatures (at 600°C) that resulted in the substrate being converted to an electrically conductive graphene-based film. Carbon hybrid films have unusually high mechanical strengths and elasticities, and they can be detached from the mother substrates via selective etching of the underlying sacrificial layers (in this case, the SiO_2 layer was etched using hydrofluoric acid solution), which can be readily transferred onto arbitrary substrates. Interestingly, these carbon hybrid films have ohmic electrical contacts throughout the junctions in the CNT/metal-catalyst/"graphene-film" system, as useful for optoelectronic devices, such as fieldemission displays (FEDs). Integration of these hybrids into flexible FEDs with excellent field-emission performance is described later in the chapter.

In addition to carbon hybrid films, inorganic nanocrystal growth on 2D graphene has also demonstrated by several groups. For example, Yoon et al. have shown that vertically aligned single-crystalline Co_5Ge_7 nanorods can be grown on a very thin HOPG layer (Fig. 6.13a–c) [49], while Kim et al. demonstrated vertically aligned ZnO nanorods on few-layer graphene sheets (Fig. 6.13d, e) [50]. In both cases, several interesting features were reported. First, as a result of small lattice mismatch, both Co_5Ge_7 and ZnO nanorods grew epitaxially with preferred orientation along

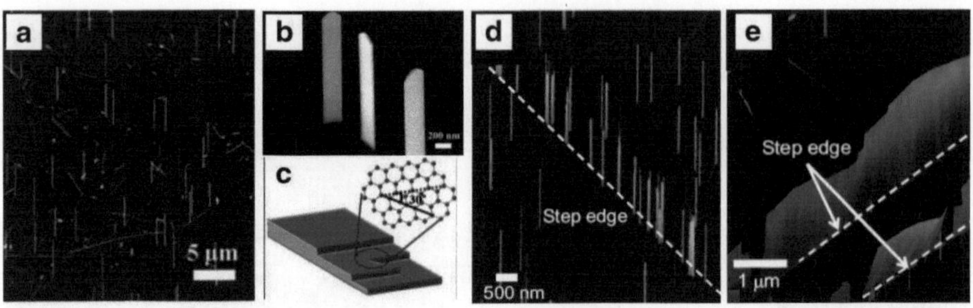

Fig. 6.13 (**a**) SEM image of vertically aligned Co_5 Ge_7nanorods grown on HOPG. (**b**) High-magnification SEM image of vertical Co_5Ge_7 nanobelts grown along the step edges of graphene. (**c**) Schematic illustration of a graphite sheet showing armchair (*solid line*) and zigzag (*dashed line*) edges (adapted from [49]). (**d**, **e**) SEM images of ZnO nanostructures near the step edges of graphene (adapted from [50])

the [100] and [0001] directions, respectively. For instance, the (100) plane of Co_5Ge_7 and the hexagonal basal plane of HOPG have a good epitaxial relationship; the c-axis lattice constant of Co_5Ge_7 ($c = 5.814$ Å) was nearly four times that of the nearest-neighbor C–C distance of the hexagonal basal plane of HOPG ($4 \times d = 4 \times 1.421$ Å $= 5.684$ Å), while the b-axis lattice constant of Co_5Ge_7 ($b = 7.641$ Å) was nearly three times the a-axis lattice constant of HOPG ($3 \times a = 3 \times 2.46$ Å $= 7.38$ Å). Second, step edges of graphene layers affected the position and morphology of the nanostructures. As shown in Fig. 6.13, Co_5Ge_7 nanobelts (rather than nanorods), ZnO nanorods aligned in a row, and ZnO nanowalls were formed on the step edges of graphene layers. This observation indicated that step edges of graphene may serve as nucleation centers for nanocrystal growth, thereby providing an opportunity to control the position and morphology of nanocrystals on graphene.

Alternatively, Lee et al. adopted a similar approach to synthesize ZnO nanorod–graphene (ZnO NRs–Gr) hybrid structures [51], although in this case ZnO seeding layers beneath or upon graphene play an important role in the hydrothermal ZnO nanorod growth in an aqueous solution (Fig. 6.14). Thus, depending on the crystallinity of the ZnO seeding layer (i.e., either polycrystal or single crystal), (1) flower-like ZnO nanorod bundles in which the ZnO nanorods were close-packed and oriented in the radial direction (Fig. 6.14a) or (2) vertically aligned, hexagonal-faceted ZnO rods (Fig. 6.14b) were formed. In addition, using holes in the graphene as a growth mask or using seed patterns on graphene, a regular array of ZnO nanorods were patterned. In this way, the NRs–Gr hybrid structure enables the construction of 3D nanostructures and the imposition of multifunctionalities, which ensures a wide spectrum of applications. First, the hybrid structures exhibited outstanding electrical conductivity and optical transparency, comparable to those of graphene (Fig. 6.14c). Second, new optical functions, i.e., UV light emission inherited from the ZnO nanorods, were introduced into the hybrid structures (Fig. 6.14d). Third, light scattering by the ordered array of ZnO nanorods on

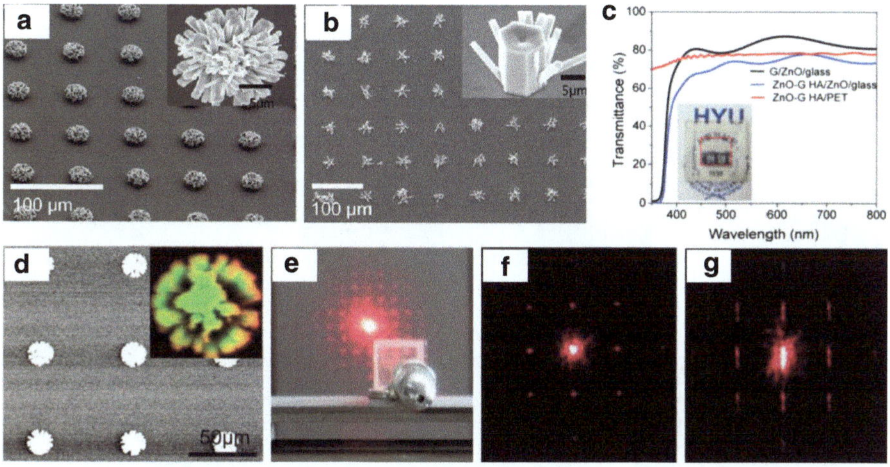

Fig. 6.14 (**a**, **b**) SEM images of ZnO NRs–Gr hybrid structures on a textured ZnO layer and ZnO epilayer as seeding layers. (**c**) Optical transmittance spectra of the ZnO NRs–Gr hybrid structure on the ZnO-coated glass substrate. *Inset*: photograph of the hybrid structure. (**d**) CL image of the hybrid structure showing distinct light emission from the array of ZnO nanorods (*inset*). (**e**) Photograph of the experimental setup for the optical diffraction of a ZnO NRs–Gr hybrid structure. Up to five orders of diffraction spots were clearly observed. Optical diffraction patterns for the hybrid structure on a plastic substrate (**f**) before bending and (**g**) after bending down to a radius curvature of 4 mm (adapted from [51])

transparent graphene yielded optical diffraction patterns for an incident laser beam. Light interaction via the nanostructured ZnO can increase the light coupling into and out of the active region of optoelectronic devices (Fig. 6.14e). Finally, the hybrid structures exhibit excellent mechanical flexibility and structural stability for a bending radius as low as $\sim$ 4 mm (Fig. 6.14e–g).

6.4.2 2D Graphene on Vertical 1D Nanostructures

The second type of 1D–2D hybrid can be fabricated by mounting 2D graphene sheets directly on the top of a 1D nanostructure, where graphene forms a freely suspended structure (Fig. 6.11b). The approach to achieve such architectures exploits the fact that free-standing graphene sheets can be readily obtained using several methods, and they can be transferred onto arbitrary targets, including nonplanar 3D structures. Due to this achievement, graphene can be applicable in exceptional circumstances where the use of TCOs is seriously limited due to their brittleness, high processing temperature required for direct deposition, and difficulty in coupling to nonplanar 3D device platforms.

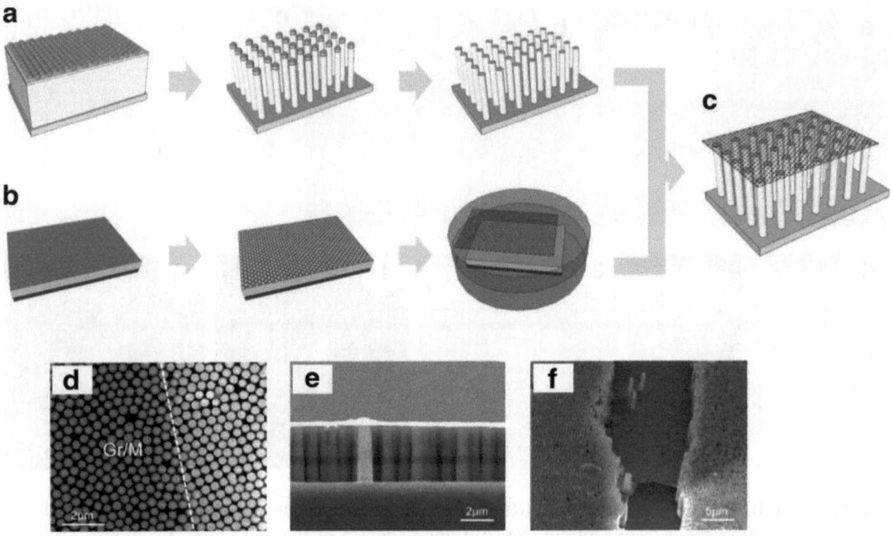

Fig. 6.15 (**a–c**) Schematic illustration of the key steps for fabricating pillar–graphene hNA including (**a**) fabrication of a vertical pillar array via dry etching with the use of silica spheres as an etching mask, (**b**) fabrication of a free-standing graphene sheet, and (**c**) transfer of the graphene sheet to the pillar array. (**d–f**) SEM images show graphene sheets suspended over a large area with the only support of the vertically aligned pillar arrays (adapted from [55])

Representative type II 1D–2D hybrids combine air-gap 1D pillar arrays with a 2D graphene sheet as the top electrode (Fig. 6.15) [55]. In this structure, a vertical array of pillars of uniform diameter and length were achieved by combining a colloidal nanosphere assembly and deep dry etching techniques (Fig. 6.15a). In parallel, a large-scale, few-layer graphene film was grown on Ni-coated SiO_2/Si substrates using CVD, with methane (CH_4) as a carbon source [26]. The graphene sheet was detached from the substrate by etching the underlying sacrificial layer (e.g., SiO_2 and Ni layers) (Fig. 6.15b). Due to the hydrophobic surfaces of the graphene sheets, the free-standing graphene sheet floated in an aqueous solution was then mounted onto the pillar arrays (Fig. 6.15c). As shown in Fig. 6.15d–f, the graphene sheet was suspended over a large area of pillar array. Tilted-view SEM images revealed that the graphene sheet contacted only the tips of the pillars. Due to the outstanding mechanical flexibility and strength of the graphene sheet, it was tightly stretched over $\sim$5– to 7 – μm-wide valleys (Fig. 6.15f).

Other examples are shown in Fig. 6.16a, b, where graphene layers were mounted on vertically grown Si nanorods and ZnO nanorods [52], respectively. Although the dry etching approach produces a well-defined regular array of pillars, those as-grown nanowires/rods lack uniformity in terms of dimension, positioning, orientation, and density of 1D nanocrystals. In this way, bottom-up synthesis of nanorods versus a pillar array fabricated from a planar wafer with a top-down etching process yielded some deviation in the nanorod length and vertical alignment,

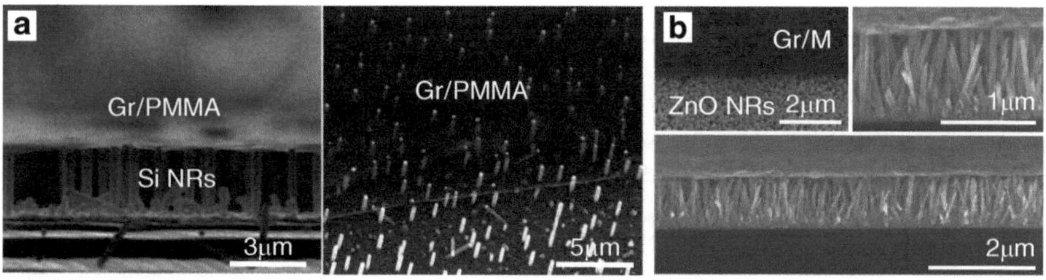

Fig. 6.16 (**a**) SEM images of Si nanowire–graphene hybrid architectures. (**b**) SEM images of ZnO nanorod–graphene/metal hybrid architectures (adapted from [52])

and thus their top surfaces did not lie in a flat horizontal plane. In addition, the existence of hemispherical metal particles at the nanorod tips and/or the small diameter leads to a significant decrease in the contact area between nanorods and graphene, whereas the average spacing between adjacent nanorods was significantly greater than the diameters of the nanorods supporting the graphene. Even in these harsh conditions, exceptionally robust and flexible graphene layers were suspended over a large area with the support of 1D nanostructures.

The successful incorporation of graphene into the 1D nanostructure (nanorod/pillar) array addresses several key features. First, graphene sheets provide efficient current spreading and injection into the active regions of the 1D nanostructures, with minimal optical absorption or reflection. Meanwhile, to ensure the high performance and reliability in the device operation, it is essential to control the electrical contact properties. Direct contact of graphene to the inorganic semiconductors may produce potential barriers at their heterojunctions, degrading device performance and reliable operation. One strategy for improving contact characteristics involves incorporation of very thin metal layers at the junction with appropriate thermal annealing. Figure 6.17a shows the I–V characteristic curves of the ZnO nanorod–graphene/metal (ZnO NRs–Gr/M) hNA before and after thermal annealing, which reveals that the I–V characteristics became linear with an abrupt 25-fold current increase. It is also noteworthy that, even after metal deposition, the graphene/metal sheets still exhibited good optical transmittance (Fig. 6.17b) [67, 68].

Second, due to the layout involving air gaps between the nanorods and robust mechanical properties of the graphene sheet, the 1D–2D hNAs offered an enhanced ability to accommodate certain levels of deformation without fracture. Figure 6.17c displays this characteristic of ZnO NRs–Gr/MhNAs. The hNA was stable without mechanical or electrical failure for bending radii less than $\sim$ 0.8 cm (corresponding to a tensile strain of $\sim$ 1.3%) under up to 100 repeated bending cycles [51].

Third, periodic arrays of 1D structures suspended in air have a much lower optical reflectance than do planar structures, even after hybridizing with graphene or a graphene/metal sheet. As shown in Fig. 6.17d, while the planar epilayer has an average reflectance value of $\sim$40% in the 400–900 nm spectral range, the pillar array exhibited an average value of $\sim$17%, demonstrating that the pillar array can effectively suppress optical reflection at the air/semiconductor interface. In this way,

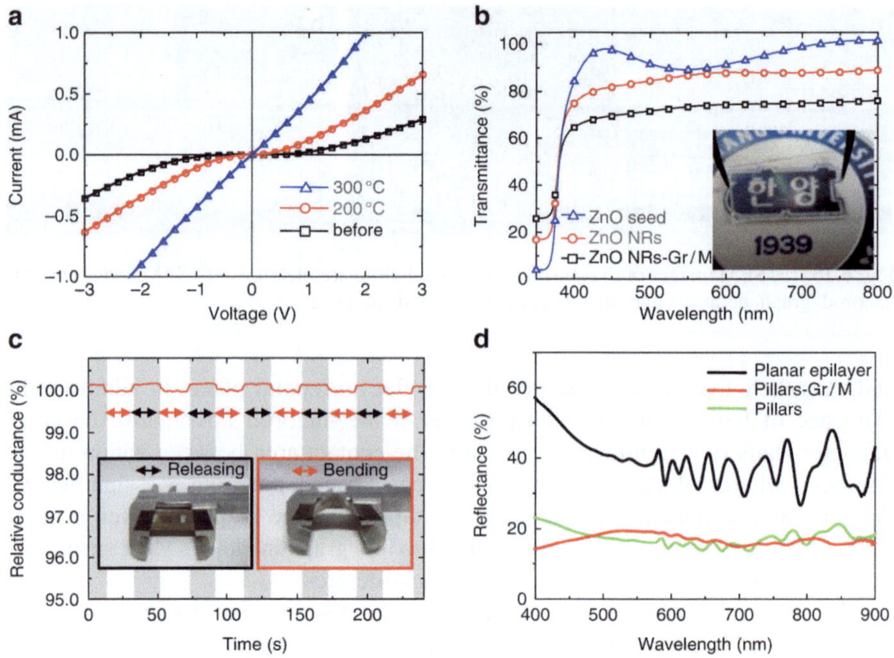

Fig. 6.17 (**a**) I–V characteristic curves of a ZnO NRs–Gr/M hNA measured before and after thermal annealing. (**b**) Optical transmittance spectra of ZnO seed, ZnO nanorods, and ZnO NRs–Gr/M hNA. (**c**) Variation in resistance changes in the device accompanying the repeated bending–unbending cyclic test. *Inset*: photographs of the ZnONRs–Gr/M hNA in the release and bending states (adapted from [52]). (**d**) Reflectance spectra of a planar epilayer (*black*), a pillar array (*green*), and the Gr/M sheet on a pillar array (*red*)

the 1D–2D hNAs provide enhanced light extraction and reduced light reflection at the air–dielectric interfaces, which is essential to achieve high efficiency in optoelectronic devices.

6.4.3 Multistage Hybrid Nanoarchitectures: Pillared Graphene

As previously discussed, multistage hNAs can be constructed by alternating growth/fabrication of a 1D nanostructure array on a graphene surface and transfer of the free-standing graphene onto a preexisting 1D nanostructure array (Fig. 6.11d). In actuality, the development of this novel 3D NA has rarely been achieved and has only been studied theoretically [69].

Based on a multiscale theoretical investigation, Georgios et al. proposed a 3D carbon nanostructure consisting of parallel graphene sheets, supported by CNTs placed vertically on graphene planes. In this structure, the pore sizes and

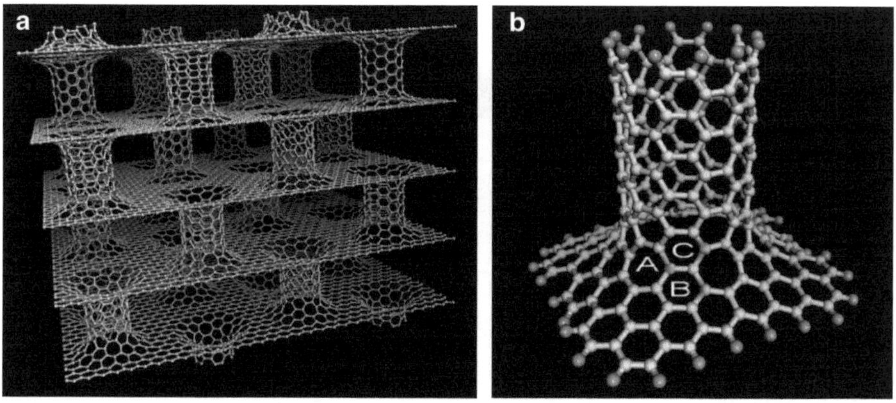

Fig. 6.18 Multistage carbon hybrid nanostructures. (**a**) A novel 3D network nanostructure proposed for enhanced hydrogen storage. (**b**) Optimized cluster of (6,6) CNTs and a graphene sheet (adapted from [69])

surface areas were tuned by controlling the CNT diameter, intertube distance, and graphene interlayer distance. Figure 6.18 shows the simulation result for the structure consisting of graphene layers with a 1.2-nm interlayer distance, stabilized by (6,6) armchair SWNTs with an intertube distance of 1.5 nm. Calculations performed on this hybrid material revealed that it is very promising for hydrogen storage [70–72]. In particular, if this material is doped with lithium cations, the storage capacity, i.e., volumetric hydrogen uptake, is increased up to 41 g H_2/L under ambient conditions, which nearly satisfies the DOE volumetric target for mobile applications.

6.4.4 *Application of 1D–2D Hybrids for Electronics and Optoelectronics*

6.4.4.1 Gas Sensors: Nanorod–Graphene Hybrid Gas Sensors

Chemical interactions of gas molecules at the surface of metal oxides involve the immobilization of conduction electrons in the near-surface region or the release of electrons back into the crystal, thereby leading to change in the electric conductivity (Fig. 6.19a). Thus, pore sizes, surface areas, and grain sizes of sensor materials are important to achieve high performance, suggesting that 1D nanorods/wires might be a suitable material. However, if a single nanorod sensor was used, a sophisticated measuring system is necessary for collecting very weak electrical signals. Instead, percolating nanowire networks [73] or laterally grown nanorod arrays [74] have been used; however, these approaches still lack reliability and controllability. As an alternative to these laterally deposited or networked nanowire sensors, vertical

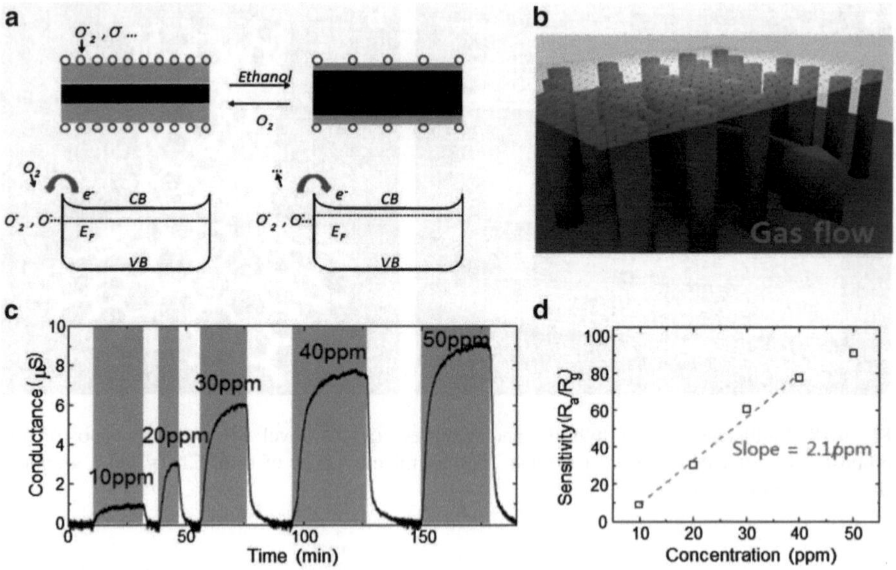

Fig. 6.19 Schematic representation of chemical reactions occurring at the surface of a semiconducting (n-type) gas sensor, according to a standard model. (**b**) Schematic of the ZnO nanorod–graphene hybrid gas sensor. (**c**) Continuous conductance changes in the sensor at different concentrations of ethanol gas at 300°C. (**d**) Plot of sensitivity versus ethanol concentration (adapted from [52])

nanorod arrays may overcome these problems and offer an attractive platform for device application. They can be easily configured to vertical device platforms useful for device scaling and integration of the largest number density in monolayered structures [75–77]. However, practical use of the 1D nanostructure in sensor applications requires electrical wiring to the individual nanorod tips without filling the interspaces between the nanorods with other supporting materials.

ZnO NRs–Gr hNAs (which correspond to type II 1D–2D hybrids, i.e., 2D graphene on vertical 1D nanorods) have been suggested to address this problem [52]. As shown in Fig. 6.16b, graphene is mechanically compliant and stretchable for transfer onto 1D nanorods, and it adheres well to the top surfaces of nanorods and behaves as a top window electrode. Indeed, the unique layout of a gas sensor composed of this material resulted in pores in the single crystal nanorod channels that allow for easy and fast gas transport (Fig. 6.19b), thereby enabling a ppm level detection of the target gas vapor. If the gas sensitivity of the sensor is defined as $S = R_a/R_g$, where R_a and R_g are the resistances of the sensor recorded in air and in the presence of ethanol gas, respectively, the sensitivity of this device was as high as ~9–10 ppm ethanol and ~90–50 pp methanol (Fig. 6.19c, d), which are significantly greater than those of previous gas sensors based on 1D ZnO nanostructures. In addition, these 1D–2D hybrids exhibit good optical transparency and mechanical deformability (Fig. 6.17b, c), offering an additional opportunity to develop mechanically flexible and/or transparent sensors.

6.4.4.2 Field-Emission Displays: Carbon Hybrid Films

Field emission involves the extraction of electrons from the surface of a material subjected to a strong electric field due to quantum mechanical tunneling [77]. Since all field-emission sources rely on field enhancement due to sharp tips/protrusions, 1D nanostructures vertically grown on conducting substrates offer challenging opportunities for FEDs. Encouraged by this idea, electron field emission from 1D nanostructures has been studied extensively. Particularly, due to small diameter, high electrical and thermal conductivity, and chemical stability, CNTs have become excellent electron emitter materials. However, most studies performed on CNTs have been performed on thick and rigid substrates, and thus the development of flexible field-emission materials with a sufficient field-emission current and electrical and mechanical robustness in an extremely deformed geometry has rarely been achieved.

To achieve this goal, carbon hybrid film made of vertically aligned CNTs grown on a graphene substrate (which corresponds to type I 1D–2D hybrids, i.e., vertical 1D nanostructures on 2D graphene) was proposed (see also Sect. 6.4.1). The use of graphene as an underlying substrate for CNT growth provides several key features. First, graphene with excellent flexibility and stretchability can be readily transferred to arbitrary structures (plastic substrates, nonplanar structures, etc.). Second, graphene has high-temperature stability that is suitable for CNT growth. Third, graphene has a similar work function to that of CNTs, and thus it is easier to ensure ohmic contacts at all junctions for efficient input electrical current. Fourth, graphene film ensures robust mechanical and electrical contacts with CNTs even in a highly deformed state. As a result, the flexible carbon hybrid film demonstrated excellent field-emission properties, even under highly deformed states (Fig. 6.20). This high-performance flexible carbon field emitter is potentially useful for diverse, flexible field-emission devices [53].

6.4.4.3 Photonic Application of 1D–2D Hybrids: Graphene-Based Nanostructured LEDs

Graphene-based hybrid LED systems combine graphene with high-performance inorganic semiconductors, such as GaAs, ZnO, and GaN, to exploit the key advantages of each component [54, 55]. To date, two approaches have been independently proposed to extend these concepts to novel hybrid LEDs; one group demonstrated this by constructing LEDs that exploit a graphene layer as an underlying substrate and bottom electrode for inorganic ZnO nanowalls/GaN epilayers (corresponding to the type I 1D–2D hybrid) [54], and the other exploits arrays of GaAs-based nanorods

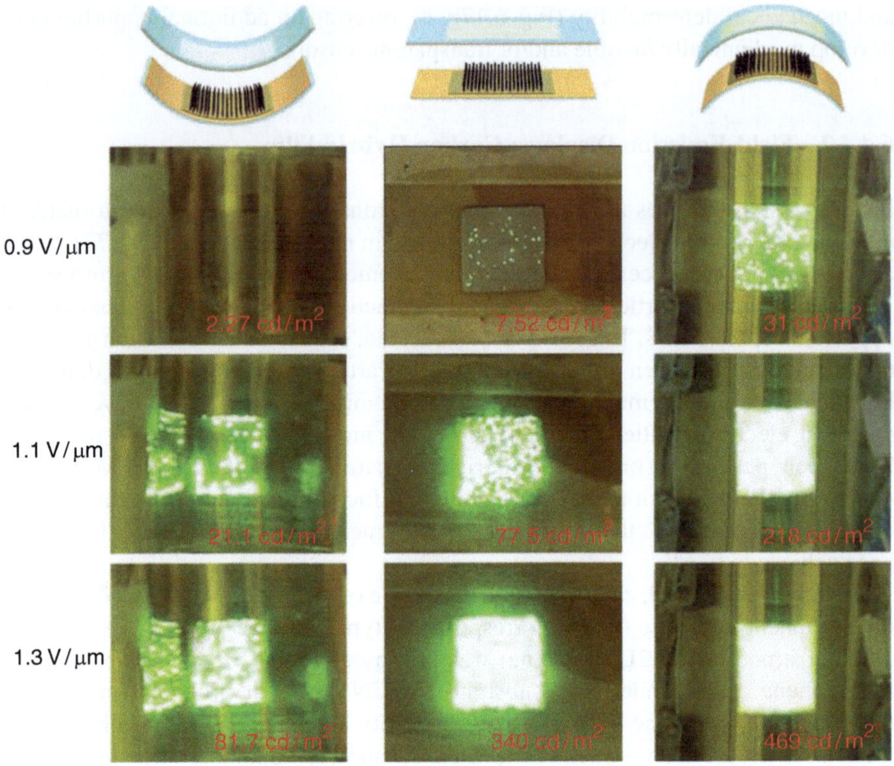

Fig. 6.20 Photographs and corresponding luminance values of the carbon hybrid field emission with concave, flat, and convex morphologies (adapted from [53])

with spanning sheets of graphene as transparent electrodes (corresponding to the type II 1D–2D hybrid) [55].

The basic strategy for LED preparation is using the first approach described in Fig. 6.21. In this approach, difficulty in growing high-quality, epitaxial GaN layers on graphene layers was resolved using high-density, vertically aligned ZnO nanowalls as an intermediate layer [78]. A mirror-smooth, epitaxial GaN thin film is difficult to achieve on pristine graphene layers, presumably because of the lack of chemical reactivity. Meanwhile, as mentioned in Sect. 6.4.1, ZnO nanorods/nanowalls grew epitaxially with preferred orientation along the [0001] direction on graphene surfaces, which were then used as a buffer layer for epitaxial GaN growth. Furthermore, these hybrid structures could be easily transferred onto foreign substrates such as glass, metal, or plastic. As shown in Fig. 6.21e, f, even after transfer onto other substrates, the hybrid LEDs exhibited relatively high performance with strong light emission comparable to those of the as-fabricated LED.

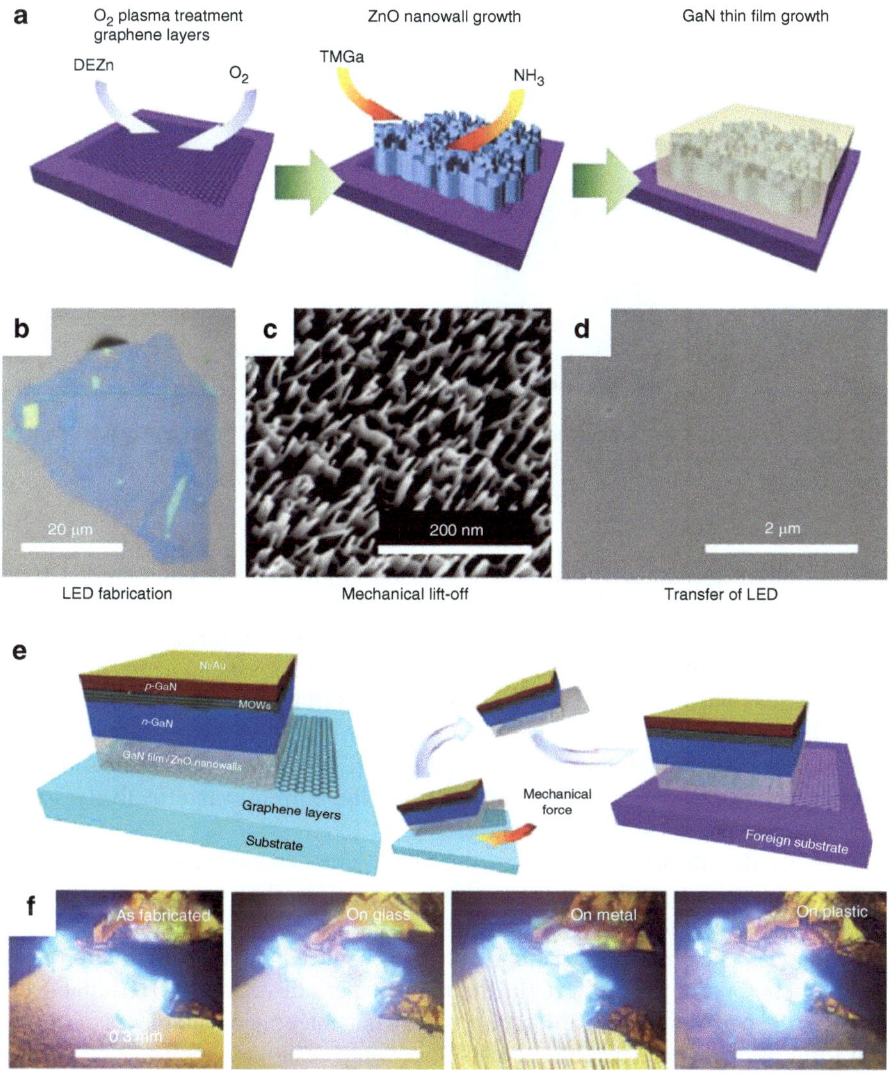

Fig. 6.21 Schematic illustrations of the fabrication and transfer processes for graphene-based hybrid GaN LEDs. (**a**) Fabrication of epitaxial GaN thin films. (**b**) Optical microscopic image of graphene layers. (**c**) SEM image of ZnO nanowalls on graphene layers. (**d**) SEM image of GaN thin film grown on the graphene/ZnO nanowalls. (**e**) Fabrication and transfer processes for thin-film LEDs. (**f**) Optical images of light emissions from the as-fabricated LED on the original substrate and transferred LEDs on the foreign metal, glass, and plastic substrates (adapted from [54])

A second type of hybrid LED combines vertical arrays of 1D pillars with spinning sheets of graphene as top transparent electrodes [55]. In this approach, graphene sheets coated with very thin metal layers exhibit good mechanical and electrical properties, and the ability to mount in a freely suspended configuration

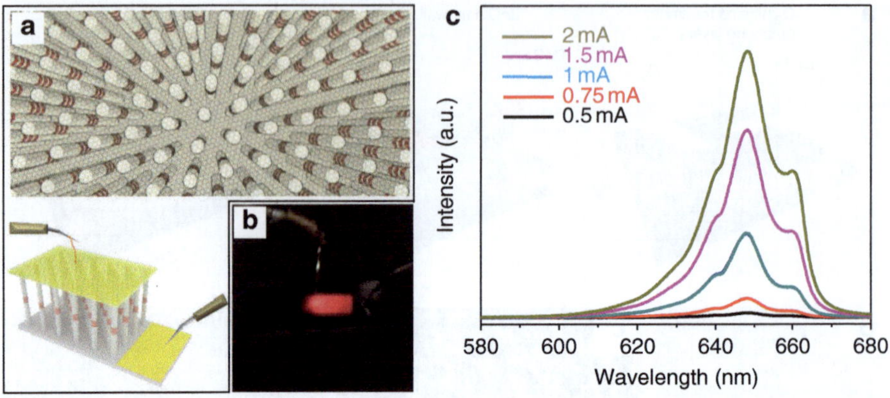

Fig. 6.22 GaAs LED. (**a**) Schematic illustration and (**b**) a photograph of the light emission from pillar–graphene hybrid LEDs. (**c**) EL spectra of PSL–Gr/M LEDs recorded at various currents (adapted from [55])

to the pillar-superlattice (PSL) arrays as a top window electrode (Fig. 6.15 and 6.22a). By ensuring good ohmic contact with a thermally annealed, very thin metal intermediate layer, efficient current injection to pillars was achieved, which enabled bright electroluminescence (EL) under forward bias. As schematically illustrated in Fig. 6.22a, when electrical contact with the Gr/M window layer was made using a $\sim$0.3 – mm-thick Au wire attached to a probe tip, a bright red emission with a dominant peak at 638 nm was observed over a large area ($\sim$ 0.5 cm $\times$ 0.3 cm) (Fig. 6.22b, c). This observation indicates that the current was effectively spread over the suspending Gr/M window layer.

Compared with planar structures, this pillar–graphene hybrid layout may provide many advantages for optoelectronics. First, 1D pillar nanostructures in our hybrid structures can effectively relieve the strain at the heterointerface and also accommodate flexural deformation. Second, enhanced light extraction/reflection and the photonic crystal effect (in the case of periodic array of 1D pillars) can improve the performance. Third, such arrays, when integrated onto pieces of plastic, can form mechanically flexible devices that would be impossible to achieve using conventional semiconductor wafer technologies. Fourth, graphene has excellent electrical/thermal conductivity as well as being mechanically flexible, thus it can effectively transfer carriers into pillars, release heat, and accommodate flexible bending.

6.5 Conclusions

In this chapter, we reviewed the latest research progress on hybrid semiconductor nanostructures using graphene-based 2D sheets as emerging materials for future electronic and photonic applications. Graphene is a promising alternative to conventional transparent electrode materials, exhibits unique properties including good electrical/thermal conductivity, mechanical strength and elasticity, and optical transparency, and may be a promising component for hybrid systems. Various types of graphene-based hybrid semiconductor nanostructures can be achieved, and they were categorized into three distinct subgroups according to the dimensionalities of the individual components: (1) 0D–2D hybrids exploiting graphene as an excellent substrate to host active nanoparticles; (2) 1D–2D hybrids incorporating a graphene layer as the underlying bottom electrode and/or spanning top window electrode; and (3) 2D–2D hybrid lamellar composites in which graphene is generally used as an anode or a cathode for organic semiconductor devices. Particularly, hybridization of a 1D structure with 2D graphene enabled sophisticated 3D architectures. Construction of 1D–2D hybrids was demonstrated mainly through the growth/fabrication of a 1D nanostructure array on a graphene surface or transfer of free-standing graphene onto a 1D nanostructure array.

Despite its very short history, graphene has already revealed a wide array of potential applications. Those discussed in this chapter are only a preview as the full potential of hybrid nanostructures is yet to be determined. For example, graphene-based hybrid devices would be promising alternatives to planar architectures where flexible and stretchable properties are required. In addition, we believe that the basic hybrid concept of exploiting graphene as robust, transparent components could provide a new strategy for the electrical interconnection of devices in many areas of electronics, optoelectronics, microelectromechanical systems and photovoltaics. Conceptually, multistage hNAs can be constructed by alternating the approaches introduced here, and they may provide new insight into the design of a new class of electronics and photonics with optimal performance and multifunctionality.

References

1. A. Makisima, Ceramic Japan **39**, 90 (2004)
2. M. Nanko, Azojomo 6, DOI: 10.2240/azojomo0288, 2009
3. Y. Chujo, KONA. **25**, 255 (2007)
4. A.K. Geim, Science **324**, 1530 (2009)
5. R. Yan, D. Gargas, P. Yang, Nat. Photonics **3**, 569 (2009)
6. A.K. Geim, K.S. Novoselov, Nat. Mater. **6**, 183 (2007)
7. K.S. Novoselov, A.K. Geim, S.V. Morozov, D. Jiang, Y. Zhang, S.V. Dubonos, I.V. Grigorieva, A.A. Firsov, Science **306**, 666 (2004)
8. K.S. Novoselov, A.K. Geim, S.V. Morozov, D. Jiang, M.I. Katsnelson, I.V. Grigorieva, S.V. Dubonos, A.A. Firsov, Nature **438**, 197 (2005)
9. Y. Zhang, Y.-W. Tan, H.L. Stormer, P. Kim, Nature, **438**, 201 (2005)
10. P.R. Wallace, Phys. Rev. **71**, 622 (1947)
11. A.H.C. Neto, F. Guinea, N.M.R. Peres, K.S. Novoselov, A.K. Geim, Rev. Mod. Phys. **81**, 109 (2009)
12. P. Avouris, Z. Chen, V. Perebeinos, Nat. Nanotechnol. **2**, 605 (2007)
13. R.R. Nair, P. Blake, A.N. Grigorenko, K.S. Novoselov, T.J. Booth, T. Stauber, N.M.R. Peres, A.K. Geim, Science **320**, 1308 (2008)

14. S.K. Bae, H.K. Kim, Y.B. Lee, X.F. Xu, J.S. Park, Y. Zheng, J.K. Balakrishnan, T. Lei, H.R Kim, Y.I. Song, Y.J. Kim, K.S. Kim, B.O. Zyilmaz, J.H. Ahn, B.H. Hong, S. Iijima, Nat. Nanotechnol. **5**, 574, (2010)
15. X. Li, Y. Zhu, W. Cai, M. Borysiak, B. Han, D. Chen, R.D. Piner, L. Colombo, R.S. Ruoff, Nano Lett. **9**, 4359 (2009)
16. A. Fasolino, J.H. Los, M.I. Katsnelson, Nat. Mater. **6**, 858 (2007)
17. I.W. Frank, D.M. Tanenbaum, A.M. Van Der Zande, P.L. McEuen, J. Vac. Sci. Technol. B. **25**, 2558 (2007)
18. M. Poot H.S.J. Van Der Zant, Appl. Phys. Lett. **92** 063111, (2008)
19. X. Lu, M. Yu, H. Huang R.S. Ruoff, Nanotechnology **10**, 269, (1999)
20. Z.H. Ni, H.M. Wang, J. Kasim, H.M. Fan, T. Yu, Y.H. Wu, Y.P. Feng, Z.X. Shen, Nano Lett. **7**, 2758 (2007)
21. P.K. Ang, S. Wang, Q. Bao, J.T.L. Thong, K.P. Loh, Acs Nano. **3**, 3587 (2009)
22. A.G. Cano-Marquez, F.J. Rodríguez-Macıas, J. Campos-Delgado, C.G. Espinosa-Gonzalez, F. Tristan-Lopez, D. Ramírez-Gonzalez, D.A. Cullen, D.J. Smith, M. Terrones, Y.I. Vega-Cantu, Nano Lett. **9**, 1527 (2009)
23. D.V. Kosynkin, A.L. Higginbotham, A. Sinitskii, J.R. Lomeda, A. Dimiev, Nature **458**, 872 (2009)
24. D. Errandonea, Nat. Mater. **8**, 170, (2009)
25. J. Hass, F. Varchon, J.E. Millan-Otoya, M. Sprinkle, N. Sharma, W.A. de Heer, C. Berger, P.N. First, L. Magaud, E.H. Conrad, Phys. Rev. Lett. **100**, 125504, (2008)
26. K.S. Kim, Y. Zhao, H. Jang, S.Y. Lee, J.M. Kim, K.S. Kim, J.-H. Ahn, P. Kim, J.-Y. Choi, B.H. Hong, Nature **457**, 706 (2009)
27. X. Li, W. Cai, J. An, S. Kim, J. Nah, D. Yang, R. Piner, A. Velamakanni, I. Jung, E. Tutuc, S.K. Banerjee, L. Colombo, R.S. Ruoff, Science **324**, 1312, (2009)
28. L. Jiao, L. Zhang, X. Wang, G. Diankov, H. Dai, Nature **458**, 877 (2009)
29. A. Yu, P. Ramesh, M.E. Itkis, E. Bekyarova, R.C. Haddon, J. Phys. Chem. C **111**, 21 (2007)
30. A. Varykhalov O. Rader, Phys. Rev. B **80**, 035437 (2009)
31. H. Ueta, M. Saida, C. Nakai, Y. Yamada, M. Sasaki, S. Yamamoto, Surf. Sci. **560**, 183 (2004)
32. J. Coraux, A.T. NDiaye, C. Busse, T. Michely, Nano Lett. **8**, 565 (2008)
33. P.W. Sutter, J.-I. Flege, E.A. Sutter, Nat. Mater. **7**, 406 (2008)
34. S.S. Datta, D.R. Strachan, E.J. Mele, A.T. Charlie Johnson, Nano Lett. **9**, 7 (2009)
35. Q. Yu, J. Lian, S. Siriponglert, H. Li, Y.P. Chen, S.-S. Pei, Appl. Phys. Lett. **93**, 113103, (2008)
36. X. Li, W. Cai, L. Colombo, R.S. Ruoff, Nano Lett. **9**, 12 (2009)
37. R.G. Gordon, MRS Bull. **25**, 52 (2000)
38. V.C. Tung, L.-M. Chen, M.J. Allen, J.K. Wassei, K. Nelson, R.B. Kaner, Y. Yang., Nano Lett. **9**, 1949 (2009)
39. E. Polikarpov, J.S. Swensen, N. Chopra, F. So, A.B. Padmaperuma, Appl. Phys. Lett. **94**, 223304 (2009)
40. J.-Y. Lee, S.T. Connor, Y. Cui, P. Peumans, Nano Lett. **8**, 2, (2008)
41. A. Patra, Y.H. Wijsboom, S.S. Zade, M. Li, Y. Sheynin, G. Leitus, M. Bendikov, J. Am. Chem. Soc. **130**, 6734, (2008)
42. D. Zhang, K. Ryu, X. Liu, E. Polikarpov, J. Ly, M.E. Tompson, C. Zhou, Nano Lett. **6**, 1880 (2006)
43. J.H. Chen, C. Jang, S. Xiao, M. Ishigami, M.S. Fuhrer, Nat. Nanotech. **3**, 206 (2008)
44. X. Wang, L. Zhi, K. Mullen, Nano Lett. **8**, 323 (2008)
45. P. Matyba, H. Yamaguchi, G. Eda, M. Chhowalla, L. Edman, N.D. Robinson, ACS Nano **4**, 637 (2010)
46. H. Wang, L.-F. Cui, Y. Yang, H.S. Casalongue, J.T. Robinson, Y. Liang, Y. Cui, H. Dai, J. Am. Chem. Soc. **132**, 13978 (2010)
47. H.Y. Yang, D.I. Son, T.W. Kim, J.M. Lee, W.I. Park, Org. Electron. **11**, 1313 (2010)
48. D.H. Lee, J.E. Kim, T.H. Han, J.W. Hwang, S. Jeon, S.-Y. Choi, S.H. Hong, W.J. Lee, R.S. Ruoff, S.O. Kim, Adv. Mater. **22**, 1247 (2010)
49. H. Yoon, K. Seo, N. Bagkar, J. In, J. Park, J. Kim, B. Kim, Adv. Mater. **21**, 4979, (2009)

50. Y.-J. Kim, J.-H. Lee, G.-C. Yi, Appl. Phys. Lett. **95**, 213101 (2009)
51. J.M. Lee, Y.B. Pyun, J. Yi, J.W. Choung, W.I. Park, J. Phys. Chem. C **113**, 19134 (2009)
52. J.S. Yi, J.M. Lee, W.I. Park, DOI:10.1016/j.snb.2010.12.033
53. D.H. Lee, J.A. Lee, W.J. Lee, S.O. Kim, Small **7** 95 (2011)
54. K. Chung, C.-H. Lee, G.-C. Yi, Science **330**, 655 (2010)
55. J.M. Lee, J.W. Choung, J. Yi, D.H. Lee, M. Samal, D.K. Yi, C.-H. Lee, G.-C. Yi, U. Paik, J.A. Rogers, W.I. Park, Nano Lett. **10**, 2783 (2010)
56. S.-I. Na, S.-S. Kim, J. Jo, D.-Y. Kim, Adv. Mater. **20**, 4061 (2008)
57. G. Eda, G. Fanchini, M. Chhowalla, Nat. Nanotech. **3**, 270 (2008)
58. G. Eda, Y.-Y. Lin, S. Miller, C.-W. Chen, W.-F. Su, M. Chhowalla, Appl. Phys. Lett. **92**, 233305 (2008)
59. S. Gilje, S. Han, M. Wang, K.L. Wang, R.B. Kaner, Nano Lett. **7**, 3394 (2007)
60. C. Gomez-Navarro, R.T. Weitz, A.M. Bittner, M. Scolari, A. Mews M. Burghard, K. Kern, Nano Lett. **7**, 3499 (2007)
61. D. Pasero, N. Reeves, A.R. West, J. Power Sources **141**, 156 (2005)
62. W.I. Park G.-C. Yi, Adv. Mater. **16**, 87 (2004)
63. J. Noborisaka, J. Motohisa, S. Hara, T. Fukui, Appl. Phys. Lett. **87**, 093109 (2005)
64. C. Thelander, P. Agarwal, S. Brongersma, J. Eymery, L.F. Feiner, A. Forchel, M. Scheffler, W. Riess, B.J. Ohlsson, U. Gösele, L. Samuelson, Mater. Today **9**, 28 (2006)
65. Y.B. Tang, Z.H. Chen, H.S. Song, C.S. Lee, H.T. Cong, H.M. Cheng, W.J. Zhang, I. Bello, S.T. Lee, Nano Lett. **8**, 4191 (2008)
66. W.I. Park G.-C. Yi, Appl. Phys. Lett. **82**, 4358 (2003)
67. K.-J. Kong, Y. Choi, B.-H. Ryu, J.-O. Lee, H. Chang, Mater. Sci. Eng. C **26**, 1207 (2006)
68. S. Malola, H. Häkkinen, P. Koskinen, Appl. Phys. Lett. **94**, 043106 (2009)
69. G.K. Dimitrakakis, E. Tylianakis, G.E. Froudakis, Nano Lett. **8**, 10 (2008)
70. A.C. Dillon, K.M. Jones, T.A. Bekkedahl, C.H. Kiang, D.S. Bethune, M.J. Heben, Nature **386**, 377 (1997)
71. C. Liu, Y.Y. Fan, M. Liu, H.T. Cong, H.M. Cheng, M.S. Dresselhaus, Science **286**, 1127 (1999)
72. Y. Ye, C.C. Ahn, C. Witham, B. Fultz, J. Liu, A.G. Rinzler, D. Colbert, K.A. Smith R.E. Smalley, Appl. Phys. Lett. **74**, 16 (1999)
73. P. Feng, Q. Wan, T.H. Wang, Appl. Phys. Lett. **87**, 213111 (2005)
74. T.-J. Hsueh, C.-L. Hsu, S.-J. Chang, I.-C. Chen, Sensor Actuat. B-Chem. **126**, 473 (2007)
75. A. Modi, N. Koratkar, E. Lass, B. Wei, P.M. Ajayan, Nature **424**, 171 (2003)
76. J. Goldberger, A.I. Hochbaum, R. Fan, P. Yang, Nano Lett. **6**, 973 (2006)
77. R.H. Fowler L. Nordheim, Proc. R. Soc. London, Ser. A **119**, 173, (1928)
78. D.P. Bour, N.M. Nickel, C.G.V.d. Walle, M.S. Kneissl, B.S. Krusor, P. Mei, N.M. Johnson, Appl. Phys. Lett. **76**, 16 (2000)

Chapter 7
Microstructural Properties of Nanostructures

Sang-Wook Han

Abstract The physical and chemical properties of nanostructures are different from their bulks. X-ray absorption fine structure (XAFS) studies of ZnO and GaN nanostructures demonstrate that the structures of nanostructures are intermediate structures of molecule and bulk. The structural distortion and surface properties of nanostructures should be counted for understanding the physical and chemical property changes of nanoparticles.

7.1 Introduction

Physical properties of a material are mainly determined by elements consisting in the material and bonds of the elements. Different bonds of same elements can make different physical properties of materials, graphite and diamond, for example. When particle size becomes in nanometer scale, the physical and chemical properties of particles can be very different from their bulks. When the particle size is comparable to Bohr radius, the charges including electrons and holes are confined in their behaviour. It results in discrete energy and widened band gap. It is called "quantum confinement". The surface-to-volume ratio of a nanoparticle is much larger than that of its bulk, so that the nanoparticle has more ratio of dangling bonds on the surface. The more dangling bonds provide the more sensitive to chemical bonds and electrical environmental changes. Furthermore, the large surface-to-volume ratio in a nanoparticle causes a structural distortion which can also change the physical and chemical properties of nanoparticles.

The quantum confinement effect has been widely studied theoretically and experimentally [1–6]. The quantum confinement effect was observed in a blue

S.-W. Han (✉)
Department of Physics Education, Chonbuk National University, 567 Baekje-daero, Deokjin-gu, Jeonju-si, Jeollabuk-do 561-756, Korea
e-mail: shan@jbnu.ac.kr

G.-C. Yi (ed.), *Semiconductor Nanostructures for Optoelectronic Devices*,
NanoScience and Technology, DOI 10.1007/978-3-642-22480-5_7,

shift from photoluminescence and Raman scattering measurements. The blue shift was observed due to quantum confinement effect from various semiconducting nanostructures, including ZnO, Si, $Ga_{1-x}Al_xN$, ZnS, ZnSe, GaN, GaAs, InAs, CdS, and CdTe [1, 2, 7–17]. The magnetic properties of nanoparticles are also different from those of bulks. Previous studies reported that $CeAl_2$ and $CePt_{2+x}$ nanoparticles displayed a nonmagnetic, Kondo ground state above an antiferromagnetic transition instead of Kondo behaviour when their particle size becomes comparable to the nanometer scale [18]. The suppression of the magnetic ground state in the $CeAl_2$ and $CePt_{2+x}$ nanoparticles is due to the inability of the nanoparticles to support spin waves, as related to spin fluctuations in these materials [18].

The physical property changes of nanostructures due to their sizes have been observed, compared with their bulk counterparts. However, the structural property changes in nanoparticles, including bond-length changes, disorder, and oxide, were also observed from nanoparticles [19,20]. These structural changes in nanoparticles should be counted for understanding the physical and chemical property changes of nanoparticles. The quantum confinement effect is generally determined by a blue shift in photoluminescence (PL) and Raman measurements. However, the blue shifts in PL and Raman spectra are contributed by the quantum confinement and lattice contraction because the energy band gap can be engineered by controlling the lattice constants of crystals. Previous studies of $Zn_{1-x}Mg_xO$ [21,22] and $Zn_{1-x}Cd_xO$ [23] reported that the band gaps were reversely proportional to the lattice constants. This means that the quantum confinement cannot be determined by the blue shift in PL or Raman measurements alone. A lattice constant change in the scale of 0.01 Å affects the blue shift of the PL and Raman measurements. However, it is very difficult to determine a small amount of a lattice change in nanoparticles.

In general, nanoparticles are formed by several to hundred thousand atoms and have an intermediate structure between a molecule and bulk. It is very difficult to determine their true structures [24]. In order to determine the structural properties of nanoparticles, various characterization techniques are used. Each technique has strong and weak points. X-ray diffraction (XRD) is a canonical tool to determine crystalline structural properties. XRD detects the average distance between atomic planes in long-range orderings. XRD cannot detect directly atomic bond lengths and atomic species. Moreover, XRD is not very useful for nanoparticle studies because nanoparticles do not have enough scattering sources and include large amount of disorder, although XRD with pair-distribution-function (PDF) analysis is better than a conventional XRD [24]. PDF analysis is based on XRD. In PDF analysis, after the scattering background is removed from XRD data measured in q-space, X-ray momentum transfer, the XRD data are Fourier transformed to r-space and fitted to a model. PDF analysis is sensitive to intermediate range orderings, 10–20 Å [24].

Scanning electron microscopy (SEM) can detect the macroscopic shape of nanoparticles but cannot show the microscopic structures. Atomic force microscopy (AFM) and scanning tunnelling microscopy (STM) can detect only the surface structures. SEM and AFM techniques limit to detect the structural properties in a subatomic scale, although they are quite useful to simply measure macro-structural properties. STM is an instrument for imaging surfaces in a 0.1 nm resolution. STM uses the concept of quantum tunnelling. The conducting tip of STM can detect the local density of states near the surface of a sample. The current crosses the tip and the surface is converted into images. STM is very useful for the study of surface but not inside nanoparticles. Furthermore, the lateral resolution of STM is approximately 1 Å which is too large to determine the structures of nanoparticles.

Transmission electron microscopy (TEM) is a local probe technique to detect atomic arrays in a specimen and widely used to detect the atomic array in nanoparticles. An electron beam is transmitted through an ultrathin specimen, interacting with electric fields due to electrons in atoms. The transmitted electrons are formed images on a fluorescence device. TEM detects the average electron density of atomic arrays along the direction of electron beam propagation. In general, TEM can detect the atomic distances in a 0.01 Å resolution. TEM is widely used as a microscopic probe in the study of crystalline nanostructures. For TEM measurements, specimens should be destroyed. TEM cannot distinguish atomic species and is not very useful for amorphous. TEM limits to detect a small amount of local distortion and structures around a specific atom.

X-ray absorption fine structure (XAFS) is a local structural probe to detect structural and chemical properties around a probe atom in a compound material. XAFS can measure the average distances of atomic shells from a probe atom. The resolution of XAFS is approximately 0.005 Å. The XAFS technique is particularly useful for compounds and doped materials. XAFS is applicable for the structural and chemical property studies of solid, liquid, and even gas. However, XAFS is an indirect measurement technique, such as XRD. The analysis of XAFS is complex. The details of XAFS are discussed later. For the determination of the structural and chemical properties of nanostructures, XAFS is quite useful because it selectively measures the local structural and chemical properties around a specific element, without depending on the crystallinity. The structural and chemical properties of nanostructures can be different from their bulks due to the large surface-to-volume ratio. Small amounts of local distortion and disorders in nanostructures cannot be detected by other techniques. Furthermore, for doped nanostructures, XAFS is a unique probe to examine the locations of doped element and local structures around the doped atoms. For nanostructures, XRD is not very useful because nanostructures do not have sufficient scattering sources. TEM can detect only limited regions and cannot distinguish atomic species. We employed XAFS techniques to examine the local structural properties of ZnO nanoparticles, ZnO nanorods, and GaN/ZnO coaxial nanorods. Unlike other techniques, XAFS can measure directly the true local structures of materials in nanostructures, which are directly related to the physical properties of nanostructures.

7.2 X-ray Absorption Fine Structure

XAFS was first studied in the 1960s and was fully understood in the 1970s by E.A. Stern and his co-workers [25]. After the first synchrotron at the Stanford Synchrotron Radiation Laboratory was built in 1973, the XAFS technique has

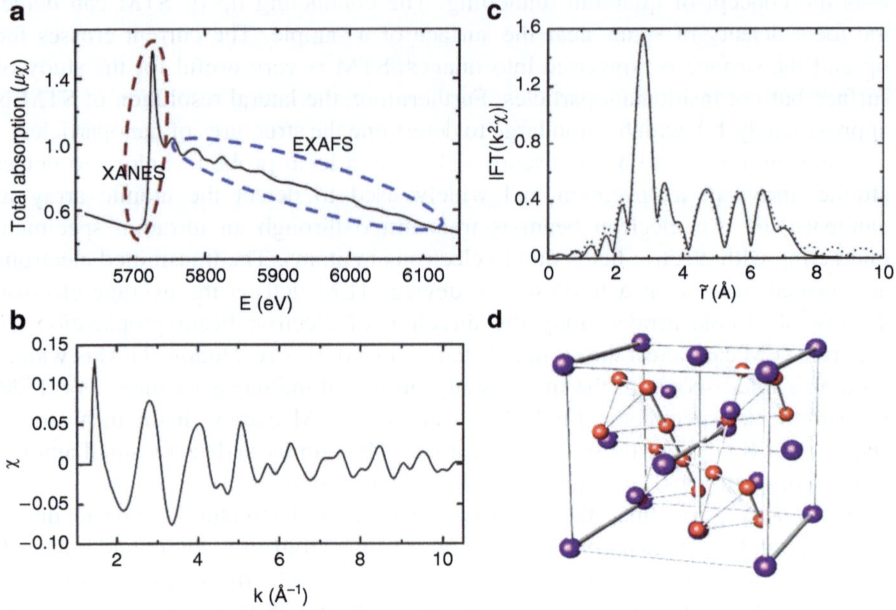

Fig. 7.1 (**a**) Total X-ray absorption from a $CeAl_2$ powder at the Ce L_{III} edge as a function of the incident X-ray energy. The *dotted ovals* indicate the regions of XANES and EXAFS. (**b**) Oscillation part (χ, EXAFS) in (**a**) as a function of the photoelectron wave number vector, k. (**c**) Magnitude of Fourier transformed EXAFS with k^2-weight as a function of the distance from a Ce atom. (**d**) Schematic diagram of the $CeAl_2$ structure

been rapidly developed to determine the local structural and chemical properties of materials. XAFS includes two parts: X-ray absorption near edge structure (XANES) and extended X-ray absorption fine structure (EXAFS), as shown in Fig. 7.1. XANES is sensitive to the energy level of an X-ray absorbing electron, chemical properties of the X-ray absorbing atom, and structural properties whereas EXAFS can describe the distance, species, and coordination number of neighbouring atoms around the X-ray absorbing atom.

When X-ray passes through a material, the X-ray is partially scattered, absorbed, and transmitted. The transmitted beam intensity is exponentially decayed with a specimen thickness and absorption coefficient. The transmitted beam intensity is obtained by integrating the absorption of each slice over the specimen thickness t, as shown in Fig. 7.2a. $I_t(\mathrm{E}) = I_0(\mathrm{E})e^{-\mu(\mathrm{E})t}$, where I_t is the transmitted beam intensity, I_0 is the incident beam intensity, μ is the absorption coefficient, and t is the specimen thickness. It is called the Beer's law. The total X-ray absorption is obtained simply by measuring the incident and transmitted beam intensities, $\mu(\mathrm{E})t = \ln[I_0(\mathrm{E})/I_t(\mathrm{E})]$.

When the bound electrons in atoms absorb X-rays and escape from the atoms, the electrons at the upper energy levels in the atoms jump into the empty lower energy levels. The transition electrons emit X-rays with the energy difference between

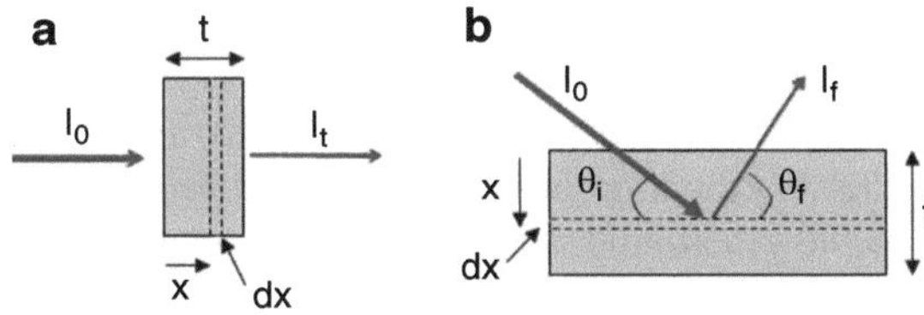

Fig. 7.2 Schematics of the transmission and fluorescence of X-rays. I_0, I_t, and I_f are the intensities of incident, transmitted, and fluorescence X-rays, t is specimen thickness and d_x is an infinitesimal thickness

the two energy levels. Therefore, XAFS can be measured with a fluorescence or transmission mode [25–27]. The total fluorescence beam intensity is obtained by integrating the fluorescence of each slice over the specimen thickness t, as shown in Fig. 7.2b

$$I_f(E) = I_0(E)\varepsilon(E)\mu(E)\int_0^t \exp\left\{-\left[\frac{\mu_T(E)}{\sin\theta_i}+\frac{\mu_T(E)}{\sin\theta_f}\right]x\right\}\frac{dx}{\sin\theta_i} \tag{7.1}$$
$$= \frac{I_0(E)\varepsilon(E)\mu_T(E)}{\mu_T(E)+\dfrac{\mu_T(E_f)\sin\theta_i}{\sin\theta_f}}\left(1-\exp\left\{-\left[\frac{\mu_T(E)}{\sin\theta_i}+\frac{\mu_T(E)}{\sin\theta_f}\right]t\right\}\right),$$

where ε is the fluorescence efficiency, $\mu_T = \mu + \mu_B$ is the total fluorescence coefficient (μ_B is the background X-ray absorption coefficient), θ_i and θ_f are the angles of the incident and fluorescence X-rays, respectively, and E_f is the fluorescence X-ray energy. In general, above the X-ray absorption edge, $\mu_T(E)$ is much larger than $\mu_T(E_f)$ and $\mu(E)$ is also larger than $\mu_B(E)$. The fluorescence intensity can be simply described, as

$$I_f(E) = I_0(E)\varepsilon(E)\frac{\mu(E)t}{\sin\theta_i}, \quad \mu(E) = \frac{I_f(E)\sin\theta_i}{I_0(E)\epsilon(E)t}. \tag{7.2}$$

The X-ray absorption coefficient is obtained by measuring the intensities of incident and fluorescence X-rays.

The X-ray absorption rate near the X-ray absorption edge is described theoretically by the Fermi's Golden rule, $| < \mathrm{f}|\hat{\mathbf{e}}\cdot\hat{\mathbf{a}}|\mathrm{i} > |^2$, where $|\mathrm{i}>$ and $|\mathrm{f}>$ are the initial and final states of the X-ray absorbing electron, $\hat{\mathbf{e}}$ is the electric field direction of the incident X-ray, and $\hat{\mathbf{a}}$ is the direction of the dipole consisting of the photoelectron and hole. The X-ray absorption coefficient is modified when the X-ray absorbing atom is surrounded by other atoms, as $\mu(E) = \mu_0(E)[1+\chi(E)]$, where μ_0 is the absorption coefficient without the surrounding atoms and χ is EXAFS determined by the surrounding atoms [25, 26]. EXAFS is described, as

$$\chi(k) = -\sum\nolimits_{j=1} 3(\widehat{e}\cdot\widehat{R}_j)^2\frac{S_0^2 N_j F_j(k,R_j)}{kR_j^2}e^{-2k^2\sigma_j^2}e^{-\frac{2R_j}{\lambda(k)}}\sin[2kR_j+\Phi_j(k,R_j)], \tag{7.3}$$

where k is the photoelectron wave number vector, $k = \sqrt{2m(E - E_0)}/\hbar$ (m electron mass, E incident X-ray energy, E_0 edge energy), S_0^2 counts the difference between with and without the hole created by the X-ray absorption, and multi-excitations, N_j is the coordination number, F_j is the backscattering amplitude of the photoelectron by the jth atomic shell, R_j is the mean distance of the jth atomic shell, σ^2 is the mean displacement of atoms in an atomic shell, $\lambda(k)$ is the mean free path of the photoelectron, and Φ_j is the overall phase shift. Since EXAFS is determined by only the neighbouring atoms without depending on specimen crystallinity, it is very powerful for determining the local structural properties of nanoparticles, amorphous, and around impurities in a material.

Figure 7.1a shows measured XAFS from a $CeAl_2$ powder with a transmission mode. EXAFS is obtained from the measured data by removing the atomic background, as shown in Fig. 7.1b. The quantitative structural properties of the specimen can be obtained by fitting the EXAFS data to the EXAFS theoretical calculation [28]. Several research groups have developed the analysis codes for XAFS data independently, including UWXAFS [29], IFEFFIT [30], WinXAS, and RSFIT. The theoretical code for the backscattering amplitude and phase shift of photoelectrons by the neighbouring atoms was developed by Rehr's group [28]. After the atomic background is removed by the AUTOBK code (part of IFEFFIT), the measured EXAFS is extracted from the raw data, as shown in Fig. 7.1b. EXAFS data in the k-space is Fourier transformed to the r-space, as shown in Fig. 7.1c. The peak positions are shifted from true bond lengths by about 0.3 Å due to the phase shift, Φ in (7.3), of the photoelectron, which is not counted in the Fourier transform. The quantitative structural properties of the measured materials can be obtained by EXAFS data fit to the EXAFS theoretical calculation [28] with an atomic model, as shown in Fig. 7.1d. In Fig. 7.1c, the dotted and solid lines are the measured data and best fit, respectively. For the fits, the FEFFIT code (part of IFEFFIT) is used.

Orientation-dependent structural properties of crystalline materials can be obtained using linearly polarized X-ray [26,31]. In (7.3), $(\hat{e} \cdot \hat{R}_j)$ is the polarization-dependent term. Polarization-dependent XAFS techniques are useful particularly for single crystals and aligned nanostructures, vertically aligned nanorods, for example. When the electric field of the incident X-ray is parallel and perpendicular to nanorod length, as shown in Fig. 7.3a b, XAFS can detect independently the structural and chemical properties in the parallel and perpendicular directions, as shown in Fig. 7.3c d, respectively. The local structural properties of nanowires can be different from their bulk, depending on their orientations because the size of their diameter is in nanometer scale whereas the length can be in micrometers with bulk properties. This means that the structural distortion in nanowires can depend on the orientation. The polarization-dependent XAFS techniques can be a powerful tool to determine the orientation-dependent local structural changes of aligned nanorods.

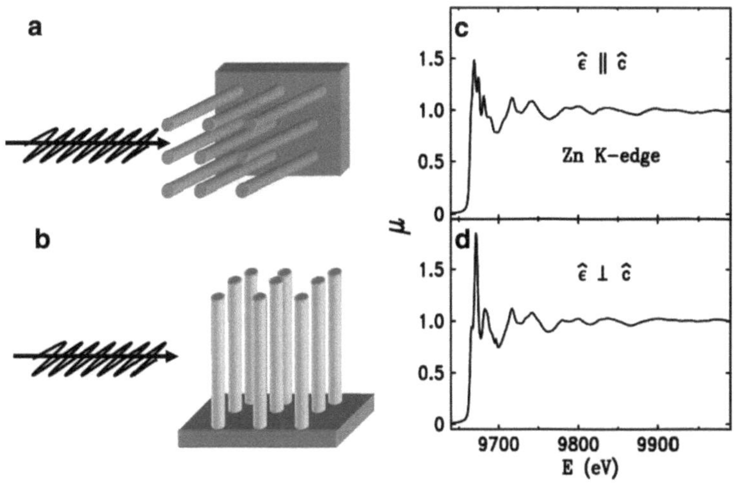

Fig. 7.3 (**a**) and (**b**) Schematics of polarization-dependent XAFS measurements with the geometry of the electric field direction of the incident X-ray parallel and perpendicular to the nanorod length, respectively. (**c**) and (**d**) Normalized X-ray absorption coefficient from ZnO nanorods at Zn *K* edge as a function of the incident X-ray energy with the two geometries, parallel and perpendicular, respectively [31]

7.3 ZnO Nanoparticles

ZnO nanoparticles with the mean size of 4.5 and 70 nm were synthesized using an aqueous solution method [32]. The high-resolution field emission TEM (FE-TEM) measurement of 4.5 nm ZnO nanoparticles demonstrates that the nanoparticles have a uniform size, as shown in Fig. 7.4a b. The FE-TEM analysis reveals that the 4.5 nm nanoparticles are an ordered structure with an average atomic distance of 3.2 Å which corresponds to the bond length of the Zn–Zn or O–O pairs in wurtzite structured ZnO crystals. However, the boundary of the 4.5 nm nanoparticles is disordered. The mean particle size of 70 nm is determined using FE-SEM measurements, as shown in the inset of Fig. 7.4c

Figure 7.4c shows temperature-dependent photoluminescence (PL) spectra from the 70 nm nanoparticles. The PL spectra demonstrate strong near-band edge emission and a weak but visible emission near 2.3 eV due to defects from the nanoparticles. In addition, the PL spectrum at low temperatures exhibits two distinct emission peaks at 3.35 and 3.41 eV, which were observed up to 200 K. No recombination peak was observed from the 4.5 nm nanoparticles due to a substantial amount of distortion and disorders in the nanoparticles. An energy band gap of ZnO powder was observed at 3.29 eV by PL at room temperature. The neutral donor-bound exciton (D^0X) of ZnO at low temperature is 3.37 eV. 3.41 eV from the ZnO nanoparticles is slightly larger than 3.37 eV from ZnO bulk. Is the blue shift caused by the quantum confinement effect? It should be noted that the mean diameter of

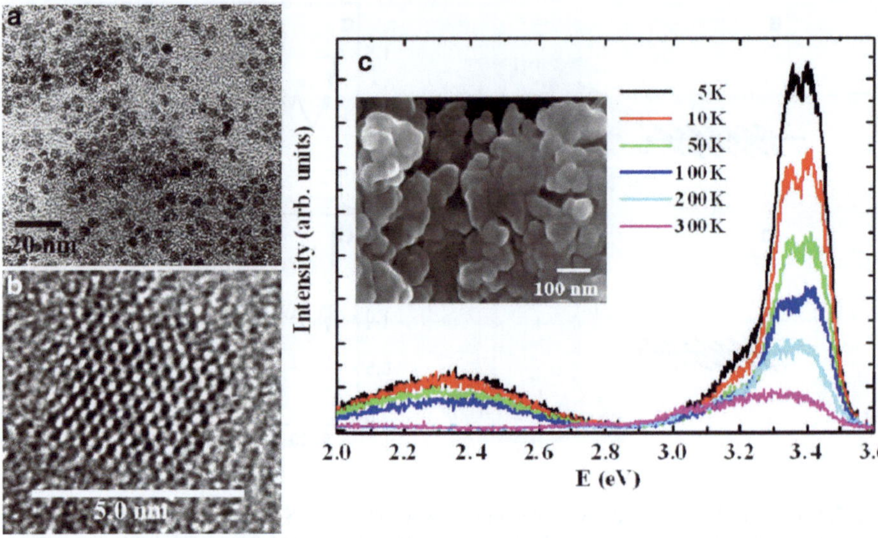

Fig. 7.4 (**a**) and (**b**) FE-TEM images of ZnO nanoparticles [19]. (**c**) PL spectra of 70 nm ZnO nanoparticles at different temperatures. The *inset* is the FE-SEM image of 70 nm ZnO nanoparticles

70 nm is much larger than the ZnO Bohr radius of 1.5 nm. The blue shift cannot be due to the quantum confinement effect.

The local structural properties of nanoparticles are examined by XAFS at 40 K, as shown in Fig. 7.5. XANES from ZnO powder and nanoparticles at Zn K edge shows gradual changes from the bulk to nanoparticles. The XANES signal of 4.5 nm ZnO nanoparticles is typically observed from ZnO nanoparticles. The XANES spectrum of the nanoparticles is often used to determine ZnO nanoparticle formed in buried layers. Figure 7.6 shows the magnitude of Fourier transformed EXAFS from ZnO powder and nanoparticles as a function of distance from a zinc atom. The local structures are changed gradually from the powder to nanoparticles. The EXAFS from the 70 nm ZnO nanoparticles shows that the 70 nm ZnO nanoparticles have an intermediate structure of ZnO powder and 4.5 nm nanoparticles. EXAFS demonstrates that the first shells (Zn–O pairs) of the three specimens are similar, whereas the second shells (Zn–Zn pairs) are quite different depending on the specimens, indicated by the intensity and position changes of the peaks. The quantitative structural properties are determined by fitting the EXAFS data to the EXAFS theory [28] using the IFEFFIT code [29, 30] and the standard analysis procedures [33].

In Fig. 7.6a, the first and the second peaks correspond to 4 oxygen and 12 zinc atoms, respectively, as the first and the second neighbouring atoms from a zinc atom in a wurtzite ($p6_3mc$) structured ZnO, as shown in the inset of Fig. 7.6c. For the fit of the ZnO powder data in Fig. 7.6a, it was begun with a fully occupied model for a wurtzite structure, shown in the inset of Fig. 7.6c. In the fit of the powder EXAFS data, the bond lengths and the mean displacement (σ^2, including thermal vibration and static disorder) of Zn–O and Zn–Zn pairs are varied. The bond lengths of the

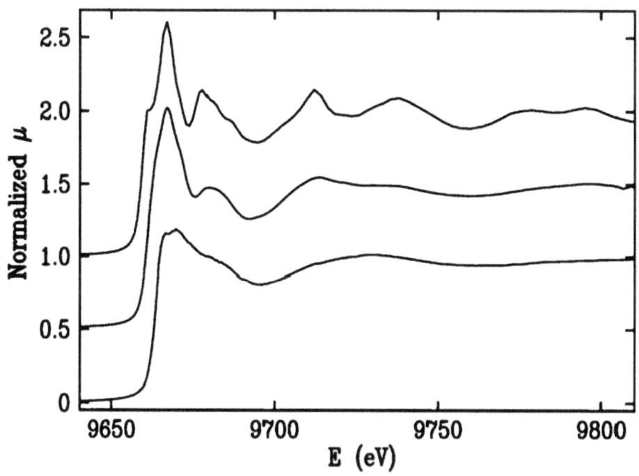

Fig. 7.5 X-ray absorption coefficient from (*top*) ZnO powder, (*middle*) 70 nm ZnO nanoparticles, and (*bottom*) 4.5 nm ZnO nanoparticles at Zn *K* edge as a function of incident X-ray energy at 40 K [19]

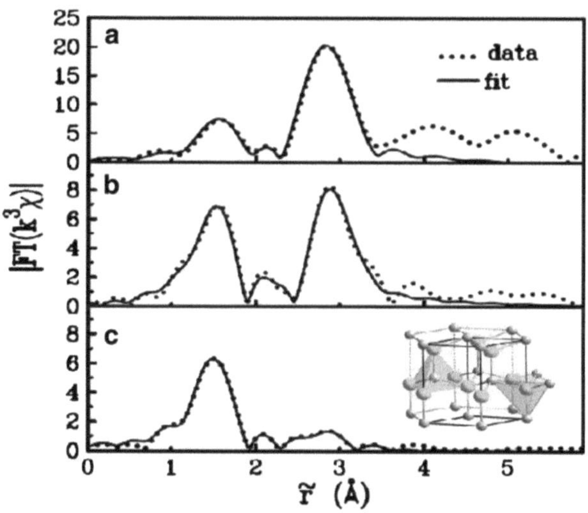

Fig. 7.6 EXAFS from ZnO (**a**) powder, (**b**) 70 nm nanoparticles, and (**c**) 4.5 nm ZnO nanoparticles [19]. The *inset* shows a wurtzite structure. Data in the range of $\tilde{r}$ = 1.2–3.5 Å are fitted

powder agree well with XRD presented in the model of Table 7.1. The best fit results are summarized in Table 7.1. No satisfactory fit to the nanoparticle EXAFS data is obtained using a fully occupied wurtzite structure model. The details of the EXAFS data fit to a distorted structure are described elsewhere [33]. A distorted wurtzite structure was used to fit the EXAFS data for the nanoparticles shown in Fig. 7.6b, c.

Table 7.1 The fit results of the EXAFS data from ZnO powder, 70 and 4.5 nm nanoparticles at 40 K. Coordination number (N), bond length (d), and displacement (σ^2). For a wurtzite ZnO crystal, one O(1) is located just above a Zn atom in the c-axis, three O(2)s at 19° from the ab-plane, six Zn(1)s at 55° from the ab-plane, and six Zn(2)s in the ab-plane. S_0^2 of 0.90(5) is used

Specimen	Zn–O(1)			Zn–O(2)			Zn–Zn(1)			Zn–Zn(2)		
	N	d (Å)	σ^2	N	d (Å)	σ^2	N	d (Å)	σ^2	N	d (Å)	σ^2
Powder	1	1.902(6)	0.003(1)	3	1.982(5)	0.003(1)	6	3.207(3)	0.004(1)	6	3.245(3)	0.004(1)
70 nm	1.0(1)	1.926(8)	0.006(1)	3.0(2)	1.980(8)	0.005(1)	6.0(4)	3.044(9)	0.012(1)	6.0(4)	3.203(5)	0.005(1)
4.5 nm	0.7(1)	1.938(5)	0.005(1)	2.1(2)	1.941(5)	0.005(1)	4.2(5)	3.00(4)	0.033(9)	4.2(5)	3.23(3)	0.026(8)

The fits of the first and second peaks in Fig. 7.6b, c show that there are approximately 30% of the vacancies in both oxygen and zinc sites of the 4.5 nm nanoparticles while no vacancy is observed in neither oxygen nor zinc site of the 70 nm nanoparticles. The vacancies in the nanoparticles suggest the presence of imperfectness in the nanoparticle crystals. There are at least two possibilities for the location of oxygen vacancies. The vacancies can spread randomly throughout the entire nanoparticles, and are placed near boundaries. The location of vacancy can be determined with a simple model calculation and comparison to the EXAFS data. For a spherical particle, the number ratio of surface atoms to total atoms can be calculated using a simple model of a volume ratio, as

$$\frac{\left[\pi R^3 - \pi (R - D)^3\right]}{\pi R^3}, \tag{7.4}$$

where R is the radius of the nanoparticle and D is the distance from the boundary. For $R = 2.25$ nm and $D = 0.32$ nm (Zn–Zn or O–O distance), approximately 37% of the atoms are placed at the boundary. If zinc atoms are the terminating atoms, approximately 37% of the Zn atoms will reside at the boundary with 1 or 2 oxygen neighbors. This means that the oxygen vacancy of the ZnO nanoparticles is approximately 20–30%. With the same model, 30–40% zinc vacancy is predicted. These model calculations correspond well to the EXAFS results. These results indicate that the vacancies on the oxygen and zinc sites are due to the boundaries of the nanoparticles but the inside defects of the nanoparticles. With the same model, approximately 3% of atoms reside at the boundary of the 70 nm nanoparticles. This means that approximately 1% oxygen and 3% zinc vacancies are present due to the boundary. Less than 5% of vacancy cannot be detected because of the resolution limit of EXAFS.

The bond lengths of the Zn–O pairs in the 4.5 nm nanoparticles are nearly the same in all directions as shown in Table 7.1. The bond length of the Zn–O(1) pairs gradually shrank, while that of the Zn–O(2) pairs expanded with decreasing the particle size. This result suggests that the four oxygen atoms around a zinc atom have a same bond length in the beginning of the ZnO crystallization. It should be noted that in a wurtzite structure of ZnO, one oxygen atom in the c-axis has a shorter bond length from a Zn than that of the other three oxygen atoms located on nearly the ab-plane. The similar bond length of the Zn–O pairs indicates an intermediate structure of a Zn–O molecule and a ZnO crystal. The bond lengths of Zn–Zn pairs in the nanoparticles are shorter than that of the ZnO powder, particularly, in the Zn–Zn(1) pairs located at about 55° from the Zn-plane. This EXAFS result demonstrates that there is a substantial amount of structural distortion in the nanoparticles. The short bond lengths of atomic pairs were also observed from $CePt_{2+x}$ and $CeAl_2$ nanoparticles [20]. In ZnS and CdS nanoparticles, structural distortions have been also observed [34].

The thermal vibration and static disorder of atoms are determined by measuring bond-length distribution. At low temperatures the bond-length distribution is generally a Gaussian distribution called as the Debye–Waller factor, $e^{-2k^2\sigma^2}$, as shown in (7.3). The temperature-dependent displacement, σ^2, contributed by the thermal vibration and static disorder of an atomic pair is described with the Einstein model, as

$$\sigma^2(T)_{\text{meas}} = \frac{\hbar h^2}{2\mu K_{\text{B}}\Theta_E}\coth\left(\frac{\Theta_E}{2T}\right) + \sigma^2_{\text{static}}, \tag{7.5}$$

where μ is the reduced mass of the atomic pairs, K_{B} is the Boltzmann's constant, and Θ_{E} is the Einstein temperature. The displacement σ^2 values of the Zn–O and the Zn–Zn pairs in the ZnO powder were approximately 0.003 Å^2. The σ^2 value at low temperatures is mainly determined by the zero-point motion and the static disorder of an atomic pair. It is noted that the XAFS data were collected at 40 K. The σ^2 value of the Zn–O pairs in the ZnO nanoparticles is about twice as large as that of the powder. For the Zn–Zn pairs, the σ^2 value of the 4.5 nm nanoparticles is about seven times larger than that of the powder. For the fits of the nanoparticle data, an asymmetric Gaussian distribution to examine a skewed pair-distance distribution was introduced. However, the asymmetric Gaussian distribution part of the displacement is negligible.

The large σ^2 value for the 4.5 nanoparticles indicates that there is significant structural disorder and distortion, particularly in the Zn–Zn pairs of the nanoparticles. For the 70 nm ZnO NPs, the σ^2 of the Zn–O pairs is similar to that of the 4.5 nm nanoparticles, while the Zn–Zn σ^2 values are much smaller than that of the 4.5 nm nanoparticles. However, the σ^2 values for the 70 nm nanoparticles are still larger than those of the powder counterpart. This indicates that there is a substantial amount of structural disorder existing in the 70 nm nanoparticles. The σ^2 value of the Zn–Zn(2) pairs located in the *ab*-plane is approximately half the Zn–Zn(1)s σ^2 located at 55° off from the *ab*-plane. This demonstrates that Zn–Zn pairs in the *ab*-plane have a tighter bond than along the *c*-axis in the nanoparticles. The σ^2 values of the three specimens demonstrate that the Zn–O pairs have a relatively tight bond whereas the Zn–Zn pairs have a loose bond in the beginning of the ZnO crystallization.

The average bond lengths of the atomic pairs, particularly Zn–Zn pairs in the 70 nm nanoparticles, are shorter than those of the ZnO powder. When these short lattice constants are compared with a previous of ZnMgO system [21, 22], the PL blue shift of the 70 nm ZnO nanoparticles can be explained by the structural distortion of the nanoparticles alone instead of by the quantum confinement effect. Distorted crystalline structures are observed from ZnS and CdS nanoparticles with a mean particle size below 5 nm [34]. The atomic arrangement of ZnS and CdS nanoparticles were examined by synchrotron XRD. In general, ZnS has a zincblende (F$\bar{4}$3m) structure. In a zincblende ZnS, a zinc atom has four sulphide atoms at a same distance as the first neighbours. Wurtzite ZnS structures are often observed but unstable, although stable wurtzite ZnS structures were observed in nanobelt shape [34]. In wurtzite ZnS, a zinc atom also has 4 neighbouring sulphide atoms as the first neighbours. Unlike a zincblende structure, one of the Zn–S bond lengths is shorter (or longer) than the others. For bulk zincblende and wurtzite, the diffraction peak positions are very different between the zincblende and wurtzite structures. However, for nanoparticles, a difference between zinc-blende and wurtzite structures is not that simple because of insufficient scattering sources and large distortion in nanoparticles [34]. XRD of ZnS and CdS nanoparticles showed that ZnS had a zinc-blende structure whereas CdS had a distorted zinc-blende structure. In bulk, ZnS and CdS have zinc-blende and wurtzite structures, respectively. As mentioned above, the difference between zinc-blende and wurtzite structures is the bond-length difference. In a molecule, the bond length of same atomic pairs is generally the same. However, the bond lengths become different when it is crystallized. In ZnS, the structures of nanoparticles and bulk are not different. Thus, the nanoparticles had a zinc-blende structure. However, for CdS, the structures of nanoparticels and bulk are different. The nanoparticle structure was an intermediate structure of molecular and bulk. ZnO is the same as CdS.

7.4 ZnO Nanorods

The physical properties of one-dimensional nanomaterials are dependent on orientations because the size and surface effects also depend on the orientations. The orientation-dependent quantum confinement effect was observed from ZnO nanorods [35] and ZnO/ZnMgO nanorod heterostructures [36]. For better understanding of the physical properties of one-dimensional nanorods, the study of orientation-dependent structural properties is necessary because the physical properties of one-dimensional nanowires can depend on their orientations. Vertically well-aligned ZnO nanorods with mean diameters of 13 and 37 nm were synthesized using a catalyst-free metal-organic chemical-vapour deposition (MOCVD) process [37]. The quantum confinement effect was observed from ZnO nanorods with a diameter of less than 20 nm [35]. SEM measurements show that the nanorods were well aligned and had a uniform size, as shown in Fig. 7.7a. XRD measurements demonstrated that the ZnO nanorods were high-quality crystals and well aligned in the *c*-axis and in the *ab*-plane [37].

Polarization-dependent XAFS is used to describe the local structural properties of vertically aligned ZnO nanorods. At room temperature, polarization-dependent XAFS measurements at Zn *K* edge were performed with the incident X-ray electric field parallel and perpendicular to the nanorod length, as shown in Fig. 7.3c, d. XANES from the nanorods shown in Fig. 7.3c, d demonstrates the dependence of the crystal orientations due to the orientation-dependent structural properties of the nanorods. After the atomic background is removed, EXAFS data are obtained and Fourier transformed to the *r*-space, as shown in Fig. 7.7c, d. In order to compare the local structural properties of ZnO nanorods and powder, the EXAFS data of ZnO powder are first analysed, as shown in Fig. 7.7b. The EXAFS data in the *r*-space are fitted to the theoretical EXAFS calculations [28] using the IFEFFIT codes [29, 30] and standard analysis procedures [34]. For the fit of the ZnO powder data, a fully occupied model of a wurtzite structure is used. The satisfactory fit is obtained, as shown in Fig. 7.7b and the fit results are summarized in Table 7.2. The

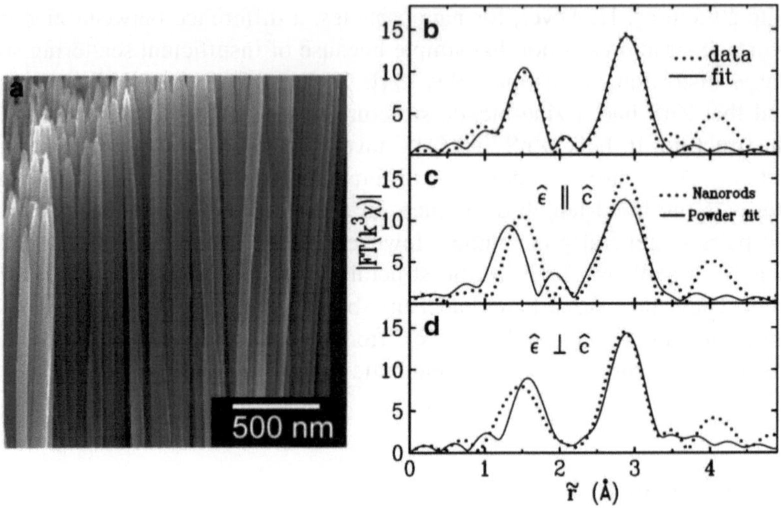

Fig. 7.7 (**a**) SEM image from ZnO nanorods [37]. Magnitude of Fourier transformed EXAFS measured at room temperature, from (**b**) ZnO powder and ZnO nanorods with an average diameter of 13 nm for the X-ray electric field direction ($\hat{\varepsilon}$) (**c**) parallel and (**d**) perpendicular to the nanorod length [31]. For the Fourier transform, the EXAFS data in the k-range of 2.5–10.5Å^{-1} are used and a Hanning window with a windowsill width of 0.5Å^{-1} is used. Data in the range of $\tilde{r} = 1.2$–3.5Å are fitted. In (**c**) and (**d**), *dotted lines* are the measured data from ZnO nanorods and *solid lines* are the theoretical calculations with structural information obtained from the ZnO powder fit in (**b**)

lattice constants of a, b = 3.245(6) Å and c = 5.206(5) Å are determined by the best fits. The lattice constants correspond to XRD results shown in Table 7.2. From the best fits, the crystalline symmetry z of oxygen=0.366 is determined. The displacements (σ^2) of Zn–O and Zn–Zn pairs of the ZnO powder are 0.003(1) Å^2 and 0.008(1) Å^2, respectively. In Fig. 7.7c d, dotted lines are the measured EXAFS from ZnO nanorods while the solid lines are the calculated EXAFS with the structural information of the ZnO powder fit in Fig. 7.7a. It is obvious that the structural properties of ZnO nanorods and powder are very different. The EXAFS from ZnO nanorods demonstrates that the Zn–O bond length of the nanorods in the c-axis is longer than that of ZnO powder whereas the Zn–O bond length of the nanorods in the ab-plane is shorter than that of ZnO powder, as shown in Fig. 7.7c, d.

To accurately determine the orientation-dependent structural properties of the nanorods, the two sets of EXAFS data in Fig. 7.7c, d are *simultaneously* fitted with the same parameters. The results of the best fit are given in Table 7.2. From a probe Zn atom, the distance of O(1) located just above the Zn atom along the c-axis is ~0.05 longer, while the distance of three O(2)s located about 19° off from the Zn ab-plane is approximately 0.025 Å shorter, compared with those bond lengths of the bulk counterpart. These results are similar to the bond-length distortions of the Zn–Zn(1) pairs located at ~55° off from the ab-plane and the Zn–Zn(2) pairs located

Table 7.2 Coordination number (N), bond length (d) and displacement (σ^2) of ZnO bulk and nanorods. S_0^2 of 0.9 is determined from the fit of the powder data and fixed for the fits of the nanorod data. For the model calculations, a fully occupied wurtzite structure with $a, b = 3.245$ Å, $c = 5.205$ Å and the crystalline symmetry z of oxygen=0.366 are used

Specimen	Zn–O(1)			Zn–O(2)			Zn–Zn(1)			Zn–Zn(2)		
	N	d (Å)	σ^2	N	d (Å)	σ^2	N	d (Å)	σ^2	N	D (Å)	σ^2
Powder	1	1.903(9)	0.003(1)	3	1.981(18)	0.003(1)	6	3.206(10)	0.008(2)	6	3.246(13)	0.008(2)
37 nm	1.0(1)	1.966(6)	0.0028(8)	2.7(3)	1.950(9)	0.0052(15)	6.0(2)	3.231(4)	0.0090(14)	5.7(5)	3.195(6)	0.0089(15)
13 nm	1.0(1)	1.947(8)	0.0037(7)	3.1(4)	1.965(12)	0.0078(17)	6.2(3)	3.236(4)	0.0084(15)	6.2(6)	3.214(6)	0.0087(16)

in the *ab*-plane, as shown in Table 7.2. Based on these fits, the lattice constants are estimated to be $a = b = 3.216$ (3.195) Å and $c = 5.305$ (5.307) Å for the ZnO nanorods with a diameter of 13 (37) nm. The EXAFS results demonstrate that the lattice constants of ZnO nanorods, a and b, are shrunken by about 0.04 Å, while c is elongated by 0.1 Å, compared with those of the bulk counterpart.

The displacements (σ^2) of all pairs except the Zn–O(2) pairs in both nanorod samples are comparable with those of the bulk counterpart, implying that there is no extra structural disorder in the bond lengths of the nanorods. The ratio of atomic pairs in the boundary to the total number of the pairs in a disk can be simply estimated with a simple model modified from (7.4), as $[\pi R^2 - \pi(R - D)^2]/\pi R^2$, where R is the radius of the disk and D is the distance from the boundary. For ZnO nanorods, the only first and second atomic layers ($D = 0.65$ nm) from the boundary may be effectively affected by the incomplete bonds at the boundary; about 19% of Zn–Zn pairs in the nanorods with a diameter of 13 nm have an extra disorder. However, from the polarization-dependent EXAFS measurements, neither extra disorder nor vacancy at the Zn sites in all directions is observed, compared with the bulk counterpart. This demonstrates that the Zn atoms are well ordered in the *ab*-plane even near the boundary, as well as along the *c*-axis, and that the terminating atoms at the boundary are oxygen atoms. As the O(2)s are the terminating atoms at the boundary, the ratio of the Zn–O(2) pairs located near the boundary to the total Zn–O(2) pairs is inverse to the diameter of the nanorods. Therefore, more disorders exist in the bond length of the Zn–O(2) pairs for the nanorods with a smaller diameter.

The quantum confinement effect was observed from ZnO nanorods with the mean diameter less than 20 nm [35]. Since the nanorods length was in micrometer scale, the quantum confinement effect was mainly determined by the electrons moving in the *ab*-plane of ZnO nanorods. EXAFS results show that the bond lengths of atomic pairs in the *ab*-plane of the 13 nm nanorods are slightly larger than those of the 37 nm nanorods. If the structural displacement affects the PL peak position, a red shift will be expected from the 13 nm nanorods due to the bondlength elongation. The EXAFS results reveal that the blue shift in PL from the 13 nm nanorods is caused by the quantum confinement effect but the shrunken bond length of atomic pairs in the *ab*-plane of the ZnO nanorods.

7.5 Coaxial GaN/ZnO Nanorods

The lattice constants of bulk ZnO are $a, b = 3.245$ Å and $c = 5.206$ Å while GaN lattice constants are $a, b = 3.160$ Å and $c = 5.195$ Å. For the wurtzite structured ZnO and GaN, the energy band gap of ZnO is 3.3 eV while that of GaN is 3.4 eV at room temperature. Recently, a theoretical model calculation of GaN/ZnO coaxial nanowires showed the quantum confinement effect expected in GaN/ZnO coaxial nanowires [40]. It should be useful to examine the local structural properties of the GaN shells on ZnO nanorod cores.

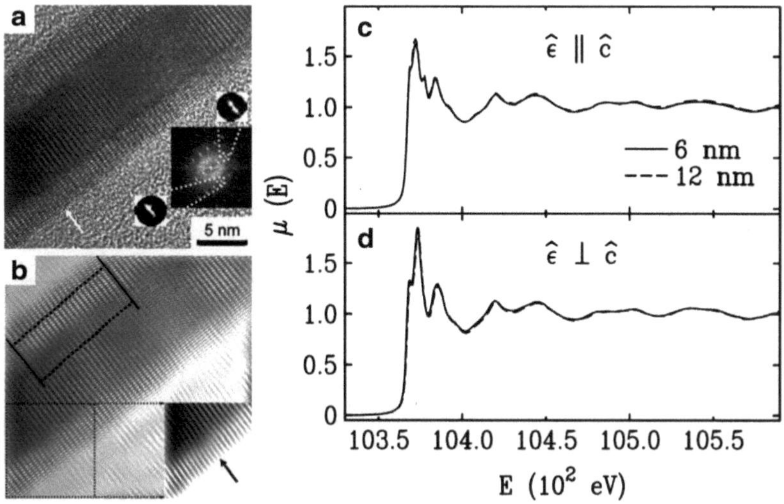

Fig. 7.8 (**a**) and (**b**) are TEM images of GaN/ZnO coaxial nanorods [38]. Normalized X-ray absorption coefficient from (**c**) and (**d**) GaN/ZnO coaxial nanorods as a function of incident X-ray energy at the Ga K-edge and 20 K [39]. For the GaN/ZnO coaxial structures, the XAFS data were collected at the X-ray electric field direction ($\hat{\varepsilon}$) (**c**) parallel and (**d**) perpendicular to the nanorod length

GaN/ZnO coaxial nanorods with different GaN shell thickness of 6 and 12 nm were synthesized using catalyst-free MOCVD [38]. TEM measurements demonstrate that both ZnO and GaN are highquality crystals and that the GaN layer are epitaxially grown on ZnO nanorods with uniform thickness, as shown in Fig. 7.8a, b. The TEM images show a dislocation of atomic layers in the GaN shells due to the lattice mismatch between ZnO and GaN. The detail local structural properties of the GaN shell are investigated using polarization-dependent XAFS measurements.

Polarization-dependent XAFS measurements from the vertically aligned GaN/ZnO coaxial nanorods are performed with the X-ray electric field parallel and perpendicular to the nanorod length [39], as shown in Fig. 7.8c,d. XANES in Fig. 7.8c,d is comparable to the polarization-dependent XANES from vertically aligned ZnO nanorods, shown in Fig. 7.3, implying that the local structural properties of GaN layer on ZnO nanorods are similar to those of ZnO nanorods. After the atomic background of the XAFS data is determined, EXAFS is obtained in the k-space and Fourier transformed into the r-space, as shown in Fig. 7.9. To minimize uncertainties only the EXAFS data in the k-range of 3.0–12.5 Å^{-1} are used for further analysis. The dotted lines in Fig. 7.9 show the magnitudes of the Fourier transformed EXAFS data from the GaN powder and GaN/ZnO coaxial nanorods. The EXAFS demonstrates that the Ga–N bond length of GaN shells in GaN/ZnO coaxial nanorods is slightly different from the powder counterpart.

The EXAFS data in the r-space are fitted to the theoretical EXAFS calculations [28] using the IFEFFIT codes [29, 30] and standard analysis procedures [33], as

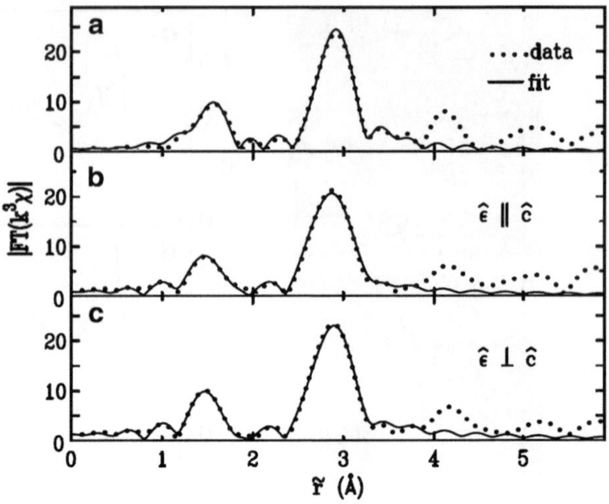

Fig. 7.9 Magnitude of Fourier transformed EXAFS with k^3-weight, from (**a**) GaN powder and GaN/ZnO coaxial nanorods with an average GaN thickness of 6 nm for the incident X-ray field parallel (**b**) and perpendicular (**c**) to the nanorods length [39]. Data in the range of $\tilde{r} = 1.2$–3.3 Å were fitted

shown in Fig. 7.9. The fits included single- and multi-scattering paths. The GaN powder data are fitted with a fully occupied model of a wurtzite structure, as shown in Fig. 7.9a. Both the zincblende and wurtzite structures have a similar tetrahedral environment that Ga(N) has 4 N(Ga) and 12 Ga(N) atoms as the first and second neighbouring atoms. For GaN with a zincblende structure, the bond lengths of the 4 Ga–N and 12 Ga–Ga pairs are $\sim$ 1.93 and 3.15 Å, respectively. However, in an ideal wurtzite phase, the Ga–N and Ga–Ga pairs have slightly different bond lengths: the bond length of one Ga–N(1) pairs located just below Ga in the c-axis is slightly larger than those of the three other Ga–N(2) pairs located at 20° off from the ab-plane, and six Ga–Ga(1) pairs located approximately 55° off from the ab-plane also have a longer bond length than the other 6 Ga–Ga(2) pairs located in the ab-plane. The fit to the EXAFS data of the GaN powder reveals that the bond lengths of the Ga–N(1) and Ga–Ga(1) pairs are about 0.025 Å longer than those of Ga–N(2) and Ga–Ga(2) pairs.

The criteria of a goodness fit of EXAFS data is (1) *R-factor* should be less than 0.02, (2) χ_ν^2 is nearly 1.0, and (3) the fit results should be physically meaningful [33]. *R-factor* and χ_ν^2 are defined, as

$$\chi_\nu^2 = \frac{N_{\text{inp}}}{N(N_{\text{inp}} - N_{\text{var}})} \sum_{j=1}^{j=N} \frac{|y_j^{\text{data}} - y_j^{\text{theory}}|^2}{\varepsilon_j^2} = \frac{\chi^2}{\nu}, \tag{7.6}$$

$$r\text{–factor} = \frac{\sum_{j=1}^{j=N} |y_j^{\text{data}} - y_j^{\text{theory}}|^2}{\sum_{j=1}^{j=N} (y_j^{\text{theory}})^2}, \tag{7.7}$$

where N_{inp} is the independent data point, N is the total number of data in the fit range, N_{var} is the number of variables, ε is the experimental error, and ν is the freedom in the fit. The EXAFS data are also fitted with a zincblende structure because GaN can have a wurtzite or zinc-blende structure. The zinc-blende model fit is also reasonably good. However, the fit values of *r-factor* and χ_{ν}^{2} with the zincblende model are about three times larger than those with the wurtzite structure. This means that the GaN powder have a wurtzite structure instead of a zincblende structure. The best fit results are summarized in Table 7.3

The two sets of EXAFS data measured at the electric field of the incident X-ray parallel and perpendicular to the c-axis can independently determine the bond lengths and their disorders of Ga–N(1), Ga–N(2), Ga–Ga(1) and Ga–Ga(2) pairs. Both sets of polarization-dependent EXAFS data from the GaN shell of GaN/ZnO coaxial nanorods are *simultaneously* fitted with the same parameters. The orientation-dependent structural properties of the GaN shells are obtained by the best fits. Although the fit starts with a wurtzite structure model, both the wurtzite and zincblende structures are reflected to the fits, because the bond lengths of the Ga–N(1), Ga–N(2), Ga–Ga(1) and Ga–Ga(2) pairs are independently varied in the fit.

The results of the best fit are summarized in Table 7.3. The fits reveal that the bond length of the Ga–N(1) in the GaN nanocrystals with a thickness of 12 nm is larger by 0.02 Å than that of the Ga–N(2) pairs. This is different from the Zn–O bond length of ZnO nanorods, as mentioned above. When the GaN thickness is reduced to 6 nm, the Ga–N(2) and Ga–N(1) pairs have a similar bond length with about 0.01 Å distortion, compared to the GaN bulk counterpart. These results strongly suggest that the GaN shells have a zincblende structure in the beginning of GaN growth on the ZnO nanorods, due to the structural strain from the lattice mismatch between the GaN and ZnO. For an ideal wurtzite structure of GaN, the bond length of Ga–Ga(1) pairs was about 0.03 Å longer than that of Ga–Ga(2) pairs. However, the EXAFS reveals that the Ga–Ga pairs have the same bond length 3.195 Å, independent of the directions. It should be emphasized that the bond lengths of Ga–Ga(1) and Ga–Ga(2) pairs are independently determined from the fit of the second peaks in Fig. 7.9b, c, respectively. The bond lengths suggest that the GaN shells on ZnO nanorods have a distorted wurtzite structure or a zincblende-like structure. The displacement σ^2 values of the Ga–N and Ga–Ga pairs in the GaN shells showed no extra disorder compared to those in the bulk counterpart, as shown in Table 7.3. The EXAFS reveals that neither extra disorder nor vacancy is present in the tubular GaN nanostructures. The EXAFS results further suggests that a blue shift in PL measurements [40–43] might be due to the quantum confinement effect but not a structural distortion.

7.6 ZnO Nanorods on GaN and Al_2O_3 Substrates

The local structural properties of ZnO nanorods grown on the different substrates of Al_2O_3 and GaN are examined using TEM and EXAFS measurements. The lattice mismatch between ZnO and GaN is much smaller than that of ZnO and Al_2O_3.

Table 7.3 Coordination number (N), bond length (d), and displacement (σ^2) of powder and tubular GaN were obtained with XAFS at 20 K. S_0^2 of 0.85(5) was determined from the fit to the GaN powder data and fixed in the tubular data fits. For the model calculations, the WZ structure with $a = b = 3.160$ Å, $c = 5.195$ Å and the crystalline symmetry z of nitrogen=0.375 are used

Specimen	Ga–N(1)			Ga–N(2)			Ga–Ga(1)			Ga–Ga(2)		
	N	d (Å)	σ^2	N	d (Å)	σ^2	N	d (Å)	σ^2	N	d (Å)	σ^2
Powder	1	1.947(5)	0.0032(5)	3	1.920(5)	0.0032(5)	6	3.206(6)	0.0050(6)	6	3.179(6)	0.0050(6)
12 nm	1.1(1)	1.942(8)	0.0039(7)	3.2(3)	1.918(4)	0.0036(7)	6.0(2)	3.195(3)	0.0049(3)	6.0(2)	3.195(3)	0.0048(3)
6 nm	1.1(1)	1.940(3)	0.0041(3)	3.3(3)	1.930(3)	0.0024(3)	6.1(4)	3.195(4)	0.0043(2)	6.2(4)	3.195(3)	0.0053(2)

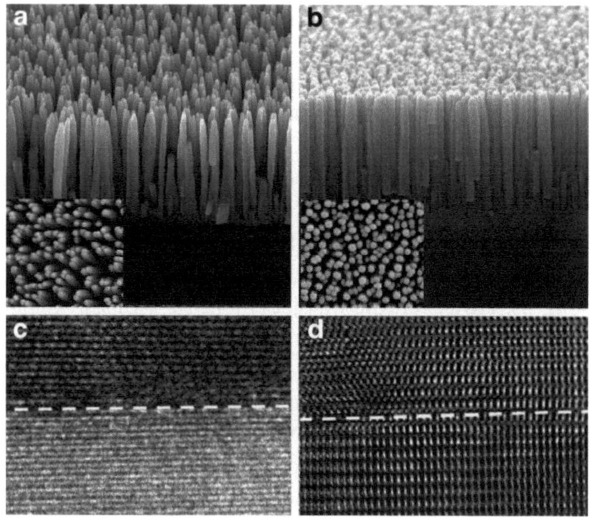

Fig. 7.10 SEM images of ZnO nanorods grown on (**a**) Al_2O_3 and (**b**) GaN substrates [44]. (**c**) and (**d**) show TEM images of the interfaces of ZnO nanorods/Al_2O_3 and ZnO nanorods/GaN, respectively. The *top* part is ZnO and the *bottom* part is the substrate. The *white-dashed lines* indicate the interfaces

Furthermore, the root-mean-square surface roughness of GaN is approximately 3.0 Å which is approximately five times smaller than that of Al_2O_3. XRD revealed that the full-widths at half maximum of ZnO (0002) diffraction peak of ZnO nanorods on GaN substrates was five times smaller than that of ZnO nanorods on Al_2O_3 substrates [44]. Moreover, ZnO nanorods on GaN were well aligned in the c-axis as well as in the ab-plane whereas they were randomly aligned in the ab-plane on Al_2O_3 substrates [44]. Figure 7.10 demonstrates the SEM images of ZnO nanorods grown on Al_2O_3 and GaN substrates. SEM shows that the mean size of nanorods is 70 and 50 nm on Al_2O_3 and GaN substrates, respectively. The vertical alignment of nanorods on GaN substrates is better than that of Al_2O_3 substrates. The top point of view of the nanorods in the inset of Fig. 7.10 shows that the density of nanorods on GaN is approximately three times larger than that on Al_2O_3 substrates. FE-TEM measurements shown in Fig. 7.10c, d demonstrate that there is disorder region near the interface of ZnO nanorods/Al_2O_3 whereas the interface of ZnO nanorods/GaN is well order. The initial structural properties of a nanorod on the GaN or Al_2O_3 substrate likely affect the entire nanorod.

The initial structural properties of the nanorods directly grown on GaN and Al_2O_3 substrates are examined using EXAFS. The nanorods are synthesized using catalyst-free MOCVD process [44]. For the ZnO nanorods on GaN, ZnO nanorods with the lengths of 0.1, 0.2, and 0.4 μm are prepared while their lengths are 0.1, 0.2, and 1.0 μm for Al_2O_3 substrates. EXAFS data are taken from ZnO nanorods with different lengths directly grown on GaN substrates at 30 K [45] and on Al_2O_3 substrates [44] at room temperature. Polarization-dependent EXAFS measurements are performed on only the ZnO nanorods grown on Al_2O_3 substrates. After the atomic background is removed from the raw data, EXAFS data are obtained. For the ZnO nanorods grown on GaN and Al_2O_3 substrates, EXAFS data in the k-space of 3–12 Å^{-1} and 2.5–10.5 Å^{-1} are used, respectively, to minimize errors in the data. The magnitude of Fourier-transformed EXAFS is shown in Fig. 7.11. There is a slight difference in the peak intensities and positions, depending on the nanorod lengths in both systems. The quantitative structural information is obtained by fitting the EXAFS data to the EXAFS theoretical calculation [28]. The EXAFS data are fitted with the IFEFFIT codes [29, 30] and standard procedures [33]. The best fit results are summarized in Table 7.4.

EXAFS of ZnO nanorods grown on GaN substrates shown in Fig. 7.11a demonstrates that the bond lengths of all pairs are nearly independent of the nanorod lengths. However, the displacement (σ^2) of Zn–O pairs in the ZnO nanorods with 0.1 μm length is twice larger than those of longer ZnO nanorods, whereas the σ^2 values of Zn–Zn pairs are independent of the nanorod lengths. The structural disorder of the Zn–O pairs is due to the small strain of lattice mismatch between the ZnO nanorods and GaN substrate. EXAFS demonstrates that the small lattice mismatch between ZnO and GaN affects only the Zn–O pairs in the *ab*-plane in the initial growth. Although the surface roughness of GaN is small, 3 Å only, there is a substantial disorder in the Zn–O(1) pairs of ZnO nanorods along the *c*-axis but the disorder disappears in longer nanorods. EXAFS results show that the structural strain due to a small lattice mismatch does not propagate through the nanorods.

Figure 7.11b, c shows the magnitude of Fourier transformed EXAFS from ZnO nanorods on Al_2O_3 substrates with different lengths. Polarization-dependent EXAFS of ZnO nanorods grown on Al_2O_3 substrates shows that the bond lengths of all atomic pairs are identical without depending on the nanorod lengths. However, the displacement σ^2 of Zn–O(2) pairs in the ZnO nanorods with 0.1 μm length is substantially larger than that of the other nanorods. The σ^2 value demonstrates that the lattice mismatch between ZnO and Al_2O_3 effectively affects the bond length of Zn–O(2) pairs in the *ab*-plane. The disorder of the Zn–O(2) pairs disappears as the nanorods grow up. The σ^2 value of Zn–O(2) in both nanorods grown on GaN and Al_2O_3 substrates demonstrates that the relaxation of Zn–O pairs in the *ab*-plane is critical to form ZnO nanorods on substrates. The σ^2 values of Zn–O(1) pairs in the nanorods grown on Al_2O_3 do not change much with nanorod length. It is in contrast to the nanorods grown on GaN substrates. The σ^2 value of Zn–O(1) in ZnO nanorods on Al_2O_3 is substantially larger than that of ZnO powder shown in Table 7.2. This means that the Zn–O(1) pairs include a static disorder. The σ^2 values of the Zn–Zn pairs in ZnO nanorods on Al_2O_3 is larger than that of the nanorods on GaN because of the measurements at different temperatures. The σ^2 values of the Zn–Zn pairs on Al_2O_3 are comparable to that of ZnO powder counterpart shown in Table 7.2, indicating no extra structural disorder in the Zn–Zn pairs. The σ^2 value of Zn–O(1) indicates that the surface roughness effect of the Al_2O_3 substrate propagates through the entire nanorods and affects the structural quality of the nanorods. These EXAFS results correspond to XRD and TEM results. XRD measurements demonstrated that the full-widths at half maximum of nanorods grown on Al_2O_3 substrates are larger than that of the nanorods on GaN substrates, as mentioned above. The surface roughness effect on the nanorod growth and crystal

Table 7.4 EXAFS results of ZnO nanorods grown on (top) GaN and (bottom) Al_2O_3 substrates. Coordination number (N), bond length (d), and displacement (σ^2) of ZnO nanorods

Rod length	Zn–O(1)			Zn–O(2)			Zn–Zn(1)			Zn–Zn(2)		
	N	d (Å)	σ^2	N	d (Å)	σ^2	N	d (Å)	σ^2	N	d (Å)	σ^2
0.1	1	1.91(1)	0.006(1)	3	1.98(1)	0.006(1)	6	3.19(1)	0.003(1)	6	3.26(1)	0.003(1)
0.2	1	1.91(1)	0.003(1)	3	1.97(1)	0.003(1)	6	3.21(1)	0.004(1)	6	3.24(1)	0.004(1)
0.4	1	1.91(1)	0.003(1)	3	1.98(1)	0.003(1)	6	3.22(1)	0.004(1)	6	3.26(1)	0.003(1)
0.1	1	1.939(5)	0.0052(5)	3	2.018(7)	0.0045(6)	6	3.209(3)	0.0088(3)	6	3.255(4)	0.0085(4)
0.2	1	1.939(6)	0.0046(5)	3	2.024(7)	0.0027(6)	6	3.209(5)	0.0089(4)	6	3.259(2)	0.0090(3)
1.0	1	1.939(5)	0.0055(5)	3	2.018(6)	0.0028(6)	6	3.206(5)	0.0085(4)	6	3.257(4)	0.0085(4)

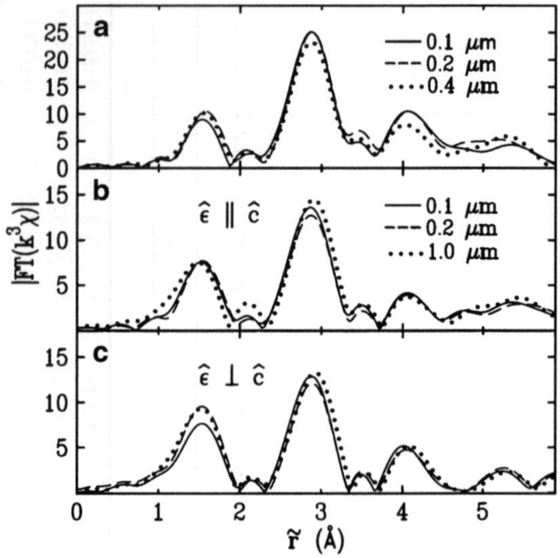

Fig. 7.11 Magnitude of Fourier transformed EXAFS with k^3-weight from ZnO nanorods grown on (**a**) GaN and (**b**, **c**) on Al_2O_3 substrates at different polarizations of X-rays [44,45]

quality was also observed from ZnO nanorods grown on ZnO homo-buffer layers [46]. The EXAFS results reveal that the strain of the Zn–O(2) pairs in the *ab*-plane due to the lattice mismatch of ZnO and substrates disappears whereas the disorder of Zn–O(1) pairs in the *c*-axis due to the surface roughness propagates through the entire nanorod when the nanorod is growing up.

Since XAFS, particularly, EXAFS, requires an X-ray absorption spectrum to obtain the structural properties of a specimen, a conventional XAFS is not very useful to examine the transient phenomena in nature, including catalysis, chemical reactions, physiological reactions, and excitons. Recently, various time-resolved XAFS (RT-XAFS) techniques have been developed to examine the structural and chemical properties of selected atoms in milliseconds to picoseconds [47–51]. Recent synchrotron X-ray sources with the X-ray pulse width of a few tenths of picoseconds provide great opportunities to examine transient behaviours in picoseconds [51].

Transient behaviours in milliseconds can be studied using quick XAFS (QXAFS) measurements [47] whereas fast reactions in microseconds can be examined using energy-dispersive XAFS (ED-XAFS) measurements [48–50] with nanosecond pulsed X-rays. However, one pulse of X-ray intensity is generally insufficient to obtain a good XAFS spectrum. Thus, XAFS data with statistically sufficient counts are obtained by accumulating signals from many pulses of X-rays. For picosecond TR-XAFS study, a femtosecond (FS) laser synchronized with pulsed X-rays is used to create excitons instantly, as shown in Fig. 7.12. A pulsed FS laser creates excitons on a specimen in advance of a selected X-ray pulse with a time-delay mode. Monochromatized and pulsed X-rays detect the structural and chemical changes of a selected atom in the specimen due to the excitons. The incident X-ray energy is selected using a double-crystal monochromator (DCM) set for the picosecond TR-XAFS measurements. In the picosecond TR-XAFS measurements, since only

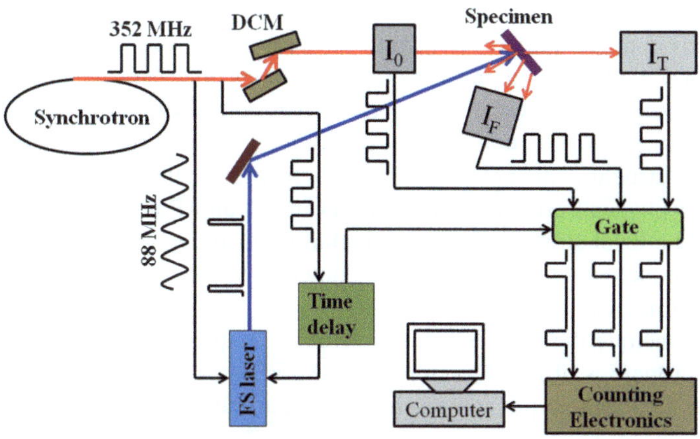

Fig. 7.12 Schematic of the experimental setup for picosecond TR-XAFS measurements

selected X-ray pulses are used, ultra-fast gate devices count the signals of incident X-rays (I_0), transmitted X-rays (I_T), and fluorescence X-rays (I_F), coming from the only selected X-ray pulses. A TR-XAFS spectrum obtained as a function of the incident X-ray energy provides the changes of local structures and chemical properties within a picosecond time resolution. Nanostructures are widely used as catalysts because the large ratio of surface-to-volume. TR-XAFS techniques can dig the new academic fields of transient phenomena, particularly, in nanostructures.

7.7 Conclusions

The structural properties of nanostructures are very important to understand their properties. However, the determination of the microstructural properties of nanostructures is quite difficult. EXAFS was used to quantitatively determine the structural properties of ZnO nanoparticles, ZnO nanorods, and GaN/ZnO coaxial nanorods. ZnO nanparticles with the mean diameter of 4.5 nm have a distorted wurtzite structure and the bond length of Zn–O pairs is shorter than that of ZnO powder counterpart. The Zn–O pairs of the ZnO nanoparticles have the similar bond length in all directions. EXAFS of ZnO nanorods with the mean diameters of 1.8 and 3.7 nm shows that the bond lengths of atomic pairs in the *c*-axis are elongated, whereas the bond lengths in the *ab*-plane are shortened. The bond lengths of the Zn–O pairs are similar in all directions in ZnO nanorods. The local structural distortion is also observed from the tubular GaN shells of GaN/ZnO coaxial nanorods. The bond lengths of Ga–N pairs are similar in the all directions. EXAFS demonstrates that the local structures of nanoparticles including ZnO nanoparticles, ZnO nanorods, and tubular GaN nanostructures, are the intermediate structures of their crystals and molecules. EXAFS measurements of ZnO nanorods grown on Al_2O_3 and GaN substrates demonstrate that the stain relaxation of the Zn–O bonds in the *ab*-plane is critical to form ZnO nanorods whereas the structural disorder of Zn–O pairs in the *c*-axis little affects the ZnO nanorod formation. EXAFS measurements show that the local structural properties of the ZnO and GaN semiconducting nanostructures are not the same as those of bulk counterparts.

The local structural difference of nanostructures from their bulk counterparts can affect other physical properties, blue shifts in PL and Raman measurements, for example. The local structural changes including disorder, distortion, vacancy, and oxidation, in nanostructures should be counted for understanding the physical and chemical properties of nanostructures.

We demonstrate that XAFS is a powerful technique to determine the true local structural properties of nanostructures. TR-XAFS will provide more opportunities to examine the transient phenomena in nature, particularly, nanostructures.

References

1. T. Takagahara K. Takeda, Phys. Rev. B **46**, 15578 (1992)
2. H.J Ryu, Z. Aksamija, D.M. Paskiewicz, S.A. Scott, M.G. Lagally, I. Knezevic, M.A. Eriksson, Phys. Rev. Lett. **105**, 256601 (2010)
3. L.T. Canham, Appl. Phys. Lett. **57**, 1046 (1990)
4. Z. Wang, Y. Fan, M. Zhao, J. Appl. Phys. **108**, 123707 (2010)
5. L. Rigutti, J. Teubert, G. Jacopin, F. Fortuna, M. Tchernycheva, A. De Luna Bugallo, F.H. Julien, F. Furtmayr, M. Stutzmann, M. Eickhoff, Phys. Rev. B **82**, 235308 (2010)
6. J. Han, T-L Chan, J.R. Chelikowsky, Phys. Rev. B **82**, 153413 (2010)
7. A.P. Alivisatos, Science **271**, 933 (1996)
8. H.K. Yadav, V. Gupta, K. Sreenivas, S.P. Singh, B. Sundarakannan, R.S. Katiyar, Phys. Rev. Lett. **97**, 85502 (2006)
9. W.-T. Hsu, K.-F. Lin, W.-F. Hsieh, Appl. Phys. Lett. **91**, 181913 (2007)
10. T.-B. Hur, Y.-H. Hwang, H.-K. Kim, Appl. Phys. Lett. **86**, 193113 (2005)
11. L. Rigutti, J. Teubert, G. Jacopin, F. Fortuna, M. Tchernycheva, A. De Luna Bugallo, F.H. Julien, F. Furtmayr, M. Stutzmann, M. Eickhoff, Phys. Rev. B **82**, 235308 (2010)
12. K. Ichino, K. Ueyama, H. Kariya, N. Suzuki, M. Kitagawa, H. Kobayashi, Appl. Phys. Lett. **74**, 3486 (1999)
13. M.C. Harris Liao, Y.H. Chang, Y.F. Chen, J.W. Hsu, J.M. Lin, W.C. Chou, Appl. Phys. Lett. **70**, 2256 (1997)
14. Y. Wu, K. Arai, T. Yao Phys. Rev. B **53**, R10485 (1996)
15. R. Trotta, L. Cavigli, L. Felisari, A. Polimeni, A. Vinattieri, M. Gurioli, M. Capizzi, F. Martelli, S. Rubini, L. Mariucci, M. Francardi, A. Gerardino, Phys. Rev. B **81**, 235327 (2010)
16. J.W. Luo, A. Franceschetti, A. Zunger, Phys. Rev. B **79**, 201301(R) (2009)
17. S. Baskoutas, A.F. Terzis, J. Appl. Phys. **99**, 013708 (2006)
18. Y.Y. Chen, Y.D. Yao, C.R. Wang, W.H. Li, C.L. Chang, T.K. Lee, T.M. Hong, J.C. Ho, S.F. Pan, Phys. Rev. Lett. **84**, 4990 (2000)
19. E.-S. Jeong, H.-J. Yu, Y.-J. Kim, Gyu-Chul Yi, Yong-Dae Choi, S.-W. Han, J. Nanosci. Nanotechnol. **10**, 3562 (2010)
20. S.-W. Han, C.-H. Booth, E.-D. Bauer, P.H. Huang, Y.Y. Chen, J.M. Lawrence, Phys. Rev. Lett. **97**, 097204 (2006)
21. W.I. Park, G-C Yi, H.M. Jang, Appl. Phys. Lett. **79**, 2022 (2001)
22. S.-H. Park, K.-B. Kim, S.-Y. Seo, S.-H. Kim, S.-W. Han, J. Electron Mater. **35**, 1680 (2006)

23. F.Z. Wang, H.P. He, Z.Z. Ye, L.P. Zhu, J. Appl. Phys. **98**, 84301 (2005)
24. S.J.L. Billinge I. Levin, Science **316**, 561 (2007)
25. D.E. Sayers, F.W. Lytle, E.A. Stern, Phys. Rev. Lett. **27**, 1204 (1971); E.A. Stern, Phys. Rev. B **10**, 3027 (1974); D.E. Sayers, E.A. Stern, F.W. Lytle, Phys. Rev. Lett. **35**, 584 (1975)
26. S.-W. Han, Int J Nanotechnol **3**, 396 (2006)
27. J.-S. Jeon, B.-H. Kim, C.-I. Park, S.-Y. Seo, C. Kwak, S.-H. Kim, S.-W. Han, Jpn. J. Appl Phys **49**, 031105 (2010)
28. A.L. Ankudinov, B. Ravel, J.J. Rehr, S.D. Conradson, Phys. Rev. B **58**, 7565 (1998); J.J. Rehr R.C. Albers, Rev. Mod. Phys. **72**, 621 (2000)
29. E.A. Stern, M. Newville, B. Ravel, Y. Yacoby, D. Haskel, Physica B **208 & 209**, 117 (1995)
30. M. Newville, J. Synchrotron Rad. **8**, 322 (2001)
31. S.-W. Han, H.-J. Yoo, S.-J. An, J. Yoo, G.-C. Yi, App. Phys. Lett. **86**, 21917 (2005)
32. L. Vayssieres, Adv. Mater. **15**, 464 (2003)
33. S.-W. Han, E.A. Stern, D. Haskel, A.D. Moodenbaugh, Phys. Rev. B **66**, 94101 (2002)
34. C. Kumpf, R.B. Neder, F. Niederdraenk, P. Luczak, A. Stahl, M. Scheuermann, S. Joshi, S.K. Kulkami, C. Barglik-Chory, C. Heske, E. Umbach, J. Chem. Phys. **123**, 224707 (2005)
35. W.I. Park, Y.H. Jun, S.W. Jung, G-C Yi, Appl. Phys. Lett. **82**, 964 (2003)
36. W.I. Park, G.-C. Yi, M. Kim, S.J. Pennycook, Adv. Mater. **15**, 526 (2003)
37. W.I. Park, D.H. Kim, S.-W. Jung, G-C Yi, Appl. Phys. Lett. **80**, 4232 (2002)
38. S.J. An, W.I. Park, G.-C. Yi, Y.-J. Kim, H.-B. Kang, Appl. Phys. Lett. **84**, 3612 (2004)
39. S.-W. Han, H.-J. Yoo, SJ An, J Yoo, G-C Yi Appl Phys Lett **88**, 111910 (2006)
40. Z. Wang, Y. Fan, M. Zhao, J. Appl. Phys. **108**, 123707 (2010)
41. T.J. Badcock, P. Dawson, M.J. Kappers, C. McAleese, J.L. Hollander, C.F. Johnston, D.V. Sridhara Rao, A.M. Sanchez, C.J. Humphreys, Appl. Phys. Lett. **93**, 101901 (2008)
42. R. Ghosh D. Basak, J. Appl. Phys. **98**, 086104 (2005)
43. H-M. Kim, D.S. Kim, D.Y. Kim, T.W. Kang, Y.-H. Cho, K.S. Chung Appl. Phys. Lett. **81**, 2193 (2002)
44. S.-H. Park, S.-Y. Seo, S.-H. Kim, S.-W. Han, Appl. Phys Lett **88**, 251903 (2006)
45. C.-H. Kwak, S.-Y. Seo, Y.-B. Lee, S.-H. Kim, S.-H. Park S.-W. Han, J Korean Phys Soc **53**, 304 (2008)
46. C. Kwak, B-H Kim, C.-I. Park, S.-H. Park, S.-Y. Seo, S.-H. Kim, S.-W. Han, J. Nanosci. Nanotechnol. **10**, 912 (2010)
47. A.J. Dent, Top. Catal. **18**, 27 (2002)
48. A. Yamaguchi, T. Shido, Y. Inada, T. Kogure, K. Asakura, M. Nomura, Y. Iwasawa, Catal. Lett. **68**, 139 (2000)
49. Y. Iwasawa, J. Catal. **216**, 165 (2003)
50. R.M. van der Veen, C. Bressler, C.J. Milne, V.-T. Pham, A.E. Nahhas, F.A. Lima, W. Gawelda, C.N. Borca, R. Abela, M. Chergui, J. Phys: Conf. Ser. **190**, 012054 (2009)
51. E.A. Stern, D.L. Brewe, K.M. Beck, S.M. Heald, Y. Feng, J. Phys: Conf. Series **190**, 012053 (2009)

Chapter 8
Luminescence Characterizations of Semiconductor Nanostructures

Jinkyoung Yoo

Abstract Exploring nanoscience confronts novel demands on the understanding of nanomaterial properties. To fulfill the demands, characterization methods for nanostructures should be sensitive to small amount of materials and have high resolution. Luminescence spectroscopy and microscopy can satisfy the requirements of nanoscale characterization although those have been used in semiconductors research for several decades. This chapter concentrates on the study of optical properties and related phenomena in semiconductor one-dimensional nanostructures using luminescence characterizations.

8.1 Introduction

One-dimensional (1D) semiconductor nanostructures such as nanowires (NWs) and nanorods (NRs) have been the subject of much attention over the past decade, with studies from moving beyond scientific exploration to utilizing nanostructures in device applications. Exciting recent developments such as fabrication of nanoheterostructures and doping in nanostructures have enabled researchers to tailor the properties of nanostructures for diverse photonic device applications, for instance, light-emitting diodes (LEDs), photovoltaic cells, and photodetectors, of which carrier generation, recombination, and separation are basic operation principles. The tailoring 1D nanostructures strongly influence the properties closely related to photonic device performance. However, fundamental understanding of physical properties of 1D semiconductor nanostructures is still lacking due to difficulties of characterization of nanomaterials.

J. Yoo (✉)
Center for Integrated Nanotechnologies, Los Alamos National Laboratory, Los Alamos, NM 87545, USA
e-mail: trip297@gmail.com

G.-C. Yi (ed.), *Semiconductor Nanostructures for Optoelectronic Devices*, NanoScience and Technology, DOI 10.1007/978-3-642-22480-5_8,

In general, characterization is based on the observation of materials' response to external perturbations such as applying electrical bias, illuminating light, changing temperature, and so on. Accordingly, optical measurement can be described as the study of the interaction between material and radiated energy in the form of light. The interaction between material and light can be observed in various phenomena such as absorption, luminescence, reflection, refraction, and scattering. Among diverse optical phenomena, luminescence, radiation of light by external perturbation, is the subject of luminescence characterization. Luminescence characterization is nondestructive, nonintrusive, and sensitive to the presence of defects or impurities in materials. Additionally, luminescence characterization can provide superior spatial, spectral, and temporal resolution. Thus, luminescence characterization is very useful to study the properties of 1D semiconductor nanostructures due to the advantages described above. In this chapter, luminescence in semiconductors is briefly introduced, and then the state-of-the-art results of the luminescence characterization of 1D semiconductor nanostructures are presented. To avoid confusion on the meaning of 1D between the dimensionality of physical properties and the shape, 1D nanostructure described in this chapter means structure with elongated shape.

8.2 Radiative Recombination in 1D Semiconductor Nanostructures

As mentioned in the introduction, luminescence characterization is a study on the emitted light from materials. To help understanding the information that can be obtained by luminescence spectroscopy, the mechanism of luminescence in semiconductors is going to be briefly explained.

When semiconductor gets an external perturbation with larger energy than its band gap, an electron in the valence band acquires suitable energy to reach to the conduction band. This excitation of an electron induces formation of a hole in the valence band. This phenomenon is called as generation. The generated electron and the hole tend to return to their ground state by giving up the excess energy through recombination. Before recombination, the generated carriers – electrons and holes – experience various processes such as diffusion, drift, scattering, exciton formation, and so on. If the generated carriers give up their energy in the form of light, the recombination is called as radiative recombination (Fig. 8.1 and Fig. 8.2). Otherwise, if the generated carriers lose the excess energy in the form of heat, the process is called as nonradiative recombination.

Luminescence measurement investigates radiative recombination in materials. There have been many kinds of luminescence measurements. Generally the method of excitation, the prerequisite of luminescence, has been used for classification of luminescence measurement. In Table 8.1 the types of luminescence and their excitation sources are summarized.

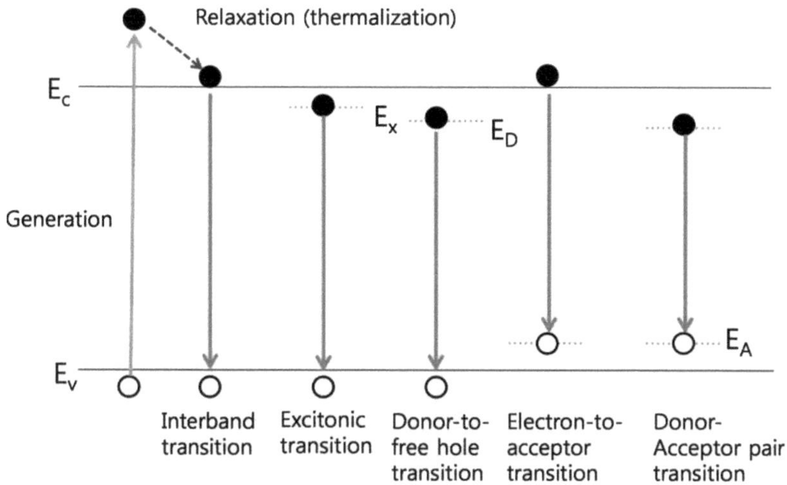

Fig. 8.1 Schematic diagram of radiative recombination processes. E_C, conduction band minimum; E_V, valence band maximum; E_X, exciton energy level; E_D: donor level; E_A, acceptor level

Table 8.1 Types of luminescence classified by excitation sources

Type	Excitation source	Type	Excitation source
Photoluminescence	Photons	Cathodoluminescence	Impact of electrons
Electroluminescence	Electric current	Chemoluminescence	Chemical reaction
Radioluminescence	Bombardment of ionizing radiation	Sonoluminescence	Sound wave
Thermoluminescence	Heat	Bioluminescence	Living organism

Luminescence is not only the basic principle of LEDs and laser diodes, but also contains essential information on electronic band structure and carrier dynamics in semiconductors. Since electronic band structure and carrier dynamics are affected by structures and defects of semiconductors, luminescence spectroscopy is very useful to study the energy levels within band gap, formed by defects and external perturbation. For instance, luminescence intensity and radiative recombination rate are very sensitive to defect states formed in band gap because radiative recombination rate, directly related to luminescence intensity at a given energy, is inversely proportional to the exponential of a given energy (Van Roosbroeck–Shockley relation) [1, 2].

Table 8.2 is a summary of essential material characteristics that can be studied by luminescence measurement for photonic devices.

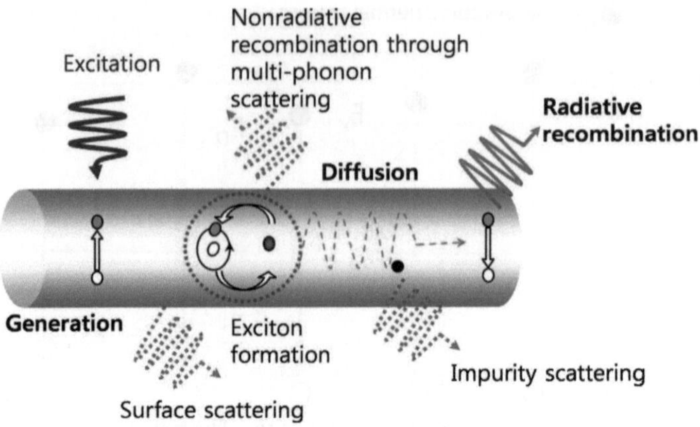

Fig. 8.2 Optical processes in 1D semiconductor nanostructures

Table 8.2 Important materials characteristics for photonic devices and luminescence measurements for investigation of the characteristics

Photonic devices	Principle	Important material characteristics	Useful luminescence measurements
Light-emitting diodes	Electroluminescence	Recombination dynamics: Recombination rate (time), surface recombination velocity	Time-resolved photoluminescence spectroscopy, electroluminescence microscopy, cathodoluminescence microscopy
Solar cells	Carrier separation in asymmetric junction, absorption	Carrier transport properties: diffusion coefficient/velocity/length	Time-resolved cathodoluminescence spectroscopy, time-resolved microphotoluminescence spectroscopy
Photodetectors	Carrier generation in asymmetric junction, absorption		

8.3 Luminescence Characterizations of 1D Semiconductor Nanostructures

8.3.1 Local Probe Techniques

The rapid development of science on 1D semiconductor nanostructures has provided novel opportunities for photonic device applications. With the rapid development, the demands on characterizations of 1D semiconductor nanostructures have become serious because the understanding on material characteristics is imperative to design and to realize photonic devices. However, conventional electrical methods such as Hall measurement and capacitance–voltage measurement to characterize semiconductors are not feasible to apply them for investigation of the properties of nanostructures due to the sizes and the shape of nanostructures. Elongated shape of 1D nanostructure prevents researchers from performing Hall measurement. The formation of electrical contact on nanostructures is not feasible, and is greatly affected by surface states of nanostructures. Thus, electrical contacts to the nanostructures make experiments be complicated. Additionally, electrical techniques are not suitable to unravel physical processes in which surface states are engaged.

Meanwhile, luminescence characterizations, inherently nondestructive and nonintrusive techniques, do not require electrical contact. Thus, luminescence characterization can be done without special preparation. Not only are luminescence characterizations nonintrusive, they also are versatile because a great number of combinations of excitation sources and detection techniques are available. The freedom of selection of excitation sources enables researchers to investigate energy levels in a wide range. Additionally, an excitation source selected circumspectly can excite selected energy level closely related to physical properties as the objective of a study. Furthermore, there have been many light detection techniques with high spatial and high temporal resolution.

For luminescence characterizations of nanostructures, the size of excitation spot is one of the most important considerations though the characterization of bundles of NWs does not require small excitation spots. To investigate the properties of single nanostructure using luminescence measurement, only single desired nanostructure should be excited selectively. Selective excitation of a single nanostructure can be obtained by using local probe techniques with spatial resolution better than 1 μm. The sub-micrometer spatial resolution can be achieved by a strong focus of excitation spot on a sample because spatial resolution is greatly affected by excitation volume and carrier generation volume. There have been many local probe techniques such as microphotoluminescence (μ-PL) spectroscopy, near-field scanning optical microscopy (NSOM), cathodoluminescensce (CL) spectoroscopy, scanning tunneling luminescence (STL) spectroscopy, and so on. Since local probe techniques have been used for study on very small objects, there have been strong demands on concurrent acquisition of morphological information and spectral information. Thus, many local probe techniques have characters of both microscopy for morphological information and spectroscopy for spectroscopic information. Among various local probe techniques, μ-PL spectroscopy and CL microscopy are most common methods. To help the understanding of local probe techniques, the principles of several local probe techniques are going to be briefly explained before presenting the reported results of local probe luminescence characterizations of 1D semiconductor nanostructures.

μ−PL spectroscopy, depicted in Fig. 8.3, is the same as PL spectroscopy except the size of excitation light spot. In PL, excitation source is light, typically laser or monochromatic beam. The incident light is absorbed in the sample, and generates

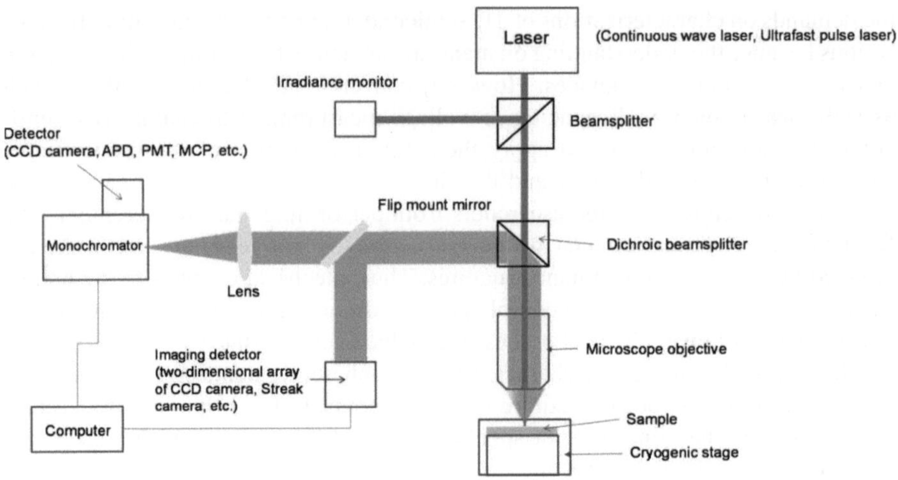

Fig. 8.3 Scheme of μ-PL measurement system. (*CCD* charge-coupled device, *APD* avalanche photodiode, *PMT* photomultiplier tube, *MCP* microchannel plate)

subsequently electron–hole pairs. If the incident light intensity is not too high or the structure of the sample does not induce carrier multiplication, each photon generates one electron–hole pair. The spatial resolution of PL can be discussed in the two categories: depth resolution and lateral resolution.

The depth resolution is determined by the penetration depth of the incident light. For direct band gap semiconductor, typical absorption coefficient is in the range of 10^4 and $10^5\,\mathrm{cm}^{-1}$ for a photon energy above the band gap. Thus, the typical penetration depth in direct band gap semiconductor is smaller than 1 μm. This value is generally shorter than the diffusion length of carriers and excitons.

μ-PL has an advantage of lateral resolution compared to conventional PL because the incident light is focused on the sample in μ-PL. The lateral resolution is limited by the minimum spot size of incident beam and by diffusion length of carriers and excitons. The minimum spot size, d, is determined by diffraction of light. According to Rayleigh criterion, one can calculate the d as follows:

$$d = \frac{1.22\lambda}{sn \sin\theta},$$

where λ is the wavelength of light, n is the refractive index of the medium in which the sample is, and θ is the half of the solid angle passing through to the focusing lens. $n \sin\theta$ is called as the numerical aperture. In practice, the d is considered to be $\lambda/2$. Typically the d is in the range of 200 and 500 nm. To focus the incident light, microscope objective with the numerical aperture of 0.4 or 0.5 is commonly used. The microscope objective can also correct the spherical aberration caused by glass window of the cryostat existing between the objective lens and the sample. The diffusion length of carriers and excitons, the other factor affecting the lateral resolution, is in the range of several nanometers and several micrometers. Thus, there is a discrepancy between the size of the focused light and the actual lateral resolution determined by the diffusion length of carriers and excitons. The discrepancy means that the spectrum of the sample can be obtained with the resolution of 1 μm even though researchers can observe the nanostructures with the size of several hundreds nm.

NSOM is another local probe technique based on optical excitation. In NSOM, the diffraction limit of the d can be overcome by proximal excitation. When the excitation of light source is very close to the sample surface, the diffraction limit is not established due to the formation of near field. To achieve near-field regime, the incident light is generally guided to a position close to the sample surface in the order of several nanometers by an optical fiber with a subwavelength-sized aperture. The distance between the aperture and the sample should be smaller than the wavelength of the light. For aperture-NSOM, the spatial resolution is given by the size of the aperture. In NSOM, since the incident light strongly interacts with the local region of the sample, there is the strong dependence of luminescence intensity on aperture-to-sample distance. This strong topological effect of near-field intensity makes the interpretation of the data obtained by NSOM very difficult. To overcome the topological effect, researchers have employed apertureless-NSOM. In apertureless-NSOM, excitation of the sample is globally occurred with large-sized beam compared to the sample, and collection of the signal in the form of the evanescent wave is done by a nanoparticle very close to the sample surface. The detailed principle of NSOM is outside the scope of this chapter.

CL microscopy is based on the excitation by electron beam. Since electron beam is the excitation source of CL, CL measurement is generally performed with electron microscope. Figure 8.4 shows the scheme of CL measurement.

In CL, electrons with the energy of several tens keV interact with the sample. The irradiation of energetic electrons on the sample generates electron–hole pairs in the sample. Unlike PL, one energetic electron in CL can generate many electron–hole pairs due to cascading effect of primary electron (PE). Figure 8.5 depicts the interaction between the incident electron and the sample. The PE injected to the sample can penetrate the sample or be reflected at the surfaces of the sample. The penetrating PEs are scattered or are deflected by the nuclei. If the PE loses its part of energy by the interaction between the PE and the electron in the sample (inelastic process), free electrons, secondary electrons (SEs), are generated in the sample. The SEs lead the generation of the other SEs through some inelastic processes. The SEs can combine with holes to form electron–hole pairs. For GaAs, a 5 keV electron generates about 700 electron–hole pairs. This multiple generation process is occurred in the generation volume determined by an average traveling distance of PEs. The generation volume by lateral spreading of the PEs is one of the most important factors limiting spatial resolution of CL.

Meanwhile, the penetration depth of the PEs depends on the energy of PEs, which is defined by the acceleration voltage of the electron beam in electron microscope. Thus, CL is capable of depth-dependent study. Fig. 8.6 shows the size

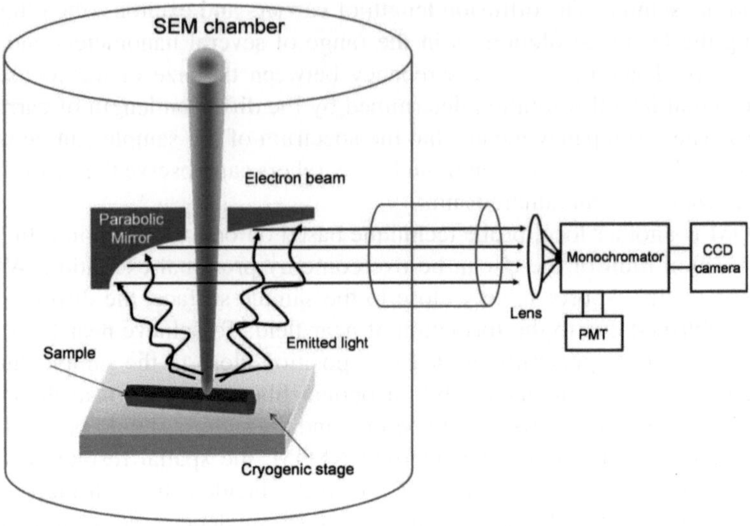

Fig. 8.4 Scheme of CL measurement

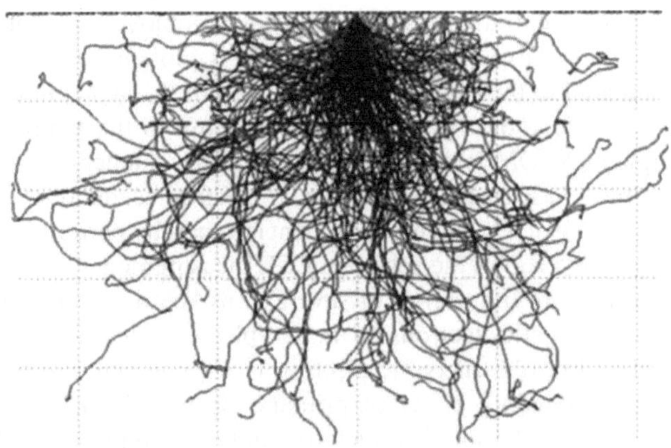

Fig. 8.5 Trajectories of 100 electrons of 20 keV in GaN calculated by the Monte Carlo method using CASINO ver. 2.42 [3]

of the generation volume and the penetration depth with different PE energies. The detailed analysis of PE interaction in the sample is described in several books. Besides the interaction between PEs and the electrons in the sample, the diffusion length of carriers and excitons is also an important limiting factor of the spatial resolution of CL.

CL has been widely used to characterize the properties of nanostructures and thin films because CL can provide concurrently spectroscopic information and morphological information with a spatial resolution of a few nanometers. The small size of the electron beam in modern electron microscope makes it possible to

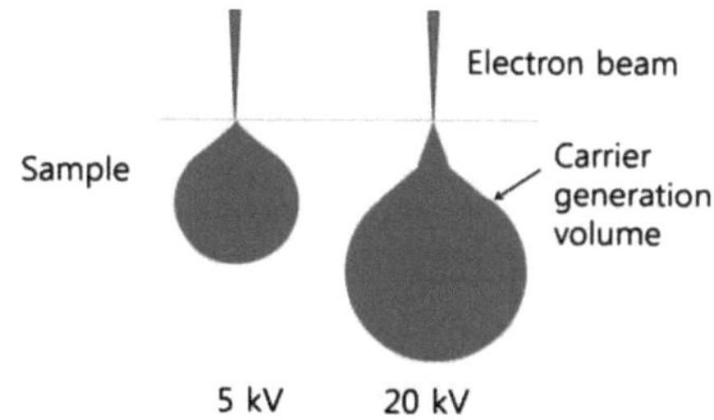

Fig. 8.6 Carrier generation volume with different acceleration voltages of electron beam

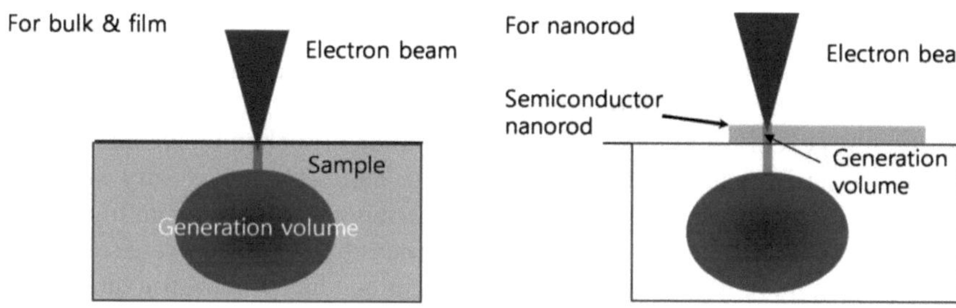

Fig. 8.7 The generation volumes for semiconductor bulk and nanorod

irradiate single nano-objects. However, as mentioned above, the excitation volume generated by cascading effect has hindered the improvement of the spatial resolution of spectroscopic study in CL. For 1D semiconductor nanostructures, the cascading effect can be avoided because the diameter of an NR or an NW is generally smaller than the penetration depth of the highly energetic electron beam. In more detail, the typical diameter of 1D semiconductor nanostructures is smaller than 100 nm, which is smaller than the penetration depth of 5 keV electron beam in almost semiconductors. Since the acceleration voltage of electron beam under the imaging mode in scanning electron microscope is generally larger than 5 kV, the almost PEs penetrate a 1D semiconductor nanostructure, and the generation volume is formed below several micrometers of a 1D semiconductor nanostructure. Thus, excitation volume in the sample is only determined by the electron beam size and the diameter of a 1D semiconductor nanostructure under the normal condition, as shown in Fig. 8.7.

STL uses low-energy carriers as excitation sources. Generally the injected carriers have energies in the range of 0.1 and 10 eV. Due to their low energy, the injected carriers can selectively excite specific states in the semiconductor. Additionally, STL provides very small excitation volume because STL measurement system is attached to scanning tunneling microscope. Several articles provide the detailed information on STL [4, 5]. The spatial resolution of spectroscopic information in STL is also governed by the diffusion length of carriers and excitons in the semiconductor.

8.3.2 *Luminescent Characteristics of Semiconductor Nanostructures*

In this section, the capability of luminescence characterization is described with several examples.

8.3.2.1 Inhomogeneity in a Single 1D Semiconductor Nanostructure

The properties of nanostructure are quite sensitive to defects due to its small volume. Although semiconductor nanostructures can be prepared in the form of single crystalline, the nanostructure can still contain many defects such as impurity atoms and surface states. Furthermore, the defects are distributed all over the nanostructures evenly or not. If the defect distribution is not homogeneous, localized physical properties can be observed. Since the nanostructure is very small, the inhomogeneity must be investigated by characterizing techniques with high spatial resolution. Local probe techniques of luminescence characterization are suitable for the research on inhomogeneity in a single 1D semiconductor nanostructure.

Figure 8.8 showed the CL results of a single ZnO nanopillar grown by the vapor–liquid–solid process using Au nanoparticles on $a - Al_2O_3$ substrate [6]. The precursor of ZnO contained small content of Ga. The diameter and the length of the ZnO pillar are 100 nm and 1 μm, respectively. The CL spectra were obtained at 10 K with the acceleration voltage of 2 kV and the beam current of about 10 pA. Under this measurement condition, the lateral resolution determined by the PEs was about 40 nm.

As shown in Fig. 8.8, the CL spectra obtained from the different excitation positions showed different spectral shapes, indicating inhomogeneous distribution of radiative recombination centers assigned to the origins of emission peaks. Each CL spectrum obtained from a spot can be deconvoluted with several sharp lines with the peak energies at 3.3593_{3-5} eV(I_{8a}), 3.3601_{1-7} eV(I_8), 3.361 eV(I_6), and 3.3636_{6-9} eV(D_oX_B). Each sharp line indicates a sharp Lorentzian component assigned to specific donor-bound exciton recombination. Since the origins of I_8 and I_{8a} can be assigned to the Ga atoms at Zn sites [7], unintentional incorporation of Ga atoms was revealed. Furthermore, the intensity ratio of I_6 to free excitonic emission (X_A), as shown in Fig. 8.9, increased as the excitation position was closer to the Al_2O_3 substrate. The spatial behavior of I_6 intensity indicates the incorporation of Al into the ZnO nanopillar because the origin of I_6 is assigned to shallow donor Al_{Zn}.

As described above, spatially resolved luminescence spectroscopy is capable of investigating the properties of single 1D semiconductor nanostructures.

8.3.2.2 Size Effect

Simple but the most distinct character of nanostructure is its small size. 1D semiconductor nanostructures have diameters that can be in the range of several

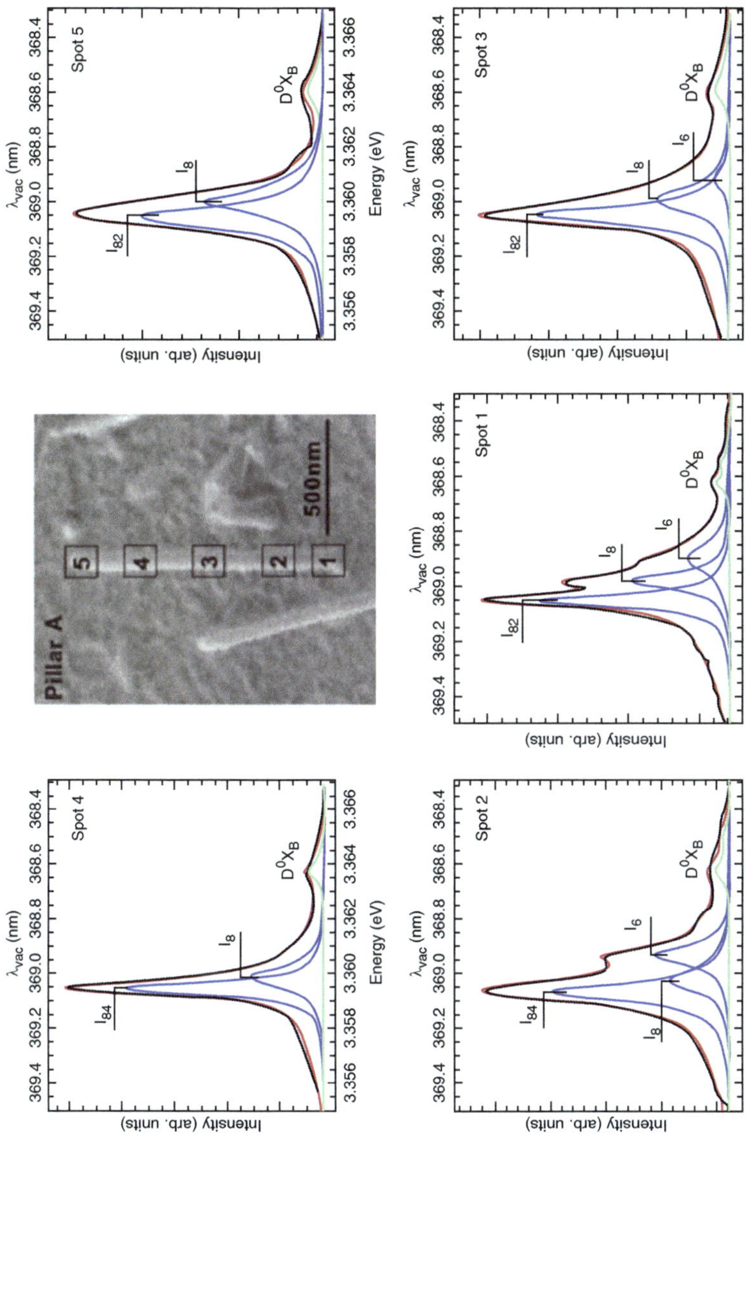

Fig. 8.8 The 10 K CL spectra from a ZnO pillar. The CL spectra were recorded for different excitation positions (spots 1–5). (Reprinted with permission from [6]. Copyright 2007 by the American Institute of Physics)

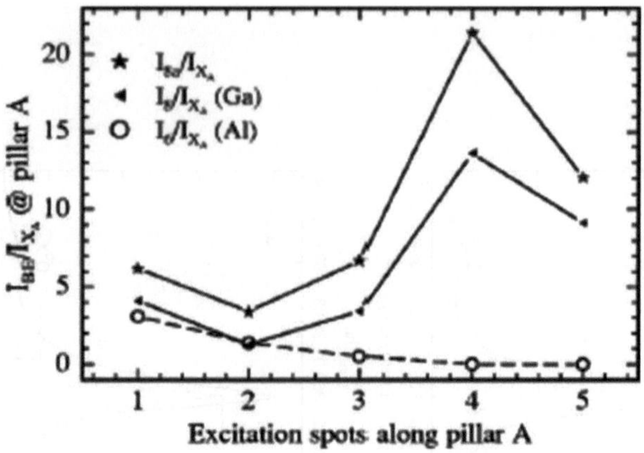

Fig. 8.9 Intensity ratio of donor-bound excitonic emission to free excitonic emission for a ZnO nanopillar. (Reprinted with permission from [6]. Copyright 2007 by the American Institute of Physics)

nanometers and several hundreds nanometers, with lengths from one hundred nanometers to several tens micrometers. The wide range of available dimensions of 1D semiconductor nanostructures brings interesting properties related to their sizes. The size dependence of luminescent characteristics can be generally classified as three regions: (1) The diameter of a 1D nanostructure is smaller than the characteristics lengths such as Fermi length, Debye length, and exciton Bohr diameter. (2) The diameter is between the characteristic length and the incident light wavelength. (3) The diameter is larger than the incident light wavelength. The case of (3) is the almost same as the bulks if the effect of surface states is excluded. Thus, this section covers only the cases of (1) and (2).

In the case of (1), quantum confinement effect is dominant. Generally the blue shift of the dominant emission energy is observed in the luminescence spectra. Figure 8.10a shows the room temperature PL spectra of InP NWs with different diameters in the range of 10 and 50 nm [8]. As shown in Fig. 8.10, the PL emission energy shifted to the higher energy side as the NW diameter decreased below 20 nm. The size-dependent blue shift of the PL emission energy can be explained by quantum confinement effect because the exciton Bohr diameter of bulk InP is 19 nm [9]. The amount of the shift is well fitted with an effective mass model (EMM) for a cylindrical potential for electrons and holes, as shown in Fig. 8.10b. The blue shift due to quantum confinement has been reported in various semiconductor NWs and NRs [10, 11].

In the regime of (2), it has been predicted that the radiative recombination rate is proportional to the diameter of the nanostructures, opposite for the (1) and (3). There have been reported the experimental results on the size-dependent radiative recombination rate in semiconductor NWs. However, Fig. 8.11 exhibits the opposite behavior to the theoretical analyses, the decrease in the recombination rate as the increase in the sizes of the NWs. The discrepancy between the theoretical prediction

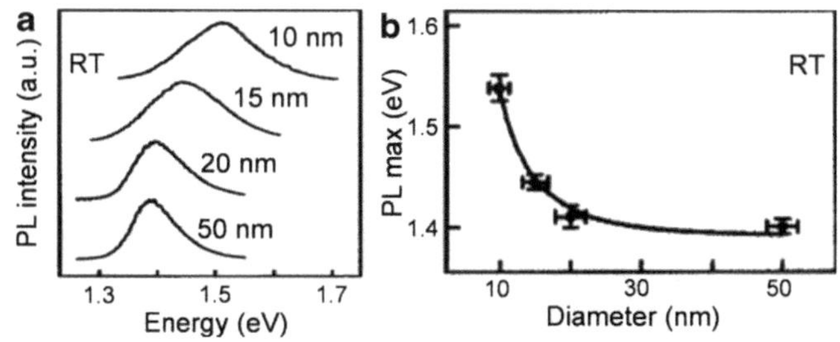

Fig. 8.10 (**a**) RT PL spectra from single InP NWs with diameters of 10, 15, 20, 50 nm. (**b**) The PL emission energies (*solid circles*) fitted with the EMM (*solid line*). (Reprinted with permission from [8]. Copyright 2002 by the American Chemical Society)

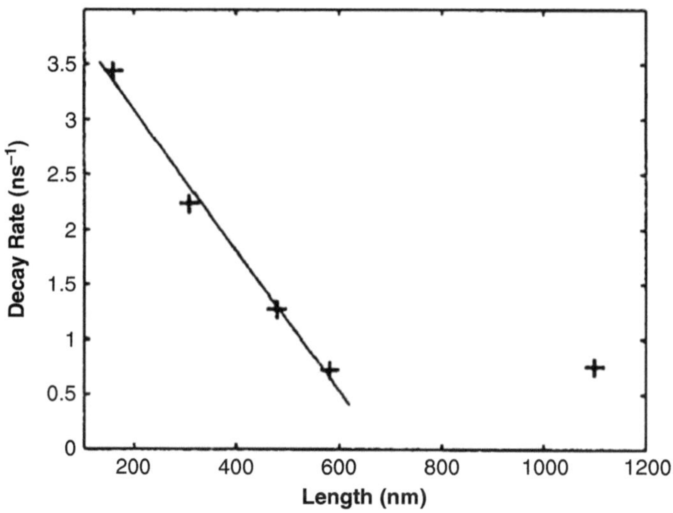

Fig. 8.11 Size-dependent PL decay rate of the ZnO NRs. (Reprinted with permission from [12]. Copyright 2003 by the American Institute of Physics)

and the experimental result has given rise to the debate on the discrimination of the effects of surface states and those of sizes on luminescent characteristics of 1D semiconductor nanostructures.

The size dependence of emission energy and recombination rate in 1D semiconductor nanostructures has opened up tuning of luminescence properties precisely.

8.3.2.3 Investigation on Surface States of 1D Semiconductor Nanostructures

The small size of nanostructure gives rise to large surface-to-volume ratio of nanostructures, which results in the dominance of surface states in nanostructures. Surface states can form energy levels within the band gap, and act as distinguished

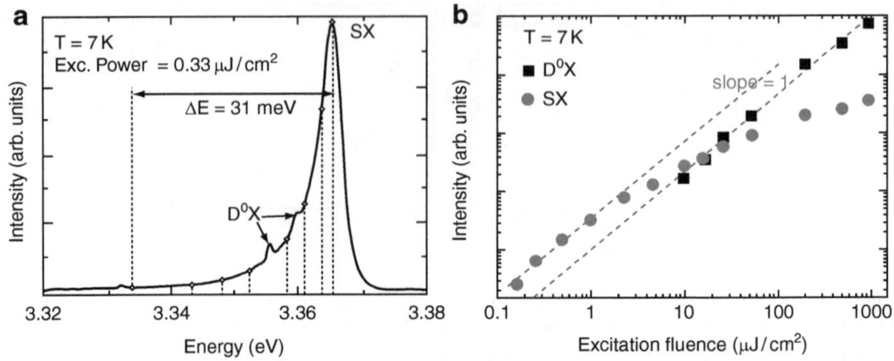

Fig. 8.12 (**a**) The 7 K PL spectrum of ZnO nanowires at the excitation intensity of 0.33 μJ/cm^2. (**b**) The peak intensities at the donor-bound excitionic emission and the SX emission as a function of excitation intensity (D_0X: donor-bound exciton). (Reprinted figure with permission from [13]. Copyright 2006 by the American Physical Society)

recombination centers. If the recombination centers originated by surface states are radiative, luminescence spectroscopy can reveal the existence and the characteristics of surface states. The fundamental understanding on the dynamics of surface-related emission can be used for the design of photonic devices based on 1D semiconductor nanostructure in which the effect of surface states can be dominant. The luminescence study on the surface states has been performed with semiconductor NWs that have the average diameters smaller than the absorption depth of the incident light.

Figure 8.12a shows the 7 K PL spectrum of ZnO NWs [13]. In Fig. 8.12a, the dominant emission was observed at 3.365_6 eV corresponding to the energy of the surface excitons (SXs) in ZnO. The dominant emission originated from the SXs indicates that ZnO NW can be a suitable platform for surface study.

Since the total number of surface states should be lower than that of volume states, it is expected that the saturation of the surface state transitions is occurred at a lower excitation intensity compared to that of the volume states. Figure 8.12b shows that a clear saturation of the intensity of SX emission is observed above 1 μJ/cm^2.

Another feature of the SX emission is rapid thermal quenching because the activation energy of the SX is 5 ± 1.5 meV much lower than the $\sim$ 16 meV as a typical value of bound exciton localization energy in ZnO [14]. The low activation energy of the SX results in feasible ionization of surface states creating nonradiative recombinationtraps of excitons and carriers. The increase in the nonradiative recombination traps gives rise to the decrease in light generation efficiency in NWs and NRs at room temperature.

The surface state-related luminescent characteristics have been observed in other semiconductor NWs, e.g., GaN NWs, InP NWs, and ZnS NWs [15–17]. The common existence of surface states in NWs and its effect on luminescence quenching have been issues in the photonic device applications because luminescence quenching and trapping carriers by surface states of semiconductor NWs are detrimental to LEDs, photovoltaic cells, and photodetectors. To overcome the quenching, surface passivation such as formation of core-shell structure and coating layers has been studied [18–20].

8.3.2.4 Impurity Characterization: Effect of Catalyst Atoms

In semiconductors, small amount of impurities in the nanomaterial can significantly change physical properties of the host materials. Reduction in the size of the materials induces the increase in the effect of impurities. Thus, it has been very important to characterize the impurities in 1D semiconductor nanostructures. In almost cases, the chemical origins of physical properties of 1D semiconductor nanostructures are similar to those of bulks and thin films. However, unlike bulks and thin films, semiconductor NWs have an additional impurity source, catalyst, because catalyst-assisted vapor–liquid–solid growth method has been widely employed to grow semiconductor NWs. Nevertheless, the study on the effect of catalyst atoms in semiconductor NWs on physical properties has rarely been reported though the incorporation of catalyst atoms in catalyzed semiconductor NWs has been confirmed [21, 22]. The difficulties in elemental analysis and electrical device fabrications are responsible for the rareness of the reports on the impurity characterizations in semiconductor NWs. With the advantage of luminescence spectroscopy, PL and CL are very useful for characterizing impurities in semiconductor NWs.

The PL spectra of GaN NWs grown by Ni catalyst-assisted methods have shown dominant emissions originated from chemical origins that have not been known in bulks and thin films. A systematic PL study suggested that the chemical origin of the peculiar emission from catalyzed GaN NWs is the Ni catalyst. Figure 8.13a shows a typical 10 K PL spectrum of Ni-catalyzed GaN NWs [23]. As shown in Fig. 8.13a, three distinct emission peaks at 3.472, 3.437, and 3.266 eV are observed in the near-band-edge emission. According to the previous results in thin films and bulks, the origins of the PL peaks at 3.472 and 3.266 eV have been attributed to neutral-donor-bound exciton and donor–acceptor-pair transitions, respectively. However, the notable emission at 3.437 eV has not been known. To study the characteristics of the PL emission at 3.437 eV, temperature-dependent (TD) PL and time-resolved (TR) PL measurements were conducted. The TDPL and TRPL studies for the emission at 3.437 eV revealed the exciton localization energy of about 17 meV and the PL decay times of 252 and 816 ps. The PL characteristic values are quite similar to those of acceptor-bound exciton in GaN. The existence of a large amount of acceptors can be explained by the unintentional incorporation of Ni atoms in Ni-catalyzed GaN NWs during the growth because the valence state of Ni in GaN is +2.

The example described above shows that luminescence characterizations can be useful for impurity characterization in semiconductor NWs, an important issue in nanoscience.

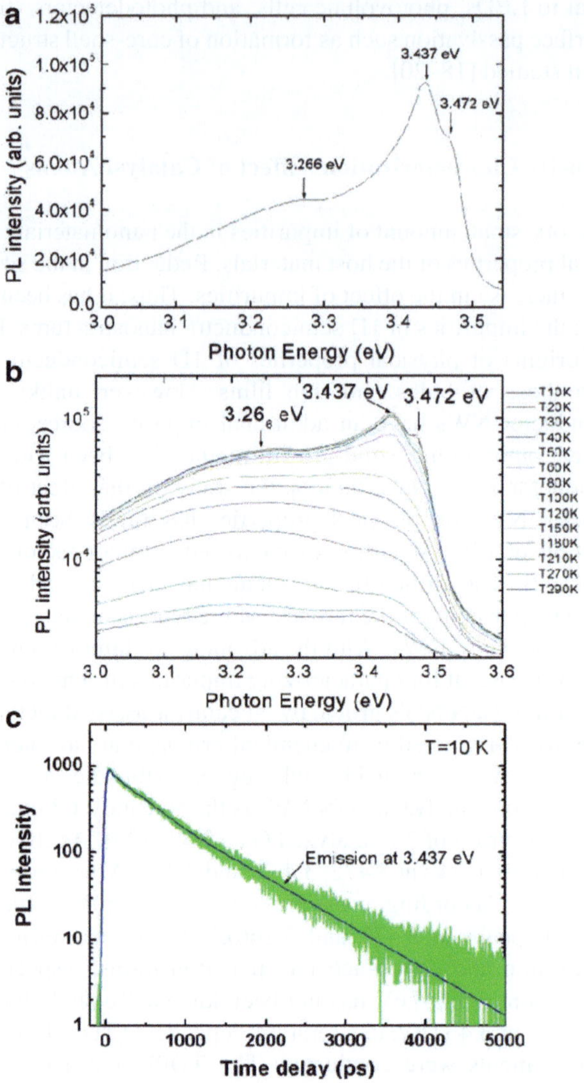

Fig. 8.13 (**a**) 10 K PL spectrum of Ni-catalyzed GaN NWs. (**b**) TDPL spectra of Ni-catalyzed GaN NWs. (**c**) 10 K TRPL spectrum measured at 3.437 eV of Ni-catalyzed GaN NWs. (Reprinted with permission from [23]. Copyright 2006 by the American Institute of Physics)

8.3.2.5 Polarization Anisotropy

Light is an electromagnetic wave consisting of electric and magnetic field components, which oscillate perpendicular to propagating direction. The nature of light generates the contrast of the strength of light–matter interaction along direction called as polarization anisotropy. In luminescence, the emission intensity can be

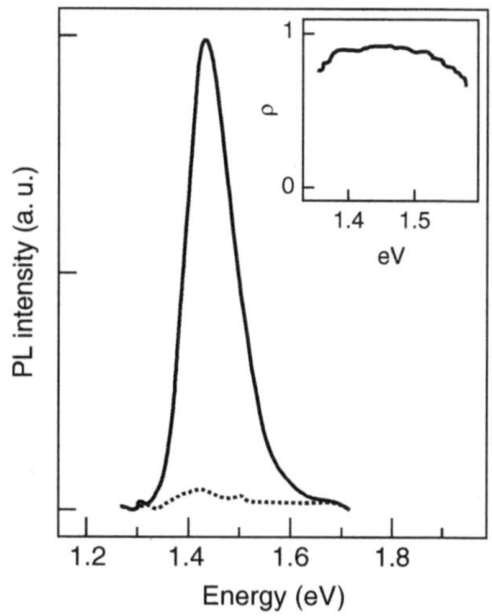

Fig. 8.14 Micro-PL spectra of a single InP NW recorded with the exciting laser aligned parallel (*solid line*) and perpendicular (*dashed line*) to the long wire axis. The *inset* shows the polarization ratio. (From [24]. Reprinted with permission from AAAS)

changed with the direction of absorption light to the orientation of the materials. The difference between the emission intensity with differently polarized incident light can be expressed in the form of polarization ratio, ρ. For luminescence of 1D nanostructures, ρ is defined as follows:

$$\rho = \frac{I_{\mathrm{P}} - I_{\perp}}{I_{\mathrm{P}} + I_{\perp}},$$

where I_{P} is the emission intensity when the electric field of the absorption light is parallel to the long axis of 1D nanostructure, and $I_{\perp}$ is the emission intensity when the electric field of the absorption light is perpendicular to the long axis of 1D nanostructure.

In the viewpoint of polarization anisotropy, 1D semiconductor nanostructure is very interesting system because of its elongated shape and small size. In the early stage of the research on 1D semiconductor nanostructures, a giant polarization anisotropy of 0.9 was reported for single InP NWs, as shown in Fig. 8.14 [24]. The giant polarization anisotropy has been reported for various 1D semiconductor nanostructures [25–29].

Two major mechanisms have been attributed for the origin of polarization anisotropy in 1D semiconductor nanostructures. One is the confinement of optical electric field due to the contrast of dielectric constants between 1D nanostructure and its environment. The other is quantum size effect.

The mechanism based on optical electric field confinement has been widely employed for the explanation of polarization anisotropy in free-standing NWs and NRs, which have larger diameters than exciton Bohr diameter. Generally the dielectric constant of semiconductor is larger than that of common environment (e.g. air or vacuum) under normal measurement condition. Table 8.3 shows the relative dielectric constants of several semiconductor materials at optical frequency. $E|c$

Table 8.3 Relative optical dielectric constants of semiconductors

Si	11.68	Ge	16.2
InAs	12.3	GaAs	10.89
InP	9.61	GaP	9.11
GaN	5.8 $(E\|c)$, 5.35 $(E\perp c)$	ZnO	3.75 $(E\|c)$, 3.70 $(E\perp c)$
* Vacuum	1	*Air	1.00059

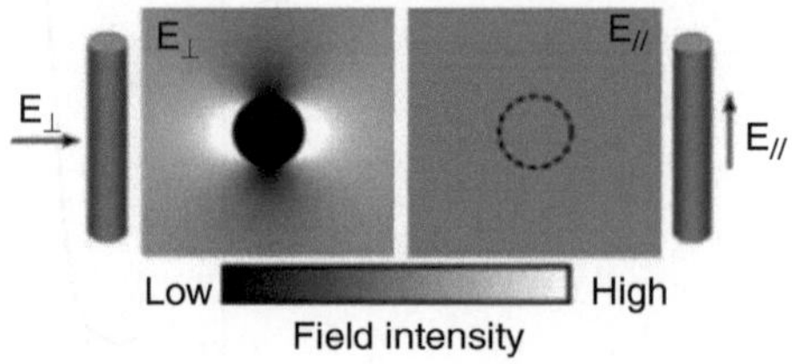

Fig. 8.15 Calculated electric field distribution near a single InP NW in a vacuum under irradiation of polarized light. In the model, an NW is considered as an infinite dielectric cylinder with much smaller diameter than the light wavelength. (From [24]. Reprinted with permission from AAAS)

The large dielectric contrast between 1D semiconductor nanostructure and surrounding results in the anisotropic electric field distribution is depicted in Fig. 8.15. As shown in Fig. 8.15, the electric field is strongly attenuated in the 1D semiconductor nanostructures for the $E_{\perp}$, whereas the electric field is homogeneously distributed for the E_{P}. Thus, the incident field parallel to the long axis of the 1D nanostructure is not reduced in the 1D nanostructure.

The dielectric contrast model predicts an interesting phenomenon, size dependence of polarization anisotropy. According to theoretical analysis, if the diameter of 1D nanostructure is smaller than the emission wavelength but much larger than the exciton Bohr diameter, the polarization ratio decreases as the diameter increases [30]. This prediction was confirmed by the polarized PL spectroscopy of single GaN NRs with different diameters, as shown in Fig. 8.16. However, the size dependence of polarization anisotropy was not observed for single InP NWs [24].

On the other hand, the quantum size effect has been considered as the main mechanism generating polarization anisotropy in quantum wires that have smaller sizes than characteristic lengths such as exciton Bohr diameter. According to the quantum size effect model, the optical transition between the valence band and conduction band of an NW is intrinsically polarized along the wire axis, and the polarized optical transition is responsible for the giant polarization anisotropy in an NW [32].

The giant polarization anisotropy is very useful for directional emission of LEDs and sensitivity of photodetectors. Fundamental understanding on polarization anisotropy in 1D semiconductor nanostructure can give valuable insight for the design of nanophotonic devices.

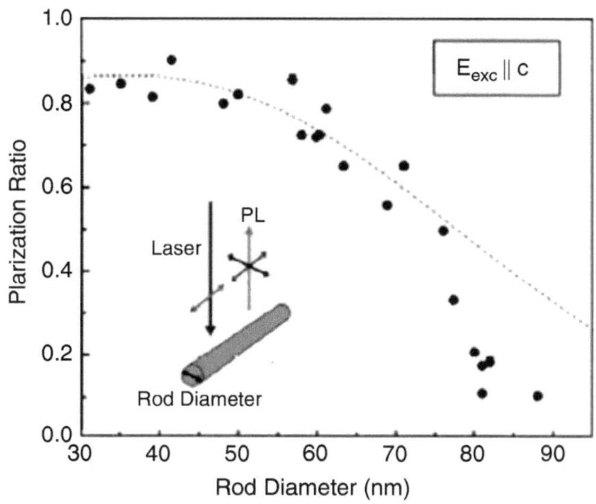

Fig. 8.16 The polarization ratio of PL intensity from single GaN NRs with different diameters. The *dashed line* is the calculated result based on the optical electric field confinement model. (Reprinted with permission from [31]. Copyright 2008 by the Optical Society of America)

8.3.2.6 Crystal Structure Dependence

In semiconductors, luminescent characteristics are determined by electronic band structure in the reciprocal space of the crystal structure. Thus, luminescence of 1D semiconductor nanostructures is affected by crystal structures. For instance, III–V semiconductors such as InP and GaAs can have two different crystal structures, zinc blende (ZB) and wurtzite (WZ). The optical transition energies of ZB and WZ semiconductors with the same stoichiometry are different [33]. Bottom-up approach of growths of 1D semiconductor nanostructures has opened novel opportunities of preparation of polytypes between WZ and ZB structures. Figure 8.17a shows the high-resolution transmission electron microscopy image (HR-TEM) of the WZ/ZB polytypism in an InP NW [34]. In Fig. 8.17a, the segments of WZ and ZB form the polytypism having staggered type-II band alignment depicted in Fig. 8.17b. Due to the type-II band alignment, the wave functions of electrons and hole are spatially separated into the ZB and the WZ segments, respectively.

The spatial separation of charge carriers results in extraordinary long electron–hole recombination lifetime at low energies because of the small overlap between the electron and hole wave functions. As the low-energy states are filled by the increase in the excitation intensity, the low-energy states are saturated, and then the electron and hole wave functions begin to penetrate the barriers. Thus, the electron–hole recombination lifetime at high energies becomes shorter than that at low energies. The transition of electron–hole recombination dynamics can be confirmed by power-dependent (PD) and time-resolved (TR) PL.

Figure 8.18a shows the normalized PL spectra of a single WZ/ZB InP homostructured NW excited using 1.59 eV laser beam with different excitation intensities

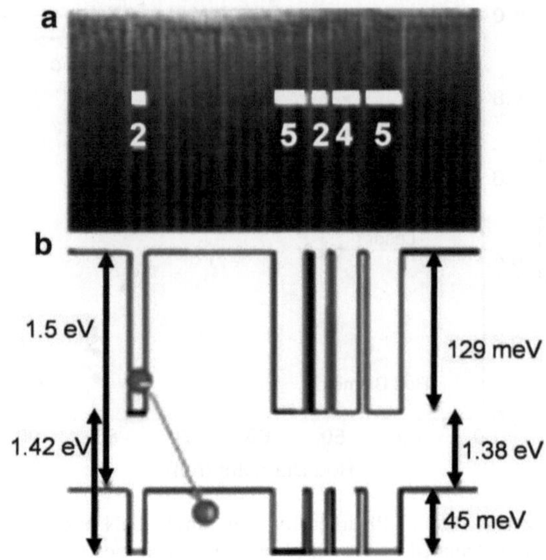

Fig. 8.17 (**a**) HR-TEM image of WZ/ZB polytypism in an InP NW. The ZB segments are identified with the *white lines*. The number indicates the number of atomic planes. (**b**) The energy band diagram of the structure presented in (**a**). The diagram shows staggered type-II band alignment. (Reprinted with permission from [34]. Copyright 2009 by the American Chemical Society)

varying by a factor of 44. The normalized PDPL spectra clearly show blue shift of the dominant emission energy. The PD blue shift is a characteristic of luminescence in polytypism NWs having type-II band alignment because the increase in the excitation intensity induces the conversion of energy transition from low-energy side to high-energy side by saturation of energy states. TRPL measurement reveals the transition of electron–hole recombination dynamics. In Fig. 8.18b, the dominant emission energy at the initial stage corresponds to the band edge emission of WZ InP. The dominant emission energy moves to lower energy side as time progresses. In the early time after laser pulse excitation, the electrons and holes in WZ segment recombine within 1 ns due to a large overlap of the electron and hole wave functions. After the filling energy states of the WZ band edge and fast decay of electron–hole pairs in the WZ segment, the electron and holes occupy energy states in the ZB and WZ, respectively. The spatial separation of electrons and holes induces much longer decay time longer than 5 ns.

Luminescent characteristics are also affected by the twinned structure in a single 1D semiconductor nanostructure [35]. The understanding on the recombination dynamics in polytypism homostructured and heterostructured 1D nanostructures can be useful for fine tuning of emission properties of 1D semiconductor nanostructures for photonic device applications.

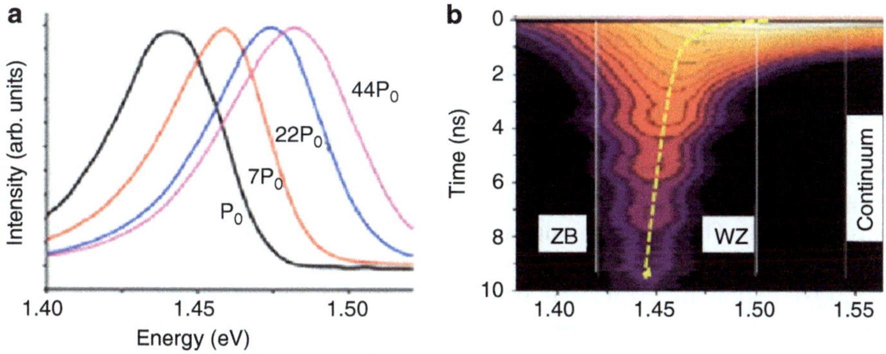

Fig. 8.18 (**a**) Normalized PL spectra of a single WZ/ZB InP NW homostructure with different excitation intensities. $P_O = 50\,\mu W$. (**b**) The evolution of TRPL spectral map with time. The *white line* labeled with WZ and ZB indicates the band gap energies of WZ and ZB structured InP, respectively. (Reprinted with permission from [34]. Copyright 2009 by the American Chemical Society)

8.3.2.7 Diffusion of Excitons and Carriers

Basically operation of a photonic device is closely related to dynamics of photoexcited carriers or excitons generated by external perturbations in the form of light or electrical bias. Thus, fundamental understanding on the dynamics of excitons and carriers in 1D semiconductor nanostructures is imperative to design photonic devices based on 1D semiconductor nanostructures. The two essential parameters of the dynamics of excitons and carriers are recombination lifetime and diffusion length. Investigation on diffusion length requires experimental techniques with high spatial resolution, whereas recombination lifetime can be directly obtained by time-resolved luminescence spectroscopy. To study diffusion length of excitons and carriers, the relation between spatial resolution of local probe technique and diffusion length of excitons and carriers can be used. In Sect. 8.3.1, it is mentioned that the spatial resolution of local probe techniques is greatly affected by diffusion length of carriers and excitons. Conversely, the spatial resolution limited by the diffusion behavior can be a measure to study exciton and carrier diffusion. If a spatially well-defined luminescent indicator is separated from the excitation position, the spatial extent of the luminescence from the indicator is proportional to the diffusion length of excitons and carriers. 1D semiconductor nanoheterostructures in which compositions are modulated along the growth axis provide a spatially well-defined luminescent indicator and the diffusion paths of excitons and carriers.

The combination of spatially resolved CL spectroscopy and nanorod quantum structure was employed to study exciton diffusion in a single MgZnO nanorod [36]. As shown in Fig. 8.19a, a $ZnO/Mg_{0.1}Zn_{0.9}O$ single-quantum-well (SQW) was formed on the top of $Mg_{0.2}Zn_{0.8}O$ nanorod. The energy band alignment in Fig. 8.19a shows that the excitons in $Mg_{0.2}Zn_{0.8}O$ nanorod stem can be flown into the $ZnO/Mg_{0.1}Zn_{0.9}O$ SQW. Because $ZnO/Mg_{0.1}Zn_{0.9}O$ SQW emits strong luminescence, the SQW can act as a suitable luminescent indicator for the presence of excitons. Exciton diffusion in a single $Mg_{0.2}Zn_{0.8}O$ nanorod can be investigated by measuring CL spectra as a function of the electron beam position along the nanorod axis with a spacing between consecutive electron beam spots. As shown in Fig. 8.19b, the excitation position-dependent CL spectra varied in the spectral shape and the SQW emission intensities. Particularly when an electron beam spot was located in the range of 220 nm from the SQW at the tip, the ZnO SQW emission was dominant. The SQW emission in the range of 220 nm resulted from the exciton diffusion in the $Mg_{0.2}Zn_{0.8}O$ nanorod because the sizes of electron beam spot and the SQW region are 50 nm and 10 nm, respectively. The exciton diffusion length can be directly estimated by the integrated CL intensity of the SQW emission as a function of the electron beam position in Fig. 8.19c, as follows:

$$I_{\mathrm{QW}}^{\mathrm{ZnO}} = I_o \exp\left(\frac{-x}{L_\mathrm{d}}\right),$$

where is L_d the exciton diffusion length, $I_{\mathrm{QW}}^{\mathrm{ZnO}}$ is the CL intensity of a ZnO SQW emission, x is the distance between electron beam position and SQW region, and I_o is a scaling factor. From the fit of CL intensity in Fig. 8.19c, the exciton diffusion length in a $Mg_{0.2}Zn_{0.8}O$ nanorod with a diameter of 120 nm was about 100 nm.

The same combination of spatially resolved CL spectroscopy and nanowire heterostructures was used for the direct measurement of carrier diffusion length in a InGaAs/GaAs nanowire heterostructures [37]. In the study, the effect of AlGaAs shell deposition on carrier diffusion length was investigated. According to the experimental results in [37], carrier diffusion length was enhanced by shell deposition with a higher band gap material.

The transport characteristic values such as diffusion length and recombination velocity give great insight into photonic device performances. Spatially resolved luminescence spectroscopy is valuable for studying transport in semiconductor nanomaterials because luminescence measurements make it possible to determine the characteristic values quantitatively.

8.4 The Limit of Luminescence Characterizations

In Sect. 8.3, it is demonstrated that luminescence characterizations are useful for study on physical properties of 1D semiconductor nanostructures. However, luminescence characterization is not an universal way to study physical properties of 1D semiconductor nanostructures.

First of all, luminescence techniques are not suitable for characterizing indirect semiconductors because radiative recombination is not an efficient process in indirect semiconductor due to momentum conservation. The luminescence measurements of indirect semiconductors such as Si and Ge have been reported in the case of nanostructures that show quantum confinement effect and strong localization effect.

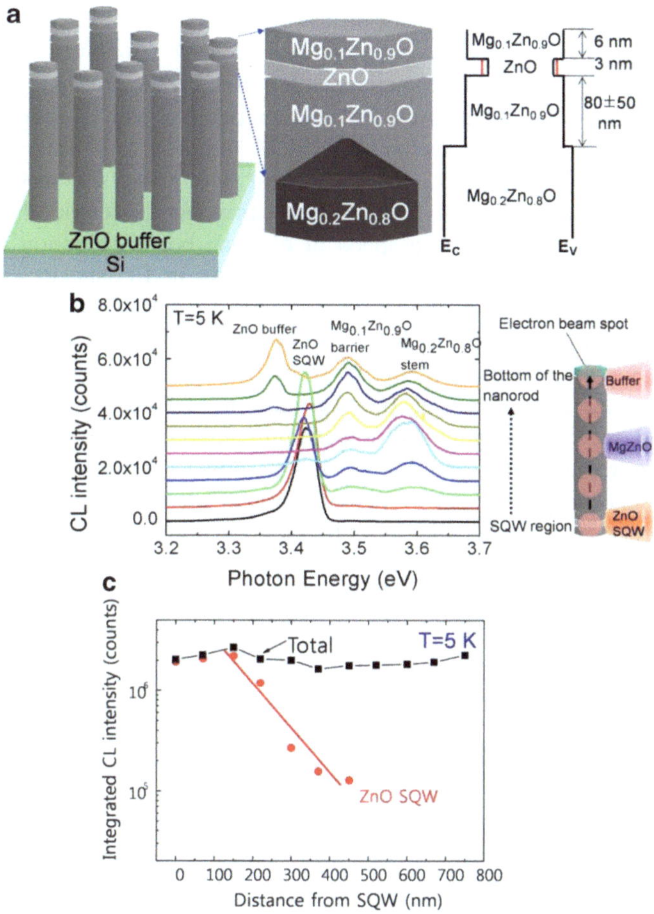

Fig. 8.19 (**a**) Schemes of $ZnO/Mg_{0.1}Zn_{0.9}O$ SQWs on the tops of $Mg_{0.2}Zn_{0.8}O$ nanorods and electronic band diagram in the nanorod quantum structure. (**b**) 5 K CL spectra of the ZnO/MgZnOnanorod SQW structure with different electron beam irradiation positions. (**c**) Integrated CL intensity of ZnO SQW emission as a function of electron beam position. (Reprinted with permission from [36]. Copyright 2008 by John Wiley and Sons Inc.)

Additionally, time-resolved luminescence spectroscopy cannot be used to measure electron and hole dynamics separately because the signal is proportional to the product of electron and hole distributions. Thus, luminescence spectroscopy is not useful to characterize minority carrier transport that is essential for the performances of photovoltaic cell and photodetector.

The great potential of luminescence characterization becomes more valuable when luminescence technique is used in conjunction with other methods such as structural characterizations and electrical characterizations.

Technically specific luminescence measurement has some limit in the measurable values. For instance, PL spectroscopy can provide high temporal resolution in the order of sub-nanosecond, but the spatial resolution of PL spectroscopy is diffraction-limited. Although NSOM is used for characterization with a spatial resolution in the order of a half of wavelength, NSOM requires very sensitive detection system because the signal in NSOM is too weak. Otherwise, CL spectroscopy can resolve a nanostructure with high spatial resolution, but the temporal resolution of CL spectroscopy is limited by a speed of electron beam deflection in the order of microseconds. The limit of temporal resolution in CL spectroscopy can be overcome by a combination of photocathode and ultrafast laser as an electron beam source. Using the electron beam source based on photocathode operated by ultrafast Ti:Sapphire laser, the temporal resolution of CL spectroscopy reached to picosecond range [38]. However, picosecond time-resolved CL spectroscopy has not been widely employed yet. Additionally, luminescence characterizations require excitation source, sample mounting system, and detection system. The capability of luminescence measurement system is generally limited by the performances of instruments as parts of a system. In the viewpoint of technical aspect, the direction of developing luminescence measurement system lies in improving detection efficiency and response speed to increase spatial and temporal resolution of luminescence measurements.

8.5 Summary

Luminescence characterizations have provided valuable information on physical properties of 1D semiconductor nanostructures using their nondestructive, nonintrusive, and sensitive characters. Especially spatially resolved and time-resolved luminescence measurements have revealed dynamics of carriers and excitons in 1D semiconductor nanostructures, which is essential to the fundamental understanding on criterion of photonic device operation. In this chapter, several examples of luminescence characterizations of 1D semiconductor nanostructures were presented. The examples showed that luminescence characterizations can be used for the investigation of effect of crystal structure, impurity incorporation, shape, and carrier (exciton) dynamics on physical properties. Although luminescence characterizations have some limitations in the criteria of measurable properties, they 0have been widely used for the characterization of 1D semiconductor nanostructures. Furthermore, the development of luminescence measurement with high spatiotemporal resolution will make it possible to study physical properties of 1D semiconductor nanostructures in the manner of improved accuracy and sensitivity.

Acknowledgements This work was supported by the Center for Integrated Nanotechnologies, a U.S. Department of Energy, Office of Basic Energy Sciences, user facility at Los Alamos National Laboratory (Contract DE-AC52–06NA25396).

References

1. W. Van Roosbroeck, W. Shockley, Phys. Rev. B **94**, 1558 (1954)
2. J.I. Pankove, *Optical Processes in Semiconductors* (Dover Publication Inc., New York, 1971)
3. http://www.gel.usherbrooke.ca/casino/download2.html
4. A. Gustafsson, M.-E. Pistol, L. Montelius, L. Samuelson, J. Appl. Phys. **84**, 1715 (1998)
5. T. Tsuruoka, S. Ushioda, J. Electron Microscopy **53**, 169 (2004)
6. M. Schirra, A. Reiser, G.M. Prinz, A. Ladenburger, K. Thonke, R. Sauer, J. Appl. Phys. **101**, 113509 (2007)
7. B.K. Meyer, H. Alves, D.M. Hofmann, W. Kriegseis, D. Forster, F. Bertram, J. Christen, A. Hoffmann, M. Straßburg, M. Dworzak, U. Haboeck, A.V. Rodina, Phys. Stat. Sol. (b) **241**, 231 (2004)
8. M.S. Gudiksen, J.F. Wang, C.M. Lieber, J. Phys. Chem. B **106**, 4036 (2002)
9. N.W. Aschcroft, N.D. Mermin, *Solid State Physics* (Harcourt College Publishers, Orlando, 1976)
10. W.I. Park, J. Yoo, G.-C. Yi, J. Korean Phys. Soc. **46**, L1067 (2005)
11. G. Zhang, K. Tateno, H. Sanada, T. Tawara, H. Gotoh, H. Nakano, Appl. Phys. Lett. **95**, 123104 (2009)
12. S.S. Hong, T. Joo, W.I. Park, Y.H. Jun, G.C. Yi, Appl. Phys. Lett. **83**, 4157 (2003)
13. L. Wischmeier, T. Voss, I. Rückmann, J. Gutowski, A.C. Mofor, A. Bakin, A. Waag, Phys. Rev. B **74**, 195333 (2006)
14. J. Grabowska, A. Meaney, K.K. Nanda, J.-P. Mosnier, M.O. Henry, J.-R. Duclére, E. McGlynn, Phys. Rev. B **71**, 115439 (2005)
15. R.A. Rosenberg, G.K. Shenoy, F. Heigl, S.-T. Lee, P.-S.G. Kim, X.-T. Zhou, T.K. Sham, Appl. Phys. Lett. **86**, 263115 (2005)
16. L.K. van Vugt, S.J. Veen, Erik P.A.M. Bakkers, J. Am. Chem. Soc. **127**, 12357 (2005)
17. O. Brandt, C. Pfüller, C. Chèze, L. Geelhaar, H. Riechert, Phys. Rev. B **81**, 045302 (2010)
18. W.I. Park, J. Yoo, D.-W. Kim, G.-C. Yi, M. Kim, J. Phys. Chem. B **110**, 1516 (2006)
19. J.P. Richters, T. Voss, D.S. Kim, R. Scholz, M. Zacharias, Nanotechnology **19**, 305202 (2008)
20. A. Bera, D. Basak, ACS Appl. Mater. Interfaces **1**, 2066 (2009)
21. D.E. Perea, J.E. Allen, S.J. May, B.W. Wessels, D.N. Seidman, L.J. Lauhon, Nano Lett. **6**, 181 (2006)
22. J.E. Allen, E.R. Hemesath, D.E. Perea, J.L. Lensch-Falk, Z.Y. Li, F. Yin, M.H. Gass, P. Weng, A.L. Bleloch, R.E. Palmer, L.J. Lauhon, Nat. Nanotechnol. **3**, 168 (2008)
23. J. Yoo, Y.-J. Hong, S.J. An, G.-C. Yi, B. Chon, T. Joo, J.-W. Kim, J.-S. Lee, Appl. Phys. Lett. **89**, 043124 (2006)
24. J.F. Wang, M.S. Gudiksen, X.F. Duan, Y. Cui, C.M. Lieber, Science **293**, 1455 (2001)
25. J.C. Johnson, H. Yan, P. Yang, R.J. Saykally, J. Phys. Chem. B **107**, 8816 (2003)
26. D. Kulik, H. Htoon, C.K. Shih, Y. Li, J. Appl. Phys. **95**, 1056 (2004)
27. D.D.D. Ma, S.T. Lee, J. Shinar, Appl. Phys. Lett. **87**, 033107 (2005)
28. J.B. Schlager, N.A. Sanford, K.A. Bertness, J.M. Barker, A. Roshko, P.T. Blanchard, Appl. Phys. Lett. **88**, 213106 (2006)
29. C.X. Shan, Z. Liu, S.K. Hark, Phys. Rev. B **74**, 153402 (2006)
30. A.V. Maslov, C.Z. Ning, Phys. Rev. B **72**, 161310R (2005)
31. H.Y. Chen, Y.C. Yang, H.W. Lin, S.C. Chang, S. Gwo, Opt. Express **16**, 13465 (2008)
32. M.P. Persson, H.Q. Xu, Phys. Rev. B **70**, 161310R (2004)
33. M. Mattila, T. Hakkarainen, M. Mulot, H. Lipsanen, Nanotechnology **17**, 1580 (2006)
34. K. Premasiri, M. Montazeri, R. Gass, L.M. Smith, H.E. Jackson, J. Yarrison-Rice, S. Paiman, Q. Gao, H. Hoe Tan, C. Jagadish, X. Zhang, J. Zou, Nano Lett. **9**, 648 (2009)
35. J. Bao, D.C. Bell, F. Capasso, J.B. Wagner, T. Mårtensson, J. Trägårdh, L. Samuelson, Nano Lett. **8**, 836 (2008)

36. J. Yoo, G.-C. Yi, L.S. Dang, Small **4**, 467 (2008)
37. J. Bolinsson, K. Mergenthaler, L. Samuelson, A. Gustafsson, J. Crys. Growth **315**, 138 (2011)
38. M. Merano, S. Sonderegger, A. Crottini, S. Collin, P. Renucci, E. Pelucchi, A. Malko, M.H. Baier, E. Kapon, B. Deveaud, J.D. Ganiere, Nature **438**, 479 (2005)

Chapter 9
Lasing Characteristics of Single and Assembled Nanowires

S.F. Yu

Abstract This chapter reviews the current achievement in realization of single-nanowire and assembled nanowire semiconductor lasers. Lasing conditions and emission mechanisms of single and assembled nanowires are also described. Furthermore, the main difficulties to obtain effective emission from nanowire semiconductor lasers under electrical excitation are discussed.

9.1 Introduction

In the early 1990s, nanowires were used as the optical gain of semiconductor lasers [1]. However, the difficulties to handle nanoscale materials limited the use of semiconductor nanowires as the optoelectronic devices. Recently, due to the successful fabrication of ZnO nanowires array on Si substrate [2], researchers have recognized the potential to realize semiconductor lasers using nanowires as the gain medium as well as the optical cavity. So far, more than few hundreds of journals have been published and the studies of nanowire lasers have no sign to be slowing down.

Nanowire lasers can be fabricated by organic and semiconductor materials. For example, photoluminescence and lasing have been demonstrated from individual and assembled organic nanowires [3–5]. In fact, semiconductor materials are the most preferred materials to grow nanowires for laser applications. This is because of their excellent optical and electrical properties that are suitable to realize electrically pumped nanolasers. ZnO is one of the most popular semiconductor materials to fabricate ultraviolet nanowire lasers [2, 6]. As ZnO material has extremely high

S.F. Yu (✉)
Department of Applied Physics, The Hong Kong Polytechnic University, Hung Hom, Kowloon, Hong Kong, P.R. China
e-mail: Siu.Fung.Yu@inet.polyu.edu.hk

G.-C. Yi (ed.), *Semiconductor Nanostructures for Optoelectronic Devices*, NanoScience and Technology, DOI 10.1007/978-3-642-22480-5_9,

excitonic gain to sustain stimulated emission at ultraviolet regime. In addition, ZnO nanostructures can be easily realized by different types of deposition methods such as vapor transport method, thermal evaporation technique, aqueous chemical method, and metal-organic chemical vapor deposition (MOCVD) with and without the use of seed layer [7]. Other wide band gap materials such as GaN [8] and ZnS [9] have also been used to fabricate nanowire lasers. Nevertheless, ZnO nanowires are more preferable for the fabrication of ultraviolet optoelectronic devices due to their ease to grow vertically aligned nanostructures on the surface of any substrates. In visible regime, CdS semiconductor may be the most appropriate material to realize nanowires as high-crystal-quality CdS exhibits high excitonic gain to support visible radiative recombination [10–12]. Furthermore, the ternary alloy CdS_xSe_{1-x} in the full composition range between CdS ($x = 1$, wavelength ~500 nm) and CdSe ($x = 0$, wavelength ~700 nm) can be obtained so that this material system is suitable for the realization of visible tunable lasers [13]. Narrow band gap GaAs- and InP-based III–V semiconductor materials can be used to fabricate infrared lasers. However, it is relatively difficult to realize nanowire lasers using these III–V material systems due to their high density of surface states [14, 15]. On the other hand, infrared nanowire lasers can be obtained from InN and GaSb [16, 17].

From the studies of nanowire lasers, lasing emission was usually collected from either individual nanowire or assembled nanowires deposited on substrates. Early studies of vertical aligned ZnO nanowire array on Si substrate have revealed that the corresponding lasing mechanism is related to the conventional Fabry–Perot resonant oscillation [2]. Similar conclusion has been given for the studies of CdS nanowire arrays and their alloy nanowire arrays [18, 19]. High-density nanowires array can also be formed inside anodic aluminum oxide (AAO) templates; however, their lasing characteristics have not been explained explicitly [20]. Dispersion of suspended nanowires in organic solvent onto substrate is another approach to study the lasing characteristics of single nanowires, but it is still arguable to explain that the corresponding lasing emission is from a single nanowire [16]. This is because many nanowires can be excited simultaneously from the substrate so that stimulated emission from individual nanowire is difficult to be identified. In fact, the observation of lasing emission from these assembled nanowires may be due to random lasing action. In this chapter, the method to recognize lasing mechanism from single and assembled nanowires lasers is discussed.

9.2 Lasing Characteristics of Single Nanowires

Single-nanowire laser exhibits lasing characteristics similar to that of a conventional Fabry–Perot cavity except the corresponding cavity length is in order of few micrometers. If the influence of surface recombination can be ignored in the nanowires, the coupling efficiency of spontaneous emission to the lasing modes, the facet reflectivity, as well as near- and far-field profiles of the nanowires are dependent on the diameter of the nanowires. Hence, it is necessary to study the lasing characteristics of nanowire lasers with the consideration of their cavity dimensions. In the following paragraphs, the critical conditions to realize stimulated emission from the nanowires are analyzed for different shape and size of the nanowires.

9.2.1 Feedback Mechanism of Single-Nanowire Lasers

Figure 9.1 shows the possible geometries of single-nanowire lasers that can be fabricated from semiconductor or organic materials. The nanowires have hexagonal, triangular, or rectangular cross-sectional areas. It is expected that only a particular set of guided modes is supported inside a nanowire with a particular geometry. Nevertheless, in order to simplify the analysis, it is possible to approximate the nanowires that have a cylindrical geometry. Hence, this structure of nanowires is used to explain the lasing characteristics of single-nanowire lasers in this section unless other specification.

There are two sets of orthogonal guided modes that can be supported inside the cylindrical nanowires. They are the transverse (along the x–y plane) and longitudinal (along the z-axis) modes. It must be noted that the transverse modes determine the effective refractive index of the nanowires as well as the near- and far-field profiles of the lasing beams. The longitudinal modes determine the lasing conditions of the nanowires. As the nanowire cavity is identical to a gain medium terminated by two reflectors, the threshold conditions for the nanowire can then be described by the balance between the round-trip gain and loss inside the cavity. Optical field propagating along the longitudinal direction is amplified and absorbed inside the nanowires due to its amplification and guiding effects. Furthermore, part of the light is reflected back into the cavity from the facets of the nanowire and the remaining light emits from the facets to the surrounding. Hence, the condition for a stationary laser oscillation can be written as:

$$r_f r_r \exp(g_{th} L' - \alpha_w L) \cdot \exp(-j 2 k_o n_{eff} L) = 1, \tag{9.1}$$

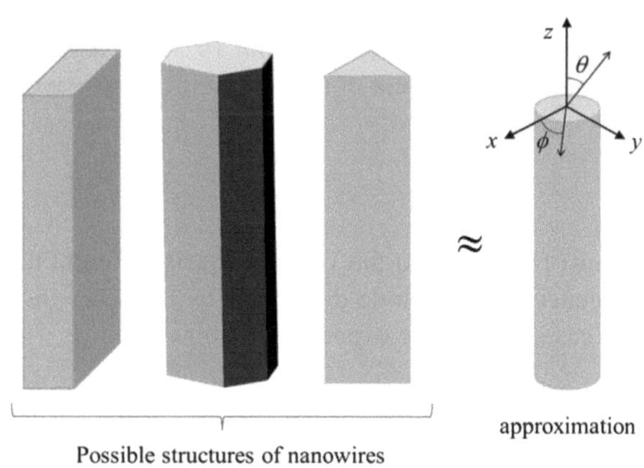

Fig. 9.1 Schematic structures of different types of nanowires

where plane wave approximation has been used for the guided modes in the nanowires. $j = \sqrt{-1}$, $k_o = 2\pi/\lambda_o$, λ_o is the resonant wavelength, n_{eff} is the dispersive effective refractive index, and r_f and r_r are the field reflectivities of the two end facets of the nanowire. α_w is the waveguide loss and L is the length of the nanowire. L' is defined as the effective traveling distance of the longitudinal modes. In principle, L' is not equal to L. This is because the guided modes can suffer from multiple reflections at the waveguide boundaries due to the large index difference between the nanowires and the surrounding so that $L' > L$ [21]. It can be shown from the real part of (1) that the threshold gain, g_{th}, of the nanowire laser can be expressed as

$$\Gamma g_{\mathrm{th}} = \alpha_w + \underbrace{L^{-1}\ln(1/r_f r_r)}_{\text{cavity loss}}, \tag{9.2}$$

where $\Gamma = L'/L$. It must be noted that due to the influence of size effect, the parameters r_f, r_r, and L' are function of the diameter of the nanowires. However, α_w is assumed to be independent on the size of the nanowires if surface recombination is ignored into the consideration. The second term on the right-hand side of (9.2) represents the cavity loss of the nanowire lasers. It can be shown that $\alpha_w <<$ cavity loss of an air-surrounded semiconductor nanowire laser with $L \sim 4\,\mu\mathrm{m}$. For example, if ZnO is the material of the nanowires, α_w should have a value less than $100\,\mathrm{cm}^{-1}$ [22]. The values of r_f and r_r are approximately equal to 0.3548 if the bulk refractive index of ZnO, which is roughly equal to 2.1, is used for the calculation. Using the above information, cavity loss is found to be about $5,000\,\mathrm{cm}^{-1}$ and this value can be higher if influence of size effect is taken into calculation of r_f and r_r. Hence, due to the short cavity length, high cavity loss is expected and it is required to have high-optical gain (i.e., in order of $10^2\,\mathrm{cm}^{-1}$) to support stimulated emission in nanowire lasers.

The round-trip phase condition of the longitudinal mode of wavelength, λ_o, can be obtained from the imaginary part of (9.1) and is given by

$$2k_o n_{\mathrm{eff}} L = 2\pi m, \tag{9.3}$$

where m is an integer. A simple expression for the mode spacing, $\Delta\lambda$, of the nanowires can be deduced from (9.3) by assuming that the phase of dielectric mirrors is independent of the resonant-cavity wavelength, which gives

$$\Delta\lambda \approx \frac{\lambda_o^2}{2n_{\mathrm{eff}}L}. \tag{9.4}$$

It must be noted that the value of n_{eff} is different to that of its bulk value due to the influence of waveguide dispersion as if occurs inside a single-mode fiber [23]. In this case, n_{eff} can be expressed as shown below

$$n_{\mathrm{eff}} = n_{\mathrm{bulk}} - \lambda_o \frac{\partial n_{\mathrm{bulk}}}{\partial \lambda_o}, \tag{9.5}$$

where n_{bulk} is the bulk refractive index, and $\partial n_{\mathrm{bulk}}/\partial\lambda_o$ is the material dispersion. In conventional Fabry–Perot lasers, $\partial n_{\mathrm{bulk}}/\partial\lambda_o$ is usually ignored for the calculation of $\Delta\lambda$. However, for the study of nanowire lasers, a negative value of $\partial n_{\mathrm{bulk}}/\partial\lambda_o$ in order of $10^{-2}\,\mathrm{nm}^{-1}$ is applied to explain the observed mode spacing from the emission spectra. Hence, $\Delta\lambda$ of nanowire lasers can be two to three times shorter than that of a conventional Fabry–Perot laser with the same value of L and gain material.

9.2.2 *Modal Characteristics of Nanowires with Different Geometries*

Modal characteristics of nanowires can be analyzed by cylindrical waveguide theory [24]. This is because the cylindrical nanowires have transverse confinement structure similar to that of the optical fibers. Hence, the modal characteristics of nanowires surrounded by air can be described by a full set of optical modes including transverse electric (TE), transverse magnetic (TM), and their hybrid modes (i.e., HE and EH modes) obtained from a cylindrical dielectric waveguide. Using this approach [25], it can be shown that the nanowires can support $\mathrm{TE_{0m}}$, $\mathrm{TM_{0m}}$, which have only three field components and no dependence on ϕ, as well as $\mathrm{HE_{nm}}$ and $\mathrm{EH_{nm}}$, which have all six field components with $\cos(n\phi)$ and $\sin(n\phi)$ dependence; see also Fig. 9.1. The index m denotes the radial dependence of the field, and n denotes angular symmetry. All modes except $\mathrm{HE_{11}}$ have low-frequency cutoffs. The cutoff frequency is [25]

$$\omega_c R/c = \begin{cases} x_{11}/\sqrt{\varepsilon-1} = 1.71 & \mathrm{HE_{12}\ and\ EH_{11}} \\ x_{01}/\sqrt{\varepsilon-1} = 1.08 \quad \text{for} & \mathrm{TE_{01}\ and\ TM_{01}}, \\ x_{02}/\sqrt{\varepsilon-1} = 2.47 & \mathrm{TE_{02}\ and\ TM_{02}} \end{cases} \tag{9.6}$$

where $\varepsilon(=6)$ is the dielectric constant of the nanowires, R is the radius of the nanowires, ω_c is the cutoff frequency, and c is the velocity of light. x_{kl} is the lth zero of the Bessel function $J_k(x)$, $k = 0, 1$. Figure 9.2 shows the dependence of the wave vector k_z along the propagation direction on the frequency ω for seven guided modes.

The penetration depth of the guided modes into the surrounding air can be understood from the transverse propagation coefficient of the guided modes, β, which can be written as

$$\beta = \sqrt{k_z^2 - \omega^2/c^2}, \tag{9.7}$$

where the penetration depth is inversely proportional to β. Figure 9.3 shows the dependencies of β on the frequency ω. As β is a real number for all the guided modes, their transverse components are decayed with the distance from the surface of the nanowire. However, $\mathrm{HE_{11}}$ mode does not have a cutoff, at low frequencies

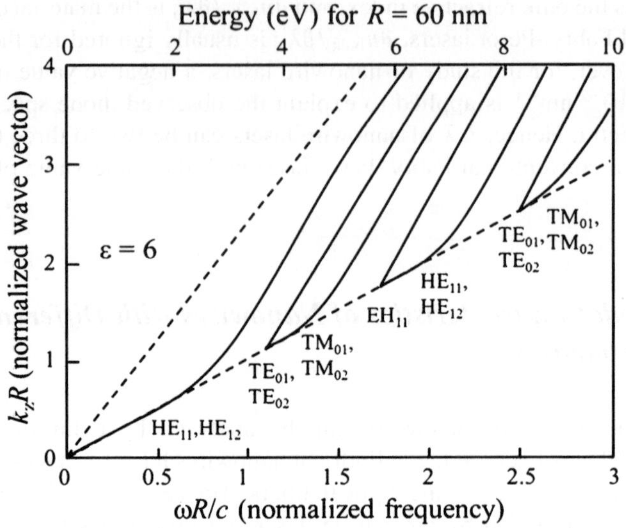

Fig. 9.2 Dispersion curves for seven guided modes of a nanowire with ($\varepsilon = 6$ surrounded by air. The left and bottom axes show the normalized propagation wave vector (k_z) and frequency (ω), respectively. The top axis shows the energy ($\bar{h}\omega$) for a fixed nanowire radius $R = 60\,\text{nm}$. The asymptotic lines are $k_z = \omega/c$ and $k_z = \sqrt{\varepsilon}\omega/c$. (Adapted with permission from [25]. ©2003, American Institute of Physics)

($\omega R/c \leq 0.7$) the field extends for a considerable distance beyond the surface, making the mode poorly confined.

As the diameter of nanowires approaches the value of lasing wavelength, it is expected that the field reflectivity, r_f and r_r, varies with diameter of the nanowires. Field reflectivity, r, can be approximated by mode matching the free-space radiation mode with a single incident and reflected guided modes at the interface of the end facet of the nanowires [26]. It can be shown that r can be expressed as

$$r \approx \int_0^\infty \frac{k_\text{I} - \xi}{k_\text{I} + \xi} \Phi_\text{I} \Phi_\text{I}^* \text{d}u \Bigg/ \int_0^\infty \Phi_\text{I} \Phi_\text{I}^* \text{d}u \,, \tag{9.8}$$

where $\xi(u) = \sqrt{k_0^2 - u^2}$, k_I is the propagation coefficient of the incident light, Φ_I is the first-order Hankel transform of the incident field, and Φ_I^* is the complex conjugate of Φ_I. As we can see, (9.8) is equivalent to the Fresnel reflectivity formula

$$r = \frac{\eta_\text{wire} - \eta_\text{air}}{\eta_\text{wire} + \eta_\text{air}}, \tag{9.9}$$

where η_wire and η_air are the wave impedance of the nanowires and air, respectively. As we expected from (9.9) that if the diameter of nanowires is reduced, its guided

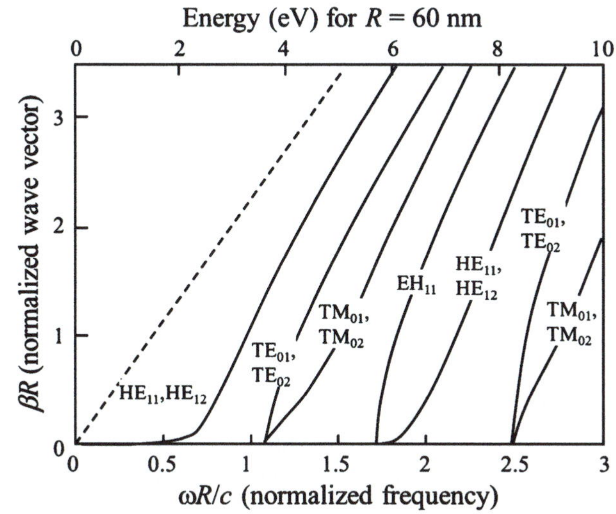

Fig. 9.3 Plot of β versus $\omega R/c$ for the modes shown in Fig. 9.2. The asymptotic line is $\beta R = (\omega R/c)\sqrt{\varepsilon - 1}$. (Adapted with permission from [25]. ©2003, American Institute of Physics)

modes will be penetrated into the surrounding air so that the corresponding η_{wire} will also be reduced. Hence, it is expected that the value of r in general reduces with the reduction of R. On the other hand, r approaches to its bulk value for a large value of R. It must be noted that (9.8) is just an approximation to the field reflectivity of nanowires; more accurate investigation requires solving Maxwell's equations using the finite-difference time-domain (FDTD) method. Figure 9.4 plots the absolute value of r versus $\omega R/c$ for the first three guided modes obtained from FDTD calculation [25]. It is observed that the value of $|r|$ increases with frequency and mode confinement for $R = 60$ nm. This is expected as the increase of $\omega R/c$ increases the confinement of guided modes so that the value of $|r|$ is also increased and a large value of $|r|$ can be reached. It is noted that $|r|$ of HE_{11} approaches its bulk value of ~0.42(if $\varepsilon = 6$ is used in the estimation) for large value of $\omega R/c$. However, the values of $|r|$ for TE_{01} and TM_{01} can be much larger than ~0.42 for large value of $\omega R/c$ due to the increase of L' of the guided modes. Nevertheless, $|r|$ is always less than unity in the frequency range of interest because of the presence of diffraction losses.

The use of cylindrical geometry to analyze and design the lasing performance of ZnS and GaN nanowire lasers may not be appropriated as ZnS and GaN have rectangular and triangular cross-sectional areas, respectively. In these cases, the polarization of the guided modes should be carefully taken into consideration. In fact, even in ZnO and CdS nanowires with hexagonal geometry, profile of guided modes may be different to that obtained from the study using cylindrical geometry [27]. Figure 9.5 plots the field distribution of the first few guided modes

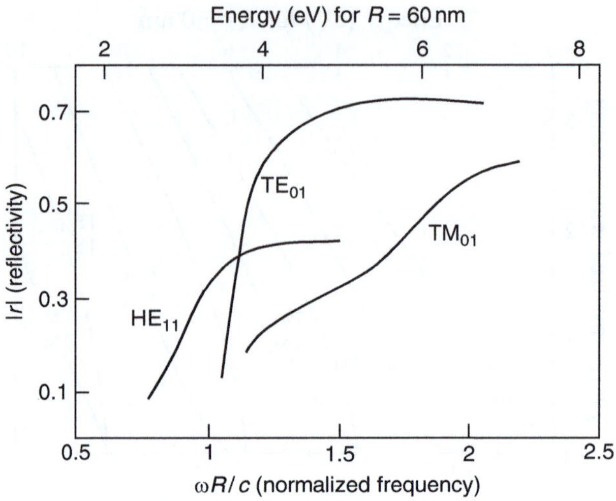

Fig. 9.4 Absolute value of the field reflection coefficient, r ($= r_f = r_r$) for the first three guided modes from the open end (top facet) of the nanowire. (Adapted with permission from [25]. ©2003, American Institute of Physics)

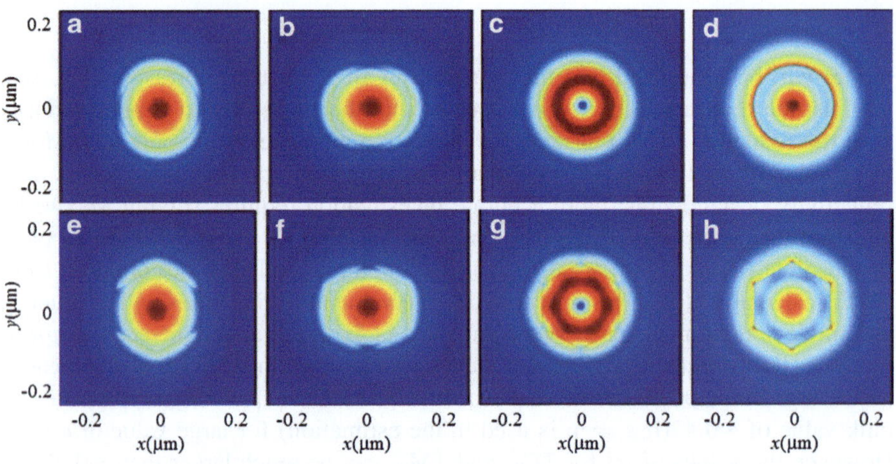

Fig. 9.5 Electric field intensity distribution of the first four guided modes of a circular nanowire with a radius $R = 100$ nm [(**a**) EH_{11}, (**b**) HE_{11}, (**c**) TE_{01}, and (**d**) TM_{01}] and a hexagonal nanowire with compatible diameter [(**e**), (**f**), (**g**), and (**h**), respectively, same names] at wavelength. (Adapted with permission from [27]. ©2009, Optical Society of America)

of circular and hexagonal nanowires. It is observed that TE_{01} and TM_{01} modes for both geometries exhibit no azimuthal field dependency and have only three field components (they are nondegenerated modes). The main differentce of modal profiles between circular and hexagonal nanowires is the HE_{11} mode. In circular nanowires, HE_{11} mode has a twofold polarization dependency. On the other hand, the hexagonal geometry of the nanowires removes this degeneracy due to the symmetry reduction of the hexagonal structure. As a result, HE_{11} polarized along the x-axis is no longer symmetric to that polarized along the y-axis. Furthermore, if we compare the field profiles in Fig. 9.5, although the TE_{01} and TM_{01} modes have the same symmetry, the modal profiles observed in hexagonal and circular nanowires are different.

The modal characteristics of nanowires also affected by the surrounding refractive index. If a ZnO nanowires of diameter equal to 170 nm is laid on a silica substrate, the polarization characteristics of the nanowires can be significantly modified. Figure 9.6 plots the profile of the two fundamental hybrid modes (EH_{11} and HE_{11}) under the influence of silica substrate [28]. The presence of the silica substrate enhances the transverse linearly polarized mode profiles. This indicated that the modal profiles of nanowires laid down on the substrate are quite different from that of the free-standing substrate.

For nanowire lasers with belt geometry, it is expected that their luminescence should be strongly linearly polarized due to the influence of high-index contrast of the dielectric rectangular waveguide [29]. Figure 9.7 shows the electric field distribution of the y and z components (i.e., E_y and E_z) in a nanobelt obtained from the FDTD method. It was assumed that the nanobelt, which was fabricated by ZnO dielectric material, has width and thickness of 150 and 40 nm, respectively. It is observed that E_y is antisymmetric only along the z-axis and E_z is antisymmetric along both y and z axes. Thus, spatial integration of E_y (E_z) will not (will) be zero, resulting in the emitted light being linearly polarized along the y-direction.

Figure 9.8 shows the modal profile of an InGaN/GaN MQW shell/GaN core nanowire [30]. It is noted that GaN-based nanowires have a triangular crosssectional area. The nanowire presented in Fig. 9.8 has an active layer surrounding the GaN core – the InGaN/GaN MQW shell provides optical gain to sustain lasing from the nanowire. The optical mode exhibits a donut shape due to the fact that high-optical

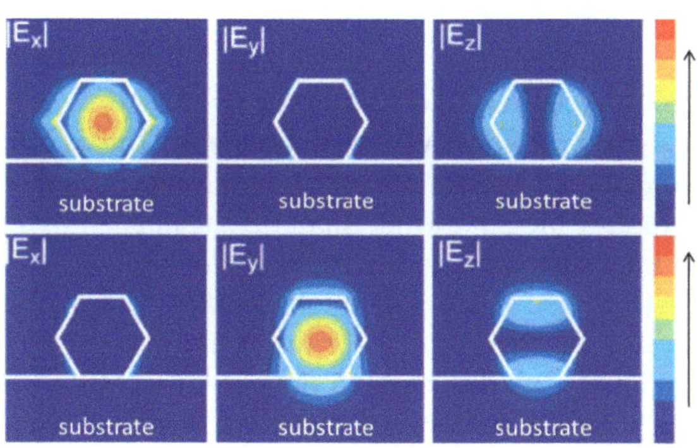

Fig. 9.6 Contour plots of normalized $|E_x|$ (E-field along the substrate), $|E_y|$ (perpendicular to the substrate), and $|E_z|$ (along the wire) for the two fundamental hybrid modes at energy equal to 3.0 eV. (Adapted with permission from [28]. ©2009, American Chemical Society)

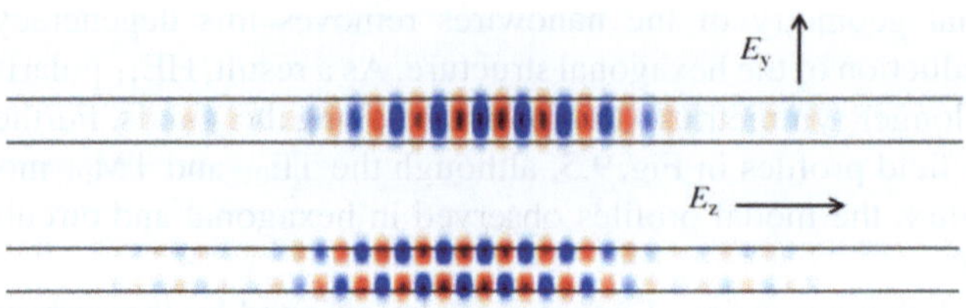

Fig. 9.7 Simulated snapshot of the *y* and *z* components of the electric field, *E*, in the nanobelt. (Adapted with permission from [29]. 2009, American Institute of Physics)

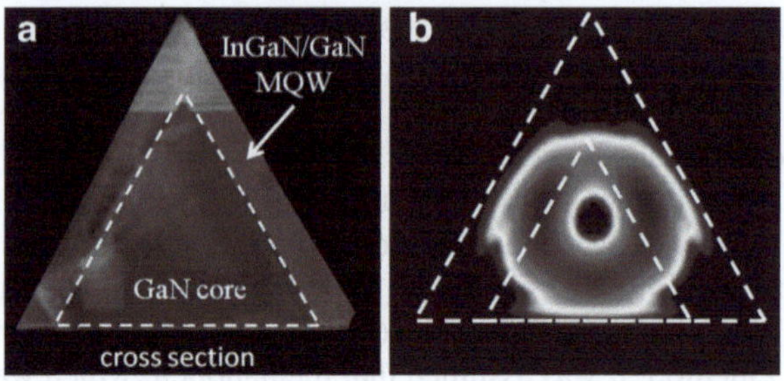

Fig. 9.8 (**a**) Scan electron microscope (SEM) image of the cross section of an InGaN/GaN MQW shell/GaN core nanowire. (**b**) FDTD simulation of the dominant laser mode of the nanowire structure. *White dashed lines* indicate the nanowire profile and the core/shell interface. (Adapted by permission from Macmillan Publishers Ltd, Nature Materials, [30]. ©2008)

gain can only be achieved inside the shell region. The formation of a hole at the center of the optical mode is to reduce the modal loss of the GaN core nanowire.

9.2.3 Near-and Far-Field Profiles

Emission profile observed from the end facets of nanowires is dependent on the corresponding near-field profile as well as their length. This is because the far-field profile is obtained from the Fourier transform of their near-field observed from the end facets. Therefore, if the nanowires support fundamental hybrid modes, which have a maximum intensity at the center of the nanowire (i.e., see Fig. 9.5e, f), the corresponding far-field also has a single-lobe profile with a peak intensity at an angle, θ, equal to 0° (see also Fig. 9.1 for the definition of θ). However, far-field profile of TE_{01} and TM_{01} modes shows a double-lobe profile with zero emission intensity at $\theta = 0^o$ due to the fact that the corresponding near-field profiles are symmetry at $\theta = 0^o$ [31]. On the other hand, short cavity length of the nanowires allows interference pattern to occur due to the coherent influence of lasing emission from the end facets of the nanowires [32]. Hence, the emission profile of the nanowires is dependent on its length.

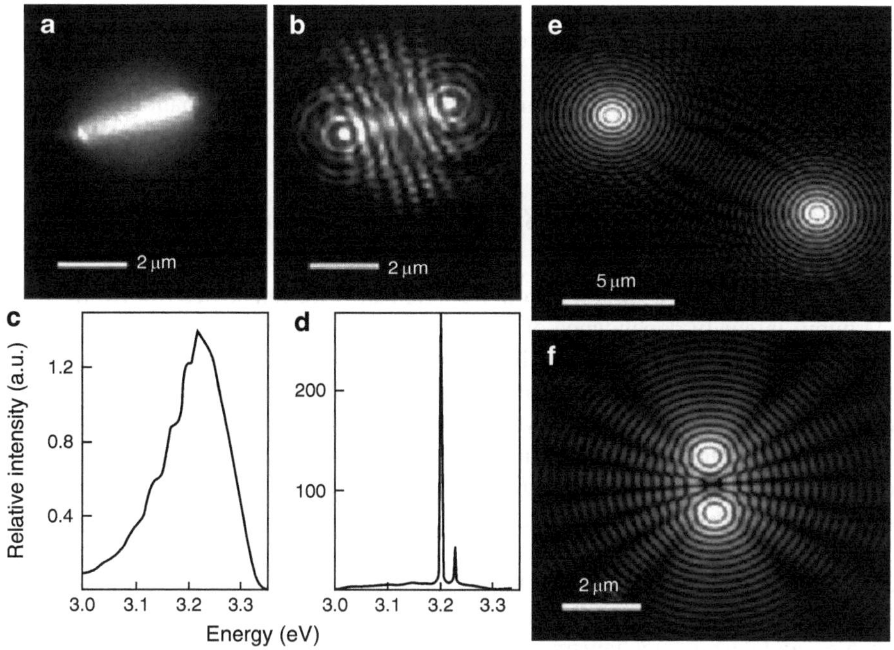

Fig. 9.9 CCD images of ZnO nanowire with different excitation intensity: (**a**) below and (**b**) well above threshold. Emission spectra taken at the excitation intensities corresponding to the images given in (**a**) and (**b**). Computer-simulated diffraction pattern for the long (**e**) and short (**f**) nanowire using spherical emission from the wire end facets. (Adapted with permission from [32]. ©2006, American Chemical Society)

Figure 9.9a, b shows the charged-coupled device (CCD) images of a ZnO nanowire (3.2 μm long and 60 nm wide) operating below and well above threshold, respectively. The accompanying emission spectra are shown in Fig. 9.9c, d. The clearly distinguishable peaks shown in Fig. 9.9d indicated that the emitted spectrum consists mainly of laser emission. The interference pattern observed in Fig. 9.9b, which depends on the emission wavelength and length of nanowires, is the indication of coherent radiation. The reason to see the interference pattern is due to the short cavity length of the nanowire so that the end facets appear to be a double-slit light source. The CCD images were also reproduced by computer simulation for two lengths (i.e., 10.2 and 2.1 μm) of nanowires with nondirection (i.e., spherical) emission from the end facets of the nanowires; see Fig. 9.9e, f. For the long nanowire, one can see that the ring pattern around the nanowire ends is nicely reproduced in the observation in Fig. 9.9b. In the calculation, incoherent luminescence was not plotted in the images. Similar interference pattern is observed from the short nanowire. This indicates that the interference patterns are originated from a superposition of emitted light from both end facets with a fixed phase difference. Full destructive interference can only be observed if the phase difference between the two emission sources is zero or an integer multiple of π. The measured interference images thus provide direct experimental evidence for a zero or fixed phase shift between the emission sources. Hence, the observation of interference pattern can be used to verify the support of lasing emission from nanowires.

9.2.4 Criteria to Achieve Stimulated Emission

Verification of lasing oscillation from a single nanowire can be determined from the (1) pump intensity dependence of the total output power – output power exhibits a superlinear increase with pump intensity above threshold, (2) excitation of sharp lasing peaks observed from the emission spectra above threshold, (3) observation of the linear dependence of $\Delta\lambda$ versus $1/L$, and (4) diffraction patterns showing coherence interference of light from the end facets of the nanowires. If these criteria can be observed, unambiguous evidence of lasing action from the nanowires can consider to be obtained. It must be noted that the far-field profile of the nanowires can only be used to identify the corresponding modal characteristics. For example, a single-lobe (double-lobe) far-field profile represents the support of fundamental hybrid (TE_{01} or TM_{01}) modes. There is no direct relationship between far-field profiles and lasing mechanism of the nanowires except that the cavity loss of the nanowires is dependent on the far-field profile of the guided modes. This is because the lasing mechanism of nanowires is equivalent to a Fabry–Perot laser; the lasing mechanism is determined by round-trip conditions of the longitudinal modes. Hence, the lasing threshold of nanowires is mostly dependent on their cavity length. We have shown that for nanowires with $L \sim 4\,\mu\text{m}$ and $r \sim 0.367$ (i.e., bulk mirror of air–ZnO interface), the corresponding cavity loss can be as large as $5,000\,\text{cm}^{-1}$. For most of the semiconductor materials, optical gain of larger than $5,000\,\text{cm}^{-1}$ is barely achieved. Therefore, if the value of L is shorter than $4\,\mu\text{m}$, the observed lasing mechanism from the nanowires should not be due to the Fabry–Perot resonant oscillation [33]. Furthermore, if the diameter of the nanowires is too small (i.e., small than $\lambda_o/2n_{\text{bulk}}$), the reflectivity of the end facets can be significantly reduced. This is due to the influence of the diffraction loss so that the resultant lasing threshold of the nanowires can be significantly increased. In this case, it may require long cavity length in order to support stimulated emission from the nanowires.

Recent study of ZnO nanowire lasers has indicated that there is a linear relationship between the diameter and length of the nanowires that can support lasing emission [34]. Figure 9.10 shows the mapping between the dimensions of nanowires and its emission characteristics under strong optical excitation. We can predict from the map that there should not have conventional Fabry–Perot resonant oscillation from ZnO nanowires with cavity length less than $3\,\mu\text{m}$. In addition, the shortest length of the ZnO nanowires that can support longitudinal resonance is limited by the diameter of the nanowires. Due to the influence of diffraction loss, cavity loss of nanowire lasers increases with the reduction of their diameter.

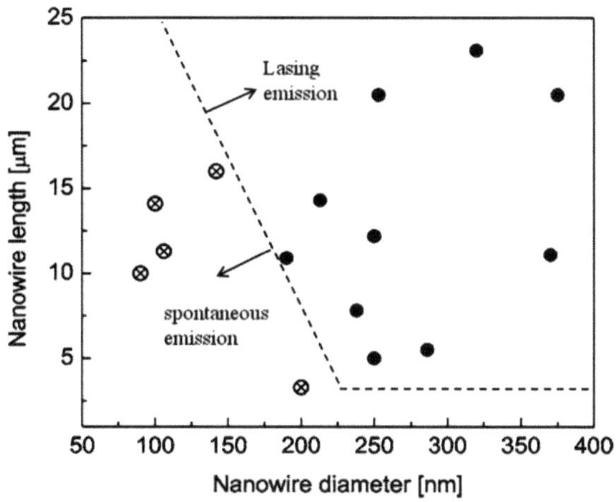

Fig. 9.10 Experimental results on lasing for nanowires of different diameter. *Filled circle* and *circled times* indicate that the ZnO nanowires lase and not lase respectively under optical excitation. (Adapted with permission from [34]. ©2008, American Institute of Physics)

Therefore, it is observed that even for long cavity ZnO nanowire, lasing may not be supported if its diameter is smaller than 100 nm. It must be noted that the value of $\lambda_o/2n_{bulk}$ inside the ZnO material is about 95 nm.

9.3 Lasing Characteristics of Assembled Nanowires

Assembled nanowires can grow vertically or horizontally onto the substrate by using appropriate seed layers. Lasing emission of these assembled nanowires has been reported under either optical [35] or electrical excitation [36]. However, lasing mechanism of the assembled nanowires is different from that of the conventional Fabry–Perot lasers. This is because assembled nanowires also support "random laser action" – no defined facets are required to sustain optical feedback. In addition, lasing characteristics of assembled nanowires are dependent on the surrounding refractive index, dimensions, and density of nanowires. Therefore, it is necessary to study the modal characteristics of the assembled nanowires. In the following paragraphs, the critical conditions to realize stimulated emission from different types of assembled nanowires as well as the corresponding lasing characteristics are investigated.

9.3.1 What is a Random Laser?

Random laser is a lasing device supported by strongly scattering media with optical gain. Figure 9.11 compares the lasing mechanism between conventional

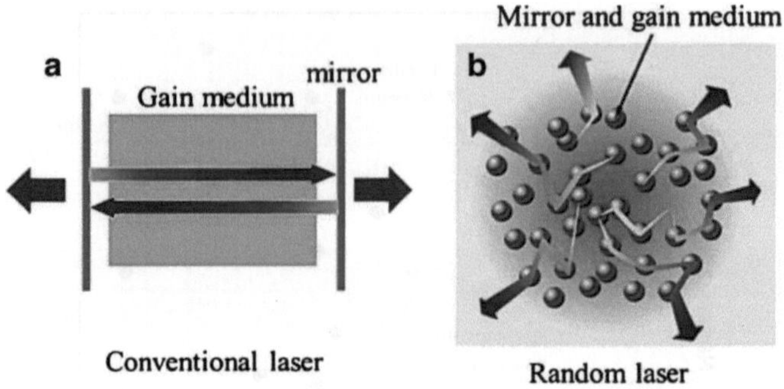

Fig. 9.11 Lasing mechanism of (**a**) conventional Fabry–Perot and (**b**) random laser

Fabry–Perot and random laser. Random lasing action is the amplification of multiple scattering photons inside a mirrorless disordered medium. Studies have shown that random lasing action can be obtained from scatterers embedded inside gain media where scattering and amplification processes are operating in separate phases. Alternative, scatterers can provide strong scattering and optical gain simultaneously to support random lasing action. The discovery of random lasers, which was originated from the work done by Letokhov [37], has led to much controversy about the physics behind the formation of laser modes. This is because although random lasers have some features in common with the conventional lasers, it is difficult to model their lasing characteristics comprehensively.

9.3.2 Feedback Mechanism of Random Lasers

Researchers have revealed that either incoherent or coherent random lasing can be excited from random media dependent on their scattering strength and free mean path of light [38]. For light suffers from multiple scattering inside a random medium, its final direction can be lost. As a result, light will not return to its original position after one round trip so that its phase information will also be missing. This type of optical feedback merely returns part of the energy to the gain medium so that it is referred to incoherent feedback; see also Fig. 9.12a. The other feature of incoherent feedback is that the emission spectrum tends to be continuous – discrete sharp peaks will not be observed from the emission spectrum. Only narrowing of emission spectrum toward the center of the amplification peak (i.e., at gain peak wavelength) is observed. Hence, the mean frequency of the lasing emission under incoherent feedback is dependent mainly on the gain peak wavelength.

Figure 9.13 shows the lasing spectra of ZnO particles, which have a mean diameter of 100 nm and density of $\sim 2.5 \times 10^{11}\,\text{cm}^{-3}$, immersed in Rhodamine 60 dye solution of concentration of 5×10^{-3} M [39]. In this case, ZnO particles (dye solution) works as scatterers (gain provider) in the random medium. When pump

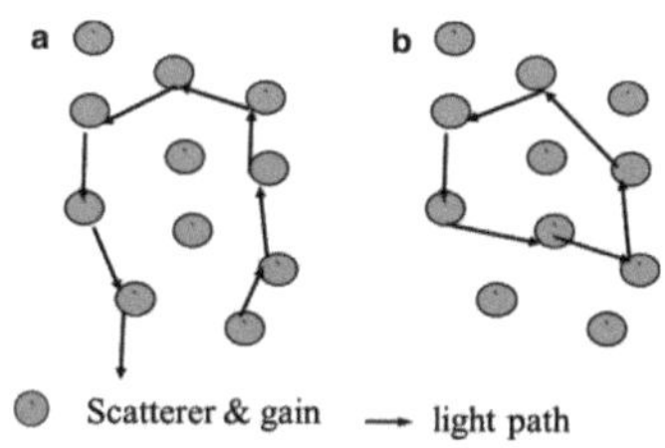

Fig. 9.12 (**a**) Incoherent and (**b**) coherent optical feedback

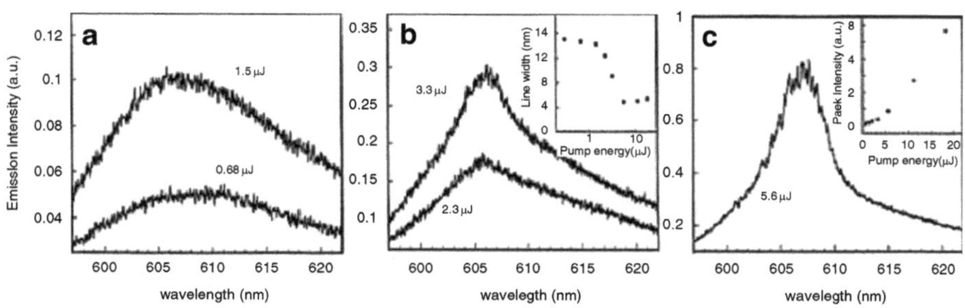

Fig. 9.13 Emission spectra when the incident pump-pulse energy is (**a**) 0.68 and 1.5, (**b**) 2.3 and 3.3, and (**c**) 5.6 μJ. The ZnO particle density is $\sim 2.5 \times 10^{11}\,\mathrm{cm}^{-3}$. The *inset* in (**b**) is the emission linewidth versus the pump-pulse energy. The *inset* in (**c**) is the emission intensity at the peak wavelength versus the pump-pulse energy. (Adapted with permission from [39]. ©2003, IEEE)

intensity reaches a threshold, P_{th}, the peak emission intensity increases linearly with respect to the pump intensity. In addition, a drastic spectral narrowing occurs at the gain peak wavelength. This process is similar to amplified spontaneous emission (ASE); however, typical ASE exhibits a gradual spectral narrowing with the excitation intensity. Observation of a sudden narrowing of spectral emission indicates that the saturation of optical gain occurs above P_{th} – the rate of photon generation by stimulated emission exceeds the photon loss rate. As linewidth narrowing comes with gain saturation, this process of light amplification by simulated emission is referred to incoherent random lasing action.

The main difference between coherent and incoherent feedback is that the coherent optical feedback preserves the phase information of the scattered light. Hence, after multiple scattering inside a random medium, the scattered light returns to its original position to form a closed-loop path of light. Furthermore, if the optical gain along the closed-loop path can overcome the loss, lasing oscillation occurs in a loop. In this case, coherent feedback is similar to a Fabry–Perot resonator that the reflected light has a multiple of π phase-shift delay. The lasing frequency of this type of random laser is determined by the phase-shift delay as well as the gain peak wavelength of the random medium. Such a laser is called coherent random laser. Figure 9.12b illustrates the explanation of coherent feedback in a random medium.

Coherent random lasing can be obtained from the same dye concentration mentioned in Fig. 9.13 with higher density of ZnO particle (i.e., $1 \times 10^{12}\,\mathrm{cm}^{-3}$)

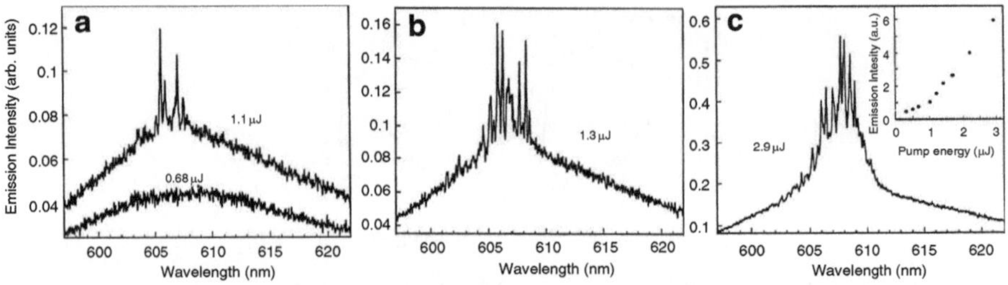

Fig. 9.14 Emission spectra when the incident pump-pulse energy is (**a**) 0.68 and 1.1, (**b**) 1.3, and (**c**) 2.9 μJ. The ZnO particle density is $\sim 1 \times 10\,\text{cm}^{-3}$. The *inset* in (**c**) shows the emission intensity versus the pump-pulse energy. (Adapted with permission from [39]. ©2003, IEEE)

[39]. Figure 9.14 shows the variation of the emission spectra with pump intensity. It is observed that sharp peaks are emerged from the emission spectra for the pump intensity excess P_{th}. Linewidth of these shape peaks is less than 0.2 nm, which is more than 50 times smaller than that of the ASE below threshold. When the pump intensity increases further, more sharp peaks appear. These discrete peaks represent the resonant modes of the random cavities formed by recurrent light scattering – closed-loop path of light.

Inside highly disordered gain media, lasing mechanism of random lasing action may be due to neither incoherent nor coherent optical feedback. In fact, localized and extended random modes, which can be excited simultaneously for the support of coherent random lasing, are used to explain the modal characteristics of highly disordered random lasers [40]. However, this type of random lasing action, which is seldom observed in assembled nanowires, is discussed in this chapter.

9.3.3 Formation of Random Cavities Using Assembled Nanowires

Recently, random lasing action has been observed from close-packed parasexiphenyl (p-6P) nanofibers deposited on mica substrates [41]. It is believed that the observed lasing action is taking place in closed-loop optical cavities realized within the tight fiber crossconnect. Furthermore, ordered ZnO nanowire arrays in AAO templates have shown coherent random lasing [20]. The ZnO nanowire arrays are densely packed ($\sim 10^{10}$ to $10^{11}\,\text{cm}^{-2}$) with a high aspect ratio up to $\sim 5 \times 10^6$ and homogeneous over a large area ($\sim 20\,\text{mm}^2$). On the other hand, randomly assembled ZnO [42] and SnO_2 [43] nanowires deposited on quartz substrate can also achieve random laser action with coherent feedback. P_{th} of less than few hundreds of kWcm^{-2} can be used to overcome the scattering loss of the random cavities. This implies that the formation of random cavities using assembled nanowires can be an effective approach to fabricate lasing devices using nanowires. The value of P_{th} can also be closed to that of the conventional Fabry–Perot single nanowire laser with similar dimensions.

On the other hand, it is reported that only ASE is observed from low-density vertically aligned ZnO nanowires [35]. This is because the gain length (scattering mean free path) is too short (long) to amplify the growth of lasing modes (to return to its original position) for the completion of closed-loop optical path. In order to modify the low-density vertically aligned ZnO nanowires to satisfy the requirement of random laser action, a low-loss slab waveguide, which allows strong transverse confinement of light inside the ZnO thin film, along the surface parallel to the sapphire substrate growth with ZnO nanowires array is proposed.

Figure 9.15a shows the schematic of the ZnO nanowires array embedded in ZnO epilayers. As we can see, MgO buffer layer of thickness ~700 nm is deposited onto 1 μm long ZnO nanowires array and followed by a ZnO film of thickness ~200 nm. The purposes of forming this slab waveguide are (1) to provide an extra gain length and (2) to reduce the scattering mean-free path (i.e., equivalent to minimize the average separation between the ZnO nanowires that occurs when light only travels in the direction parallel to the substrate). It must be noted that the refractive indices of MgO and ZnO are roughly equal to 1.76 and 2.1, respectively, at ~390 nm and the effective refractive index of the slab waveguide can be estimated to be 1.99. Hence, light scattering formed by the ZnO nanowires is maintained, which is due to the discontinuity in refractive index between the boundaries of the slab waveguide and nanowires.

Figure 9.15b shows the light–light curve and emission spectra of the modified nanowires array under optical excitation at room temperature. It is observed that TE and TM polarizations are supported inside the nanowires array. This indicates that the scattered lights are optically guided inside the slab waveguide. Furthermore, single-broad emission spectra with an FWHM of about 15 nm were observed from the emission spectra. When the excitations exceed P_{th}, sharp lasing peaks of a linewidth less than 0.4 nm are emerged from the single-broad emission spectra. A further increase in pump intensity increases the number of lasing modes as the increase in optical gain excites more cavity modes with higher losses.

The generation of single-broad emission spectra above threshold is due to the weak optical feedback from the facets of the slab waveguide. The excitation of shape peaks is in fact caused by the formation of closed-loop paths for light through recurrent scattering (i.e., coherent feedback); see the inset of Fig. 9.15b. It can be shown that the average separation between nanowires is less than 90 nm and the field reflectivity between the nanowires and slab waveguide is about 0.03. Hence, the estimated scattering mean-free path is less than the emission wavelength so that the claim of recurrent scattering is justified.

Coherent random lasing is also observed from randomly assembled nanowires [42, 43]. Figure 9.16 shows physical features and SEM images of randomly assembled SnO_2 nanowires. The SnO_2 nanowires, which have average length and width of ~10 μm and ~100 nm, respectively, are closely packed together to form randomly assembled nanowires. The randomly assembled SnO_2 nanowires are fabricated by using a vapor transport method. Figure 9.17 shows the emission spectra and light–light curve of the randomly assembled SnO_2 nanowires. It is observed that when excitation power exceeds P_{th}, sharp peaks at around 387 nm with linewidth less

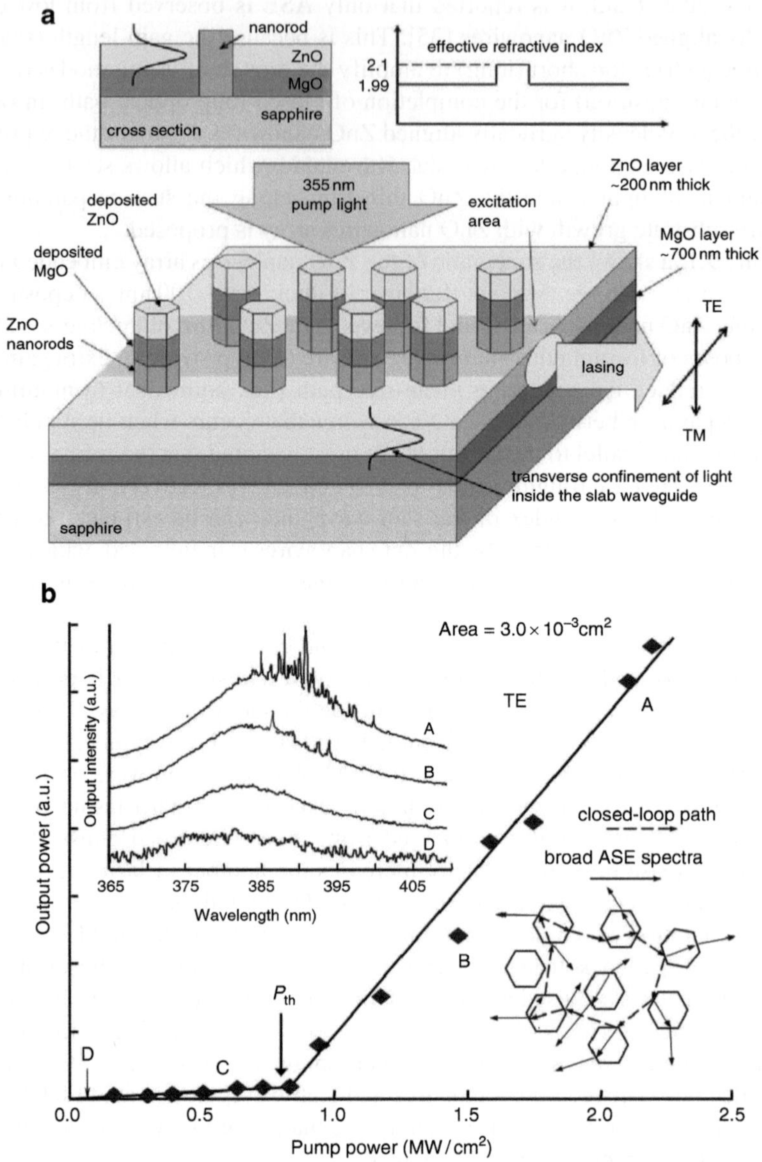

Fig. 9.15 (**a**) Schematic diagram of the ZnO nanowires array embedded in the ZnO epilayers. Bottom half of the diagram shows the arrangement of optical excitation of the sample. Top half of the diagram shows the cross section of the sample and the corresponding effective refractive index of the slab waveguide and ZnO nanowires. (**b**) Light–light curves and emission spectra of TE polarization form the ZnO nanowires array embedded in ZnO epilayers. The *inset* on the top-left-hand corner is the emission spectra at various pump intensities and that on the bottom-right-hand corner is an illustration showing the formation of closed-loop path for light through recurrent scattering (*dashed arrow*) and single-broad ASE spectra (*solid arrow*) in the sample. (Adapted with permission from [35]. ©2004, American Institute of Physics)

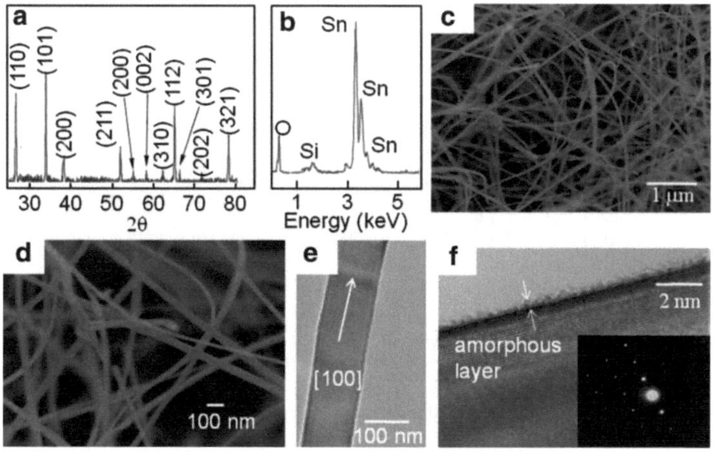

Fig. 9.16 (**a**) XRD pattern, (**b**) EDS spectrum, (**c**) low-magnification SEM image, and [(**d**) and (**e**)] high-magnification SEM image of the randomly assembled SnO_2 nanowires. (**f**) HRTEM of a SnO_2 nanowire and the *inset* shows the corresponding SAED pattern. (Adapted with permission from [43]. ©2009, American Institute of Physics)

than 0.4 nm emerged from the emission spectra. It is believed that the sharp peak represents a closed-loop path of light (i.e., see the inset of the figure), which randomly formed a cavity mode inside the randomly assembled nanowires, due to the coherent optical feedback of scattering light. However, if the density and size of the SnO_2 nanowires are reduced from the current values, closed-loop paths may not be formed.

Randomly assembled ZnO [42] and ZnO/ZnS [44] nanowires of dimensions similar to that of the SnO_2 nanowires can also be fabricated by vapor transport method. As the refractive index of SnO_2 is closed to that of ZnO and ZnO/ZnS at UV regime, it is expected that their scattering strength should be similar. In addition, ZnO and ZnO/ZnS have optical gain higher than that of SnO_2. This is because ZnO and ZnO/ZnS nanowires support ultraviolet excitonic radiative recombination, while SnO_2 is mainly dependent on the defect emission. Hence, randomly assembled ZnO and ZnO/ZnS nanowires should have longer gain length than that of SnO_2 nanowires. Random lasing action should support the randomly assembled ZnO and ZnO/ZnS nanowires. In fact, we have shown that both randomly assembled ZnO and ZnO/ZnS nanowires can support coherent random lasing action under optical excitation at room temperature. In addition, the corresponding lasing characteristics are similar to that of randomly assembled SnO_2 nanowires. Similar lasing characteristics are also reported from the randomly assembled alloy CdS_xSe_{1-x} nanowires [13]. Experiment has shown that full range of alloy composition of randomly assembled alloy CdS_xSe_{1-x} nanowires can be grown on the same substrate through a single run of deposition. In addition, the sample can demonstrate spatially tunable lasing emission from ~500 to ~700 nm. This is because different compositions of S and Se of the CdS_xSe_{1-x} randomly assembled nanowires can be formed on the

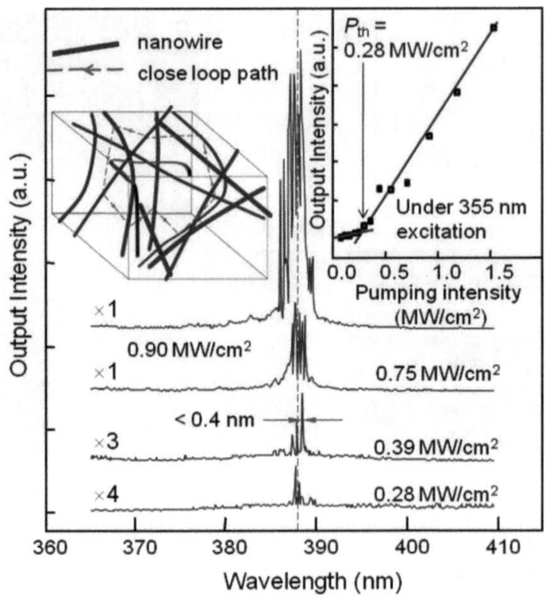

Fig. 9.17 Emission spectra of the randomly assembled SnO_2 nanowires under the excitation of 355 nm optical pumping. The corresponding light–light curve is also shown in the figure. The *inset* schematic shows the formation of closed-loop path of light inside the randomly assembled SnO_2 nanowires. (Adapted with permission from [43]. ©2009, American Institute of Physics)

substrate with spatial distribution of substrate temperature. This implies that random lasing can be easily achieved in randomly assembled nanowires than that in the alignment of vertical array.

9.3.4 Criteria to Achieve Stimulated Emission

After study the lasing characteristics of the assembled nanowires, readers may want to distinguish between random lasing action and conventional Fabry–Perot resonant oscillation. It must be noted that "gain saturation" is the main feature for all types of lasers including conventional Fabry–Perot and random lasers. This is equivalent to demonstrate the linear dependence of output power with the excitation power over a narrow emission range. However, this information is just only sufficient to verify that the gain medium can overcome the cavity loss to sustain stimulated emission. In fact, the main question is how to show the presence of random lasing action in either nanowires array or randomly assembled nanowires?

For random cavities supporting either incoherent or coherent random lasing action, we expect to see:

1. The multidirectional radiation, in which the emission spectra are different at different angle of observation.
2. The pumping threshold, P_{th}, is inversely proportional to $A_{\mathrm{th}}^{2/3}$ of the random medium where A_{th} is the excitation area at threshold.

For the media supporting coherent feedback, sharp peaks (i.e., their linewidth at least 50 times narrower than that of the spontaneous emission spectrum) emerge from the emission spectra, however

3. $\Delta\lambda$ cannot be found as the spectral separation of the sharp peaks is unevenly distributed over the emission spectrum.

If sharp peaks do not appear from the emission spectra but the criteria (1) and (2) are observed, we can conclude that the assembled nanowires support incoherent random lasing. Furthermore, if sharp peaks are observed from the emission spectra and the criteria (1), (2), and (3) are satisfied, we can confirm the presence of coherent random lasing.

How to design incoherent and coherent random lasers using assembled nanowires? For vertically aligned nanowires, coherent random lasing can be obtained from highly packed nanowires such as those embedded inside AAO templates [9, 20], or epilayers [35]. It is also possible to obtained coherent random lasing from high-density nanowires array with density similar to that of the highly disordered films [45]. However, low-density vertically aligned nanowires array will only support incoherent random lasing [35, 46]. On the other hand, it is noted that most of the randomly assembled nanowires support coherent random lasing [19, 47, 48]. This is because randomly assembled nanowires have strong three-dimensional confinement of optical modes so that the overall scattering strength is larger than that of the vertically aligned nanowires array (i.e., vertically aligned nanowires array only have two-dimensional confinement of optical modes). Hence, this may be the reason why coherent random lasing is commonly observed from random assembled nanostructures than that from vertically aligned nanowires array.

How to optimize the lasing efficiency of assembled nanowires? In fact, random laser is not an effective lasing device because the light path is not always traveled within the gain medium. Therefore, the lasing characteristics of random lasers may not be as good as the conventional Fabry–Perot lasers. Nevertheless, if we can bury the entire random cavity inside a low index surrounding (i.e., such as a buried waveguide randomly assembled nanowires laser [42]), cavity loss will then be limited to the two end facets of the random cavities and the lasing efficiency can be improved. In this case, quasi one-dimensional random cavities may exhibit compatible lasing performance to that of the conventional Fabry–Perot lasers. This may be the only possible way to achieve high-conversion-efficiency lasing using random cavities.

9.4 Single and Assembled Nanowires Laser Diodes

Most of the single nanowires and assembled nanowires are excited under optical pumping; however, this is not desirable for practical application. Therefore, the realization of electrically pumped laser actions is worth an effort. In the following paragraphs, we discuss on the recent development of single [49] and randomly assembled [36] nanowire laser diodes.

9.4.1 Single-Nanowire Electrically Driven Lasers

Electrically driven lasing requires efficient electron (n-type) and hole (p-type) injection into the cavity region. The use of GaAs-based semiconductor nanowires may be an appropriate choice as efficient injection layers of n- and p-types are available on the shelf. Nevertheless, the first electrical pumped single-nanowire laser diode was fabricated from CdS material [49]. This may be due to the difficulty of achieving high-optical-quality GaAs-based semiconductor nanowires. On the other hand, it is still a problem in developing high-mobility p-type CdS to realize CdSp–n homojunction. Therefore, p-type Si was used as the hole injection layer to form a heterojunction with a CdS nanowire for the realization of a single CdS nanowire laser [49]. This is possible because CdS nanowires exhibit n-type behavior with doping concentrations on the order of 10^{18}–10^{19} cm^{23} and electron mobilities of $\sim$100 cm^{2}V^{-1}s^{-1}. Figure 9.18a, b shows the schematic and CCD image of the proposed CdS nanowire laser diode. The average diameter and length of the nanowires is about 150 nm and 20 μm, respectively. Strong room-temperature electroluminescence (EL) is observed from the exposed CdS nanowire ends under forward bias. Figure 9.18c plots current–voltage (I–V) curve of the CdS nanowire laser diode. It is noted that the device exhibits a rectification response with turn-on voltage at around $\sim$4 V. Figure 9.18d shows the emission spectra of the nanowire laser diode at different injection currents. Narrowing of broad emission spectrum centered at $\sim$510 nm is observed for injection current larger than 200 μA. Sharp peaks with linewidth of $\sim$0.3 nm and average mode spacing, $\Delta\lambda$, of $\sim$1.8 nm are observed from the lasing spectra. This confirmed the excitation of Fabry–Perot cavity modes from the CdS nanowire. If we assumed that the refractive index of bulk CdS is 2.5 at 510 nm, the reflectivity at the end facets without considering diffraction effect is about 0.4286. Hence, the cavity loss is $\sim$850 cm^{-1} for nanowire has a cavity length of 20 μm. This indicates that the optical gain of CdS nanowire can be larger than the cavity loss to support Fabry–Perot resonant oscillation.

9.4.2 Electrically Pumped Nanowire Array Lasers

It was reported that highly disordered ZnO films with grains and voids can support random lasing action under optical and electrical excitation [50]. This is because

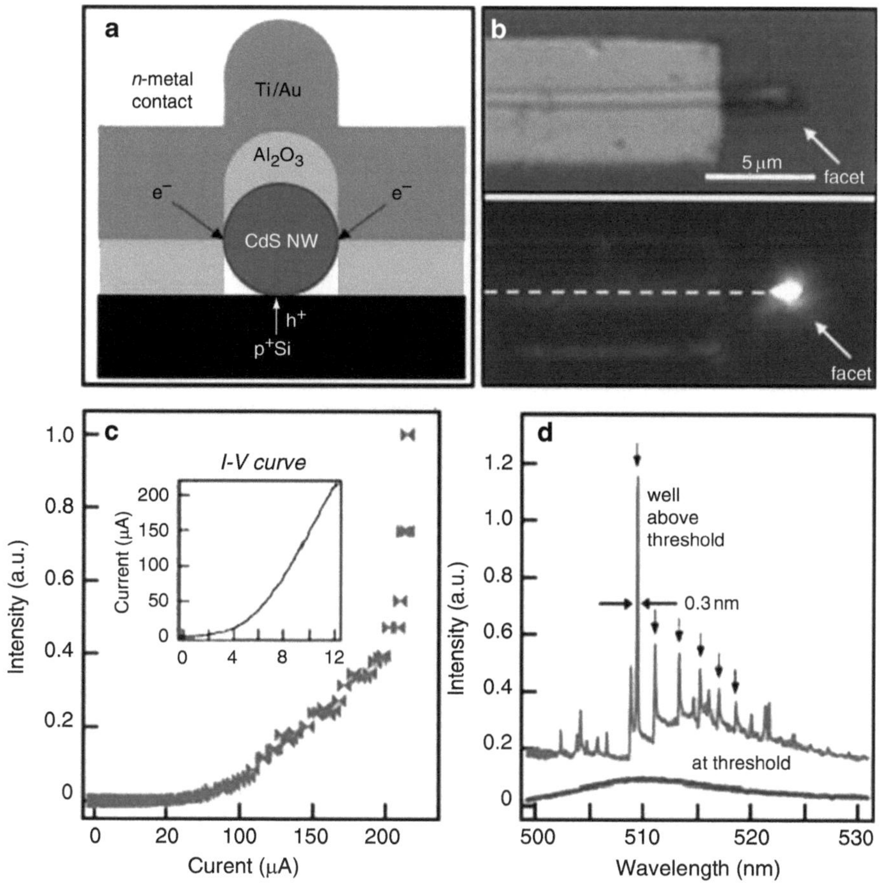

Fig. 9.18 Nanowire electrical injection laser. (**a**) Schematic cross section of the proposed CdS nanowire laser diode. (**b**) Shows (*top*) an optical image of the laser and (*bottom*) EL image recorded with an injection current of about 80 μA. (**c**) Emission intensity versus injection current. *Inset* shows the *I*–V curve of the device. (**d**) EL spectra obtained from the nanowire end with injection currents of 120 μA (*bottom*) and 210 μA (*top*). (Adapted by permission from Macmillan Publishers Ltd, Nature, after [49], ©2003)

strong optical scattering and high excitonic gain at ultraviolet wavelength can be achieved simultaneously to support stimulated emission. On the other hand, it is believed that strong optical scattering can be obtained in a medium with large variation of refractive index. Hence, it is possible that high-optical-quality ZnO nanowire array, which has controllability in spatial variation of refractive index, can achieve strong amplification of scattering light. Therefore, the use of dense ZnO nanowire array should be able to realize ultraviolet random laser diodes. Figure 9.19 shows the schematic diagram of a metal–insulator–semiconductor (MIS) random laser based on the ZnO nanowire array on Si substrate [36].

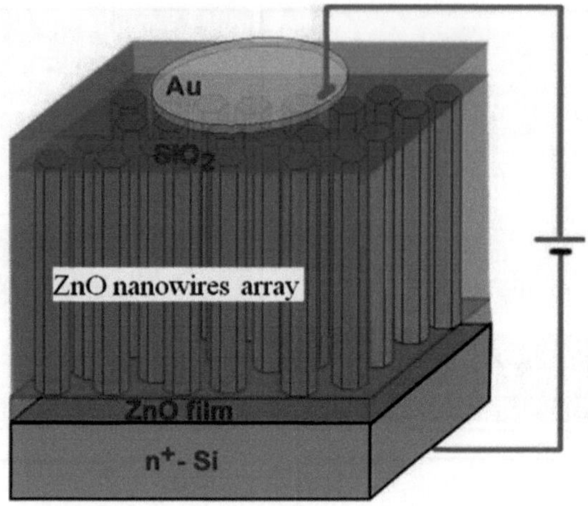

Fig. 9.19 Schematic diagram of the metal (Au)–insulator (SiO_2)–semiconductor (ZnO nanowires array) structure on Si substrate. (Adapted with permission from [36]. ©2009, Optical Society of America)

ZnO nanowire array (~2 μm long and ~40 to ~80 nm diameter) was grown on ~50 nm thick ZnO thin film seed layer deposited on a n^+-doped Si substrate as the electron injection layer. The top end facets of the ZnO nanowires were deposited with a ~90 nm thick of SiO_2 film. A ~20 and ~100 nm thick Au films were deposited on to the SiO_2 film and back of the n^+-Si substrate respectively to form the electrodes. This design avoids using p-type ZnO as the hole injection layer. This is because the thin SiO_2 film blocks electrons from escaping to the Au electrode. Hence, the injection of holes from the Au electrode can be effectively recombined with electrons inside the ZnO nanowires through the presence of thin SiO_2 film.

Figure 9.20a shows the I–V curve of the ZnO vertically aligned nanowire array random laser. It is noted that although the forward-bias current is high, the device demonstrates a rectifying behavior. It is found that the MIS device based on ZnO nanowire array can exhibit EL only under forward bias. Figure 9.20b plots the corresponding evolution of the EL spectra with the increase of forward-bias voltage. For the device operating below 4.5 V, broad spontaneous emission peak centered at around 383 nm is observed. For bias at and above 4.5 V, sharp peaks which have linewidth of less than 0.2 nm are emerged from the emission spectra. Further increase of voltage excited more sharp peaks from the emission spectra. The detected output power as a function of injection current is also shown in Fig. 9.20c. As we can see, a kink is observed at current ~95 mA and a linearly I–V relationship is observed. From these observations, it is reasonable to justify that lasing emission is obtained from the ZnO nanowire array. Furthermore, due to the randomly distribution of the ZnO nanowire array, it is believed that only random lasing action will support stimulated emission from the device. Hence, the proposed device can achieve optical gain through stimulated emission together with multiple scattering

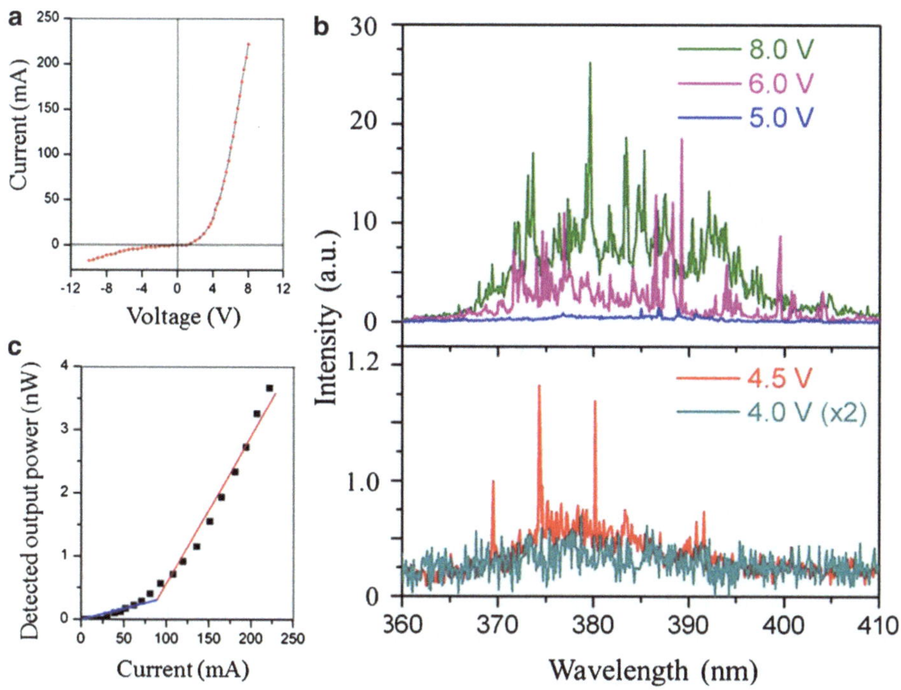

Fig. 9.20 (**a**) I–V curve of the MIS device based on ZnO nanowires array. (**b**) EL spectra of the device under different forward-bias voltages. (**c**) Detected output power as a function of the injection current. (Adapted with permission from [36]. ©2009, Optical Society of America)

under electrical excitation. The current–output power curve, appearance of kink, and the excitation of narrow sharp peaks in the emission spectra have verified the support of lasing emission from the nanowire array. From the emission spectra, it can be shown that the lasing characteristics of the ZnO nanowire array are mainly related to random lasing action. This is because $\Delta\lambda$ is nonuniformly distributed over the emission spectra. Furthermore, it can be shown that the emission spectra of the device are different at different observation angles. Hence, it is verified that lasing emission observed from the MIS ZnO nanowire array is due to coherent random lasing.

9.5 Conclusion and Discussion

This chapter reviews the lasing characteristics of single and assembled nanowire at room temperature. For single and isolated nanowires with long enough length and wide enough diameter (i.e., the corresponding cavity loss is less than $\sim$5,000 cm^{-1}), conventional Fabry–Perot lasing operation can be excited under optical or electrical excitation. However, if the cavity length is too short (<2 μm) and the diameter is too narrow ($<\lambda/2n_{\text{bulk}}$), the nanowires may not be able to sustain conventional Fabry–Perot resonant oscillation. Nevertheless, it is still possible to obtain lasing emission from these nanowires if they form assembled nanowires. In this case, random lasing action, which can be either incoherent or coherent random lasing, can be supported by the assembled nanowires. There are some similarity between conventional Fabry–Perot lasing and random lasing action: (1) the peak emission intensity increases linearly with respect to the pump intensity and (2) a drastic spectral narrowing at gain peak wavelength for the pump intensity reaches a threshold, P_{th}. In order to verify the presence of random lasing action, it must be noted that random lasing action is multidirectional and the profile of emission spectra is dependent on the observation angle. Furthermore, P_{th}, should be inversely proportional to $A_{\text{th}}^{2/3}$ of the assembled nanowires. Furthermore, the main feature of coherent random lasing is the excitation of sharp peaks (i.e., its linewidth is 50 times less than that of the ASE below) and $\Delta\lambda$ is nonuniformly distributed over the emission spectrum.

For the single and assembled nanowires to be useful, they have to be operated under electrical excitation. However, there are very few reports have demonstrated electrically pumped single or assembled nanowires lasers. The main difficulties to realize nanowire lasers are to obtain n- and p-types carrier injector layers as well as to find suitable materials to grow high-optical-quality nanowires. The first CdS single-nanowire laser diode using p^+-Si layer as the hole injection layer formed heterojunction with a CdS nanowire. This is because there is a problem in developing high-mobility p-type CdS and its alloys. Similarly, there is still a difficulty to obtain reliable p-type ZnO layer as the hole injection layer so that high-energy injection of holes through the thin dielectric layer was used to drive the n-type assembled ZnO nanowires. It seems that although CdS and ZnO nanostructures can be easily fabricated by standard vapor transport technique, the difficulty to obtain highly reliable p-type hole injector layer using these material systems prevents further development of electrically pumped nanowire lasers. On the other hand, reliable n- and p-type GaN can be obtained commercially so that it is possible to realize nanwires laser diodes using GaN. Unfortunately, this is still a challenge to grow vertically aligned GaN nanowires with extremely high-optical-quality and relatively easy size controllability. Similar problem occurs in GaAs- and InP-based semiconductor materials. On the other hand, although the development of n- and p-type III–V materials is in a mature state, the difficulty to achieve high-optical-quality nanostructures drags the fabrication of electrical pumped nanowire lasers using GaAs- and InP-based III–V materials.

There are at least two difficulties to be overcome in the future in order to simplify the fabrication of single nanowire laser diodes. The first difficulty is to realize "lattice-matched" heterojunction in order to obtain effective recombination of electrons and holes respectively inside the nanowires. Although high-crystal-quality ZnO- and CdS-based semiconductor nanowires can easily be obtained by different deposition technique, the fabrication of p-type ZnMgO- and CdSSe-based semiconductors is still in progress. We may need to wait for several years to receive p-type lattice-matched hole injection layers for ZnO and CdS nanowires. We can, however, use suitable substitution for the hole injection layer. For example, wide band gap p-type AlGaN layer, which is approximately lattice-matched to ZnO, can be used as the hole injector for the ZnO nanowires [51]. The second difficulty is to grow a single nanowire horizontally on the injection layers for the improvement of carrier injection efficiency and uniformity. Recently, single-nanowire laser was fabricated by depositing an isolated nanowire onto a substrate through dispersion. Alternatively, single nanowire can be dropped mechanically on a substrate. The drawback of these techniques is the poor interface condition – poor injection efficiency of carriers into the nanowires. Therefore, there is a need to develop some method to grow lattice-matched single nanowire horizontally onto the injection layer/substrate. Furthermore, the formation of isolation layers to separate the injection of electrons and holes into individual nanowire may be a difficult task to achieve as most of the nanowires have a cylindrical geometry. In the coming future, there will be more reports on the realization of novel electrically pumped nanowire lasers. It is also believed that the nanowire laser diodes can have the capability to operate over a wide range of visible and ultraviolet wavelength.

Acknowledgements This work was supported by The Hong Kong Polytechnic University research grant no. 1-ZV6X.

References

1. S.Y. Hu, M.S. Miller, D.B. Young, J.C. Yi, D. Leonard, A.C. Gossard, P.M. Petroff, L.A. Coldren, N. Dagli, Appl. Phys. Lett. **63**, 2015 (1993)
2. M.H. Huang, S. Mao, H. Feick, H.Q. Yan, Y.Y. Wu, H. Kind, E. Weber, R. Russo, P.D. Yang, Science **292**, 1897 (2001)
3. F. Quochi, F. Cordella, R. Orrù, J.E. Communal, P. Verzeroli, A. Mura, G. Bongiovanni, A. Andreev, H. Sitter, N.S. Sariciftci, Appl. Phys. Lett. **84**, 4454 (2004)
4. K.H. Cheng, Y.C. Zhong, B.Y. Xie, Y.Q. Dong, Y.I. Hong, J.Z. Sun, B.Z. Tang, K.S. Wong, J. Phys. Chem. C **112**, 17507 (2008)
5. D. O'Carroll, I. Lieberwirth, G. Redmond, Nat Nanotechnol. **2**, 180 (2007)
6. H.C. Hsu, C.Y. Wu, H.M. Cheng, F. Hsieh, Appl. Phys. Lett. 89, 013101 (2006)
7. U. Ozgur, Y.I. Alivov, C. Liu, A. Teke, M.A. Reshchikov, S. Dogan, V. Avrutin, S.J. Cho, H. Morkoc, J. Appl. Phys. 98, 041301 (2005)
8. J.C. Johnson, H.J. Choi, K.P. Knutsen, R.D. Schaller, P.D. Yang, R.J. Saykally, Nat Mater. **1**, 106, 2002
9. J.X. Ding, J.A. Zapien, Chen, Y. Lifshitz, S.T. Lee, X.M. Meng, Appl. Phys. Let. **85**, 2361 (2004)
10. Y.K. Liu, J.A. Zapien, C.Y. Geng, Y.Y. Shan, C.S. Lee, Y. Lifshitz, S.T. Lee, Appl. Phys. Lett. **85**, 3241 (2004)
11. R. Agarwal, C.J. Barrelet, C.M. Lieber, Nano Lett. **5**, 917 (2005)
12. A.L. Pan, R.B. Liu, Q.L. Zhang, Q. Wan, P.B. He, M. Zacharias, B.S. Zou, J. Phys. Chem. C, **111**, 14253 (2007)
13. A.L. Pan, C. Zhou, E.S.P. Leong, R.B. Liu, A.H. Chin, B.G. Zou, C.Z. Ning, Nano Lett. **9**, 784 (2009)
14. X.F. Duan, J.F. Wang, C.M. Lieber, Appl. Phys. Lett. **76**, 1116 (2000)

15. B. Hua, J. Motohisa, Y. Kobayashi, S. Hara, T. Fukui, Nano Lett. **9**, 112 (2009)
16. A.H. Chin, S. Vaddiraju, A.V. Maslov, C.Z. Ning, M.K. Sunkara, M. Meyyappan, Appl. Phys. Lett. **88**, 163115 (2006)
17. M.S. Hu, G.M. Hsu, K.H. Chen, C.J. Yu, H.C. Hsu, L.C. Chen, J.S. Hwang, L.S. Hong, Y.F. Chen, Appl. Phys. Lett. **90**, 123109 (2007)
18. B.L. Cao, Y. Jiang, C. Wang, WH. Wang, L.Z. Wang, M. Niu, J. Zhang, Y.Q. Li, S.T. Lee, Adv. Funct. Mater. **17**, 1501 (2007)
19. Y.K. Liu, J.A. Zapien, Y.Y. Shan, H. Tang, C.S. Lee, S.T. Lee, Nanotechnology **18**, 365606 (2007)
20. C.H. Liu, J.A. Zapien, Y. Yao, X.M. Meng, C.S. Lee, S.S. Fan, Y. Lifshitz, S.T. Lee, Adv. Mater. **15**, 838 (2003)
21. A.V. Maslov, C.Z. Ning, IEEE J. Quantum Elect. **40**, 1389 (2004)
22. S.F. Yu, C. Yuen, S.P. Lau, J. Fan, IEEE J. Quantum Elect. **40**, 406 (2004)
23. J. Wilson, J. Hawkes (ed.) *Optoelectronics – An Introduction* (Prentice Hall, London 1998)
24. A. Snyder, J.D. Love (ed.) *Optical Waveguide Theory* (Chapman and Hall, London, 1983)
25. A.V. Maslov, C.Z. Ning, Appl. Phys. Lett. **83**, 1237 (2003)
26. R.R. Burton, M.S. Stern, P.C. Kendall, P.N. Robson, Opt. Quant Electron. **28**, 1677 (1996)
27. A.L. Henneghien, B. Gayral, Y. Desieres, J.M. Gerard, J. Opt. Soc. Am. B. **26**, 2396 (2009)
28. H.Y. Li, S. Ruhle, R. Khedoe, A.F. Koenderink, D. Vanmaekelbergh, Nano Lett. **9**, 3515 (2009)
29. X.L. Xu, F.S.F. Brossard, D.A. Williams, D.P. Collins, M.J. Holmes, R.A. Taylor, X.T. Zhang, Appl. Phys. Lett. **94**, 231103 (2009)
30. F. Qian, Y. Li, S. Gradecak, H.G. Park, Y.J. Dong, Y. Ding, Z.L. Wang, C.M. Lieber, Nat Mater. **7**, 701 (2008)
31. M.A. Zimmler, F. Capasso, S. Muller, C. Ronning, Semicond. Sci. Technol. 25 024001 (2010)
32. L.K. van Vugt, S. Ruhle, D. Vanmaekelbergh, Nano Lett. **6**, 2707 (2006)
33. J.H. Choy, E.S. Jang, J.H. Won, J.H. Chung, D.J. Jang, Y. Kim, Adv. Mater. **12**, 1911 (2003)
34. M.A. Zimmler, J.M. Bao, F. Capasso, S. Müller, C. Ronning, Appl. Phys. Lett. **93**, 051101 (2008)
35. S.F. Yu, C. Yuen, S.P. Lau, I. Park, G.C. Yi, Appl. Phys. Lett. **84**, 3241 (2004)
36. X.Y. Ma, J. Pan, P.L. Chen, D.S. Li, H. Zhang, Y. Yang, D.R. Yang, Opt Express, **17**, 14426 (2009)
37. V.S. Letokhov, Soviet Phys. JETP-USSR, **26**, 835 (1968)
38. H. Cao, Wave in Random Media, **13**, R1 (2003)
39. H. Cao, J.Y. Xu, Y. Ling, A.L. Burin, E. Seeling, X. Liu, R.P.H. Chang, IEEE J. Sel. Top Quant Electron. **9**, 111 (2003)
40. J. Fallert, R.J.B. Dietz, J. Sartor, D. Schneider, C. Klingshirn, H. Kalt, Nat. Photonics **3**, 279 (2009)
41. F. Quochi, F. Cordella, R. Orrù, J.E. Communal, P. Verzeroli, A. Mura, G. Bongiovanni, A. Andreev, H. Sitter, N.S. Sariciftci, Appl. Phys. Lett. **84**, 4454 (2004)
42. H.Y. Yang, S.F. Yu, H.K. Liang, C. Pang, B. Yan, T. Yu, J. Appl. Phys. **106**, 043102 (2009)
43. H.Y. Yang, S.F. Yu, S.P. Lau, S.H. Tsang, G.H. Xing, T. Wu, Appl. Phys. Lett. **94**, 241121 (2009)
44. H.Y. Yang, S.F. Yu, J. Yan, L.D. Zhang, Appl. Phys. Lett. **96**, 14111 (2010)
45. S.F. Yu, C Yuen, S.P. Lau, Appl. Phys. Lett. **84**, 3244 (2004)
46. S. Hirano, N. Takeuchi, S. Shimada, K. Masuya, K. Ibe, H. Tsunakawa, M. Kuwabara, J. Appl. Phys. **98**, 094305 (2005)
47. S.H. Lee, S. Lee, J.S. Ha, H.J. Lee, J. Lee, J.Y. Lee, S.K. Hong, T. Goto, M. Cho, T. Yao, J. Phys. D: Appl. Phys. **42**, 225403 (2009)
48. Y. Jiang, J. Zhang, J.S. Jie, X.M. Meng, J.A. Zapien, S.T. Lee, Adv. Mater. **18**, 1527 (2006)
49. X.F. Duan, Y. Huang, R. Agarwal, C.M. Lieber, Nature **42**, 241 (2003)
50. H.K. Liang, S.F. Yu, H.Y. Yang, Appl. Phys. Lett. **96**, 101116 (2010)
51. W.I. Park, G.C. Yi, Adv. Mater. **16**, 87 (2004)

Chapter 10
Nanophotonic Device Application Using Semiconductor Nanorod Heterostructures

Takashi Yatsui, Gyu-Chul Yi, and Motoichi Ohtsu

Abstract "Quantitative innovation" in optical technology is required for future optical information-transmission systems, that is, increasing the integration of photonic devices by reducing their size and heat generation. Furthermore, novel applications such as optical information-processing systems are expected by realizing "qualitative innovation," meaning novel functions and operations in photonic devices that are impossible with conventional photonic devices, such as lasers, modulators, and waveguides. This chapter reviews how the "nanophotonics" provides us "qualitative innovation."

10.1 Introduction

To reduce device size and heat generation, it is still insufficient to quantize the plasma oscillation because the position of the photon is defined only in a space larger than the wavelength of light, which is the consequence of the uncertainty principle. That is, the localization of the light cannot be denned in subwavelength space. However, if a subwavelength-sized nanometric particle is used to absorb the light, it works as a photodetector, and consequently, the photon can be detected and its position is determined by the size of the particles with high spatial accuracy. This means that a local interaction between nanometric particles and photons is required to reduce the size of photonic devices beyond the diffraction limit. Furthermore, the energy transferred via this interaction must be dissipated in the

T. Yatsui (✉) · M. Ohtsu
School of Engineering, University of Tokyo, 2-11-16 Yayoi Bunkyo-ku, 113-8656 Tokyo, Japan
e-mail: yatsui@ee.t.u-tokyo.ac.jp; ohtsu@ee.t.u-tokyo.ac.jp

G.-C. Yi
National Creative Research Initiative Center for Semiconductor Nanostructures, Department of Physics and Astronomy, Seoul National University, 599 Gwanak-ro, Gwanak-gu, Seoul 151–747, Korea
e-mail: gyuchul.yi@gmail.com

G.-C. Yi (ed.), *Semiconductor Nanostructures for Optoelectronic Devices*, NanoScience and Technology, DOI 10.1007/978-3-642-22480-5_10,

nanometric particles or adjacent macroscopic materials to fix the position and magnitude of the transferred energy. Since plasmonics does not deal with this local dissipation of energy, it is irrelevant for reducing device size. Local energy transfer and its subsequent dissipation are possible only using optical near fields [1], which are the elementary surface excitations on nanometric particles, or in other words, dressed photons that carry the material energy. Novel optical technology utilizing the local interaction between nanometric particles via optical near fields is called "nanophotonics" [2]. Even if the incident propagating light exhibits classical properties, nanophotonics enables the generation of light with quantum mechanical properties by laying out the position of the nanometric particles in an appropriate manner and controlling their electronic energy state densities. As a representative device of nanophotonic device, the principles of nanophotonic AND-gate device using ZnO nanorod heterostructures are described in this section.

10.2 ZnO Nanorod Heterostructure for Nanophotonic Device

ZnO nanocrystallites are promising material for realizing nanoscale photonics device, i.e., nanophotonic devices [3] owing to their large exciton binding energy [4–6] and large oscillator strength [7]. In the nanophotonic devices, optical near field acts as a carrier to transfer the signal. Furthermore, the recent demonstration of semiconductor nanorod quantum-well structure enables us to fabricate nanometer-scale electronic and photonic devices on single nanorods because of its extremely high quality of crystallinity [8–11]. Recently, semiconductor quantum-well structures using ZnO/ZnMgO heterostructures were fabricated on the end of the nanorod and the quantum confinement effect even from the single quantum-well structures (SQWs) was successfully observed [12]. More recently, the realization of p-type ZnO opens up significant opportunities for the optoelectro device based on ZnO [13]. Near-field spectroscopy has made a remarkable contribution to investigations of the optical properties in nanocrystallite [14], and resulted in the observation of nanometer-scale optical image, such as the local density of exciton states [15]. However, reports on semiconductor quantum structure are limited to naturally formed quantum dot (QD) [15–17]. In this section, we report low-temperature near-field spectroscopy of artificially fabricated ZnO quantum-well structures (QWs) on the end of a ZnO nanorod.

10.3 Near-Field Evaluation of Isolated ZnO Nanorod Single-Quantum-Well Structure for Nanophotonic device

To confirm the promising optical properties of ZnO QDs, we performed near-field evaluation of isolated ZnO SQWs. ZnO/ZnMgO SQWs were fabricated on the ends of a ZnO stem with a mean diameter of 40 nm and a length of 1 μm.

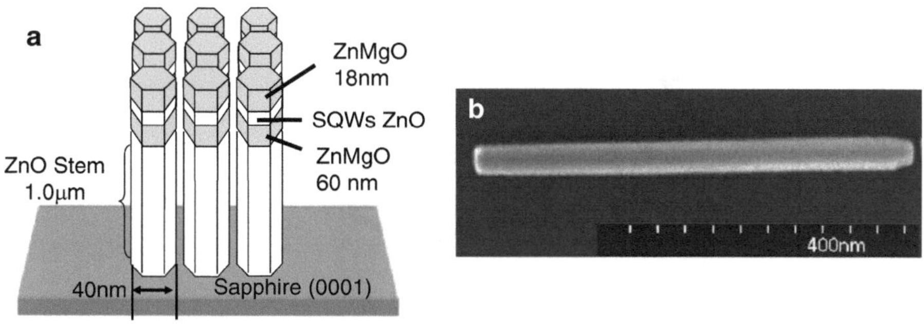

Fig. 10.1 (**a**) Schematic of ZnO/ZnMgO SQWs on the ends of ZnO nanorods. (**b**) SEM image of the dispersed ZnO/ZnMgO SQWs

They were grown vertically from the sapphire (0001) substrate using catalyst-free metal-organic vapor phase epitaxy, in which the ZnO nanorod was grown in the c orientation [11, 12]. The Mg concentration in the ZnMgO layers averaged 20 atm.%. Two samples were prepared for this study, their ZnO well layer thicknesses, L_w, were 2.5 and 3.75 nm, while the thicknesses of the ZnMgO bottom and top barrier layers in the SQWs were fixed at 60 and 18 nm, respectively (Fig. 10.1a). After growing the ZnO/ZnMgO nanorod SQWs, they were dispersed so that they were laid down on a flat sapphire substrate to isolate them from each other (Fig. 10.1b).

The far-field PL spectra were obtained using a He–Cd laser ($\lambda = 325$ nm) before dispersion of the ZnO/ZnMgO nanorod SQWs. The emission signal was collected with the achromatic lens ($f = 50$ mm). To confirm that the optical qualities of individual ZnO/ZnMgO SQWs were sufficiently high, we used a collection-mode near-field optical microscope (NOM) using a He–Cd laser ($\lambda = 325$ nm) for excitation, and a UV fiber probe with an aperture diameter of 30 nm. The excitation source was focused on a nanorod sample laid on the substrate with a spot size approximately 100 μm in diameter. The PL signal was collected with the fiber probe, and detected using a cooled charge-coupled device through a monochromator. The fiber probe was kept in close proximity to the sample surface (~5 nm) using the shear-force feedback technique. The polarization of the incident light was controlled with a $\lambda/2$ waveplate. In contrast to the naturally formed QD structure (a high monolayer island formed in a narrow quantum well), the barrier and cap layers laid on the substrate allowed the probe tip access to PL source, which reduced carrier diffusion in the ZnO SQWs and the subsequent linewidth broadening, thereby achieving a high spatial and spectral resolution. In addition to the PL measurements, absorption spectra were obtained using a halogen lamp, where the absorption was defined by the ratio $I_{\text{well}}/I_{\text{back}}$ between the signal intensities transmitted through the well layer (I_{well}) and substrate (I_{back}, 50-nm apart from the well layer) (Fig. 10.2). The absorption signal was collected with the same fiber probe with an aperture diameter of 30 nm. Since the ZnMgO layers (bottom and top barrier layers are 60 and 18 nm, respectively) are much thicker than that of the well layer (~3 nm), any difference in the transmission signals between I_{well} and I_{back} was not detected,

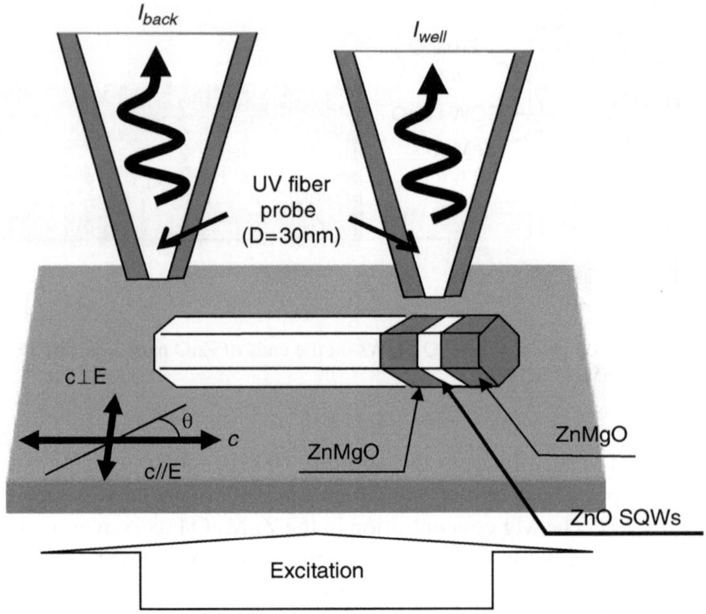

Fig. 10.2 Schematic of experimental setup for near-field PL spectroscopy. c c-axis of the ZnO stem. θ angle between the ZnO stem and the direction of incident light polarization

which resulted in no detection of the absorption peak originated from the ZnMgO layers.

As a preliminary near-field spectroscopy experiment of the ZnO SQWs, we obtained near-field PL spectra of the ZnO SQWs with $L_w = 3.75$ nm (Fig. 10.3a) obtained with polarization perpendicular to the c-axis ($\theta = 90$ in Fig. 10.2). Two typical spectra are shown, one with a single peak at 3.375 eV (NF_1) and the other with several sharp peaks around 3.375, 3.444, and 3.530 eV (NF_2), while NF_b is a background spectra (Fig. 10.3a). Several conclusions can be drawn from these spectral profiles. First, comparison with the far-field PL spectrum (FF: dashed curve in Fig. 10.3a) showed that the emission peak I_2^{ZnO} at 3.375 eV was suppressed and I_{QW} (3.444 eV) and I_{ZnMgO} (3.530 eV) were enhanced in NF_2, indicating that peaks I_2^{ZnO} and I_{znMgO} originated from the ZnO stem and ZnMgO layers, respectively. Second, since the peak position of I_{QW} was consistent with the theoretical prediction (3.430 eV) using the finite square-well potential of the quantum confinement effect in the ZnO well layer for $L_w = 3.75$ nm, we concluded that peak I_{QW} originated from the ZnO SQWs. The theoretical calculation used $0.28m_0$ and $1.8m_0$ as the effective masses of an electron and hole in ZnO, respectively, at a ratio of conduction and valance band offsets ($\Delta E_C/\Delta E_V$) of 9, and a band gap offset (ΔE_g) of 250 meV [12]. The spatial distributions of the near-field PL intensity of peaks I_2^{ZnO} and I_{QW} [Fig. 10.3b, c] supported the postulate that the blue-shifted emission was confined to the end of the ZnO stem. Third, the spectral width (3 meV) of peak I_{QW} was much narrower than those of the far-field PL spectra (40 meV).

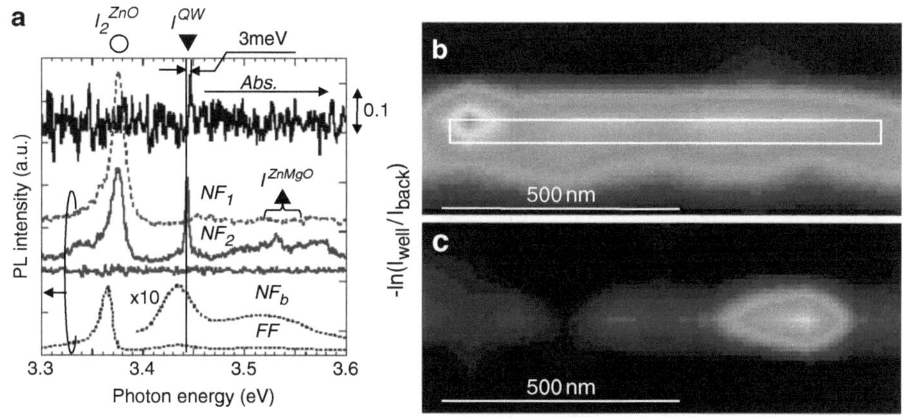

Fig. 10.3 Near-field PL and absorption spectroscopy of isolated ZnO SQWs ($L_w = 3.75$ nm) at 15 K. (**a**) NF_1, NF_2 near-field PL spectra. NF_b background noise, *Abs.* near-field absorption spectrum, *FF* far-field PL spectrum of vertically aligned ZnO SQWs ($L_w = 3.75$ nm). NOM images of isolated ZnO SQWs obtained at (**b**) 3.375 and (**c**) 3.444 eV. The *rectangle* shown in *dashed white lines* indicates the position of the ZnO stem

To estimate the homogeneous linewidth of isolated ZnO SQWs, we observed the power dependence of the near-field PL spectra [Fig. 10.4a] by varying the excitation power densities from 0.6 to 4.8 W/cm^2. The shape of each spectrum was fitted using the Lorentzian function indicated by the solid curve. Figure 10.4b, c shows the integrated PL intensity (I_{PL}) and linewidth (Δ) of the fitted Lorentzian, which increased linearly and remained constant around 3 meV, respectively. These results indicate that emission peak I_{QW} represented the emission from a single-exciton state in ZnO SQWs and that the linewidth was governed by the homogeneous broadening. Fourth, the Stokes shift of 3 meV [Fig. 10.3a] was much smaller than the reported value (50 meV) in ZnO/ZnMgO superlattices [18,19]. The small Stokes shift may result from the decreased piezoelectric polarization effect by the fully relaxed strain for the ZnO/ZnMgO nanorod quantum structures in contrast to the two-dimensional (2D) ZnO/ZnMgO heteroepitaxial multiple layers. This argument is supported by theoretical calculation of electronic structure of double barrier InAs/InP/InAs/InP/InAs nanorod heterostructures [20], concluding that any strain at heterointerfaces relaxes in nanorods within a few atomic layers in contrast to 2D pseudomorphic heteroepitaxy.

Based on these experiments, a major investigation of the optical properties of isolated ZnO SQWs was performed by analyzing the polarization-dependent PL spectrum of isolated ZnO SQWs ($L_w = 3.75$ nm). As shown in Fig. 10.5a, NF_0 is a near-field PL spectrum obtained with parallel polarization with respect to the c-axis, $\theta = 0°$, and this exhibits a new peak I_{1b}^{QW} at 3.483 eV, which is out of peak in the far-field spectrum (3.435 eV $\pm$ 20 meV). Peak I_{1a}^{QW} is the same as I_{QW} in Fig. 10.3a.

As the ZnO has valence band anisotropy owing to the wurtzite crystal structure, the operator corresponds to the Γ_5 (Γ_1) representation when the electric vector

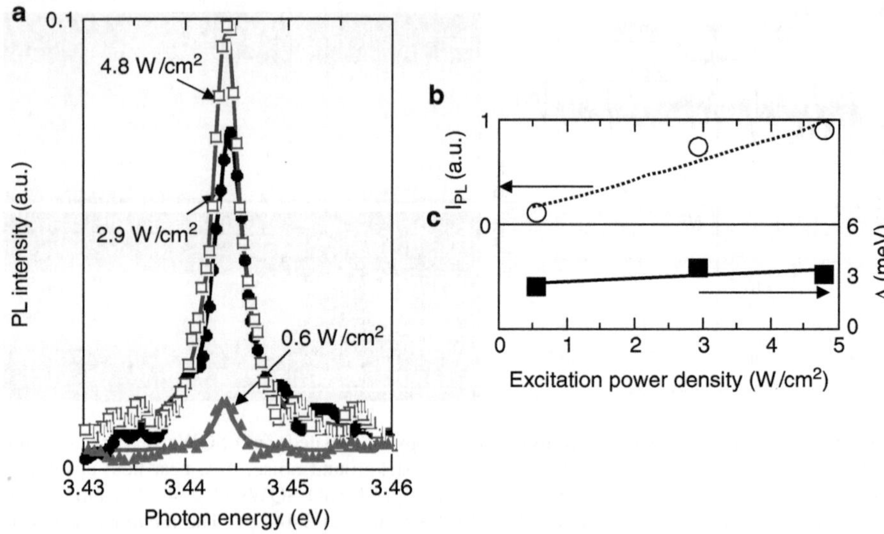

Fig. 10.4 Power-dependent near-field PL spectroscopy of isolated ZnO SQWs ($L_w = 3.75$ nm) at 15 K. (**a**) Near-field PL spectra of isolated ZnO SQWs at excitation densities ranging from 0.6 to 4.8 W/cm^2. The integrated PL intensity, I_{PL}, (**b**) homogeneous linewidths, Δ, (**c**) as a function of the excitation power density

E of the incident light is perpendicular (parallel) to the crystalline c-axis. By considering the energy difference between Γ_5 and Γ_1 in the center of the zone around 40 meV for bulk material [7, 21, 22], and the direction of the incident light polarization with respect to the c-axis, emission peaks I_{1a}^{QW} and I_{1b}^{QW} in Fig. 10.5a are allowed for the exciton from Γ_5 and Γ_1, respectively. Note that this is the first observation of a Γ_1 exciton in a PL spectrum, while the observation of emission from Γ_1 has been realized only for bulk ZnO crystals using time-resolved reflection spectroscopy [21, 22]. Since the exciton binding energy of the emission from Γ_1 (50–56 meV) [22, 23] is comparable to that from Γ_5 (60 meV), this successful observation originates from the enhancement of the exciton binding energy owing to the quantum confinement effect [6]. Furthermore, in contrast to the bulk ZnO film, our sample configuration using laid ZnO nanorod SQWs has realized π polarization ($\theta = 0°$), allowing the detection of the emission from the Γ_1 exciton. The homogeneous linewidth of emission peak I_{1a}^{QW} (Γ_5) is in the range 3–5 meV, while that of I_{1b}^{QW} (Γ_1) is 9–11 meV (Fig. 10.5b). This difference is attributed to the degeneracy of the transition of the Γ_1 exciton with continuum and to the contribution of the residual strain field, and results in sensitive dependence of the Γ_1 exciton on the strain, as reported in the GaN [24]. The solid triangles and circles in Fig. 10.5c show the respective normalized integrated PL intensity at I_{1a}^{QW} and I_{1b}^{QW}, respectively, which are in good agreement with the sine-squared and cosine-squared functions represented by the solid curves. These results indicate that

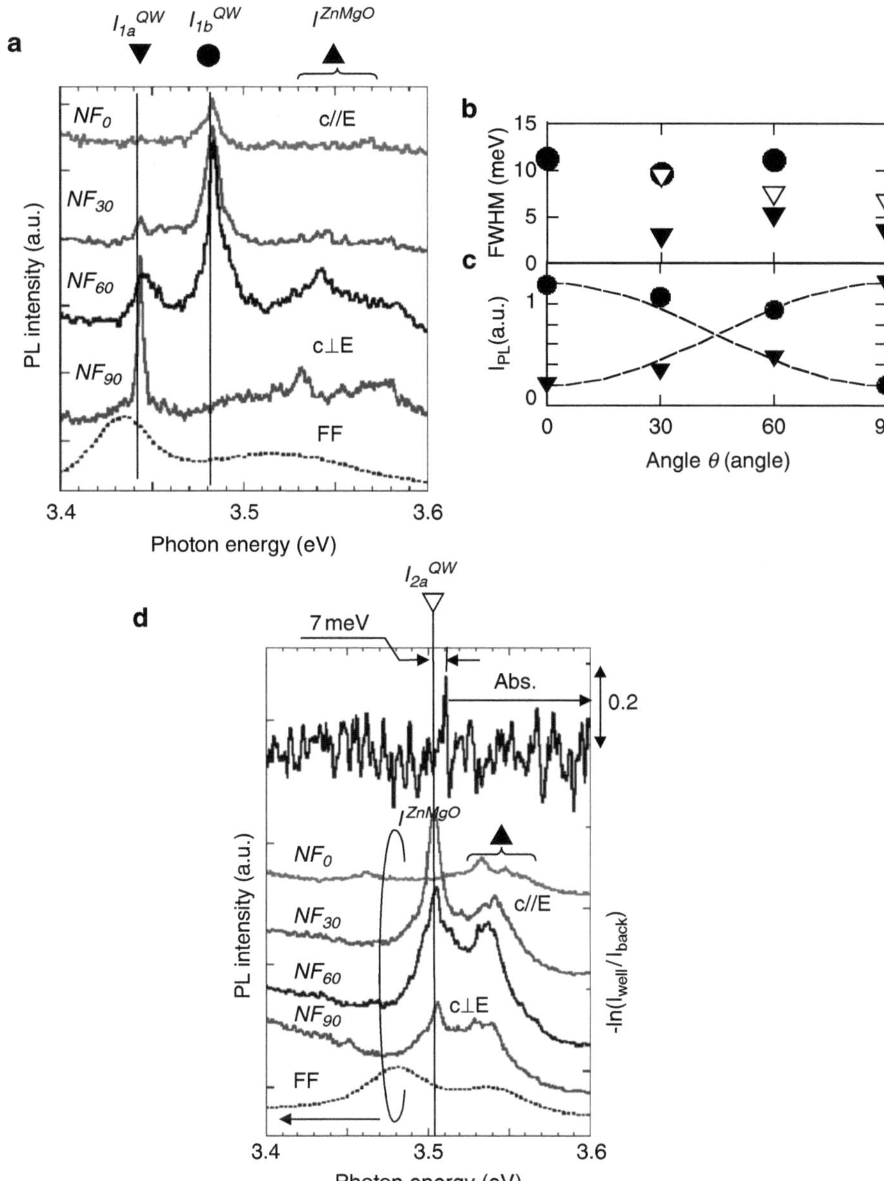

Fig. 10.5 Polarization dependence of near-field PL spectra of isolated ZnO SQWs obtained at 15 K. (**a**) NF_θ, FF near-field and far-field PL spectra of isolated ZnO SQWs ($L_w = 3.75$ nm) for $\theta = 0$, 30, 60, and 90°, (**b**) *Solid triangles and circles* are the polarization dependence of the linewidth of I_{1a}^{QW} and I_{1b}^{QW}, respectively in (**a**). *Open triangles* are the polarization dependence of linewidth of I_{1a}^{QW} in (**d**). (**c**) *Solid triangles and circles* are the integrated PL intensities, I_{PL}, of I_{1a}^{QW} and I_{1b}^{QW}, respectively, normalized to the total PL intensities ($I_{1a}^{QW} + I_{1b}^{QW}$). (**d**) NF_θ, FF near-field and far-field PL spectra of isolated *ZnO* SQWs ($L_w = 2.5$ nm). *Abs.* absorption spectrum

emission peaks I_{1a}^{QW} and I_{1b}^{QW} originate from unidirectional transition dipoles that are orthogonal to each other.

To study the linewidth broadening mechanism, Fig. 10.5d shows the polarization-dependent near-field PL spectra (NF_0–NF_{90}) and absorption spectrum obtained for isolated ZnO SQWs with a thinner well layer ($L_{\mathrm{w}} = 2.5$ nm). In NF_0–NF_{90}, the emission peaks I_{znMgO} around 3.535 eV originate from the ZnMgO layers. Emission peak I_{2a}^{QW} originates from the Γ_5 exciton in the SQWs, as was the case for I_{1a}^{QW} in Fig. 10.5a, since the position of peak I_{2a}^{QW} (3.503 eV) is consistent with the theoretical prediction (3.455 eV) using the finite square-well potential of the quantum confinement effect in the ZnO well layer. In comparison to ZnO SQWs with $L_{\mathrm{w}} = 3.75$ nm, however, emission peak I_{2a}^{QW} had a broader linewidth (7–10 meV), which is attributed to the shorter exciton dephasing time. In the nanocrystallite where the excitons are quantized, the linewidth should be determined by the exciton dephasing time. Such dephasing arises from the collisions of the excitons at the irregular surface, so that the linewidth is d^{-2} (d is the effective size of the quantum structure) [25]. The observed well-width dependence of the spectral linewidth, $3.75^{-2}/2.5^{-2} \sim 3/7$, and the Stokes shift of 7 meV [see Fig. 10.5c] larger than that for $L_{\mathrm{w}} = 3.75$ nm (3 meV) are supported by this dephasing mechanism quantitatively. Although emission peak I_{2a}^{QW} was suppressed for $\theta = 0°$, no peaks corresponding to the Γ_1 exciton in SQWs were detected owing to the reduction of the exciton binding energy, since the peak energy of Γ_1 for the ZnO SQWs with $L_{\mathrm{w}} = 2.5$ nm is comparable with that of ZnMgO.

10.4 A Nanophotonic AND-Gate Device Using ZnO Nanorod Double-Quantum-Well Structures

Using time-resolved near-field spectroscopy, we demonstrated the switching dynamics that result from controlling the optical near-field energy transfer in ZnO nanorod double-quantum-well structures (DQWs). We observed nutation of the exciton population between the resonantly coupled exciton states of DQWs, where the coupling strength of the near-field interaction decreased exponentially as the separation increased. Our results provide criteria for designing nanophotonic devices.

As reviewed in the previous section, the quantum confinement effect has been observed from SQWs structures [12, 26–28]. In this section, we review the time-resolved near-field spectroscopy to demonstrate the switching dynamics that result from controlling the optical near-field energy transfer in ZnO nanorod DQWs. We observed nutation of the population between the resonantly coupled exciton states of DQWs, where the coupling strength of the near-field interaction decreased exponentially as the separation increased [29].

To evaluate the exciton energy transfer between the resonantly coupled QWs, three samples were prepared (Fig. 10.6a): (1) SQWs with a well-layer thickness

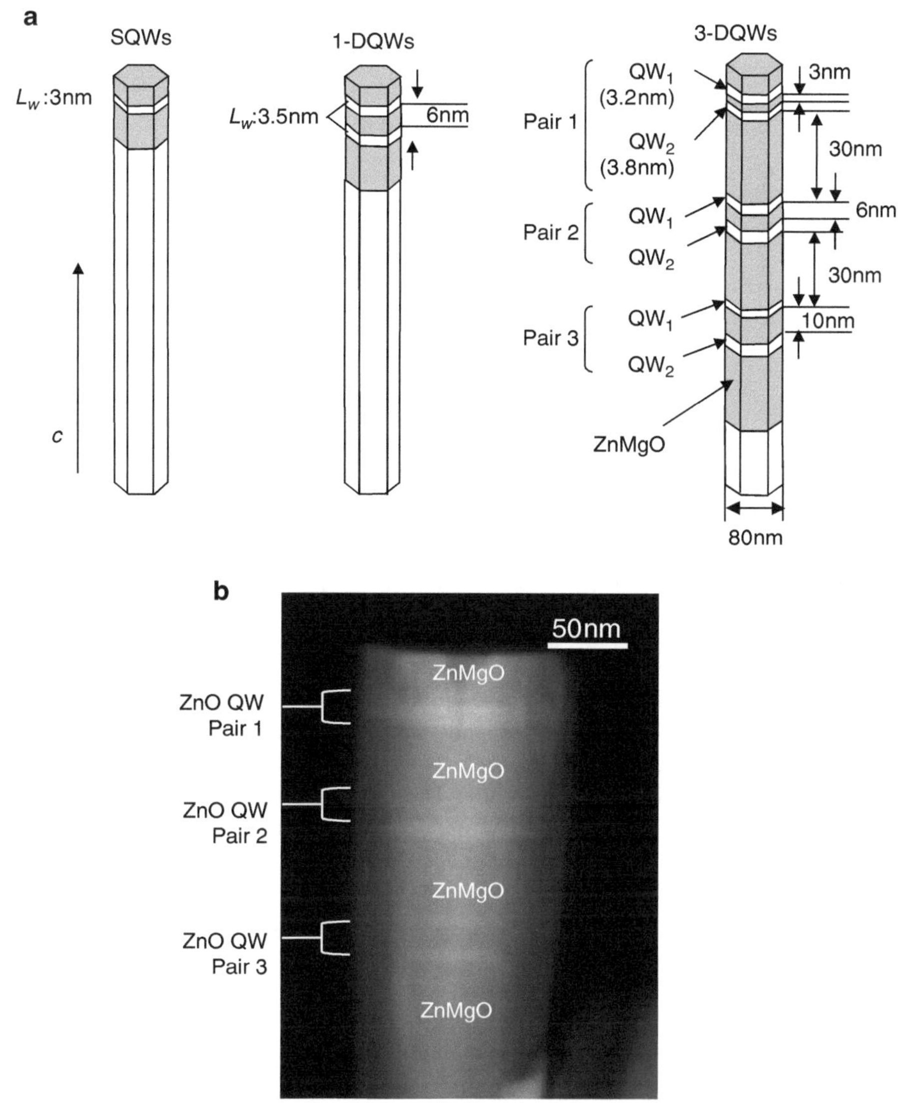

Fig. 10.6 ZnO/ZnMgO nanorod quantum-well structures, *c* *c*-axis of the ZnO stem. (**a**) Schematic of ZnO/ZnMgO SQWs, DQWs (1-DQWs), and triple pairs of DQWs (3-DQWs). (**b**) Z-contrast TEM image of 3-DQWs clearly shows the compositional variation, with the bright layers representing the ZnO well layers. *Scale bar*: 50 nm

of $L_w = 2.0$ nm (SQWs) for the comparison with the DQWs, (2) DQWs with $L_w = 3.5$ nm with 6-nm separation (1-DQWs), and (3) three pairs of DQWs with $L_w = 2.0$ nm with different separations (3, 6, and 10 nm), where each DQW was separated by 30 nm (3-DQWs). ZnO/ZnMgO quantum-well structures (QWs)

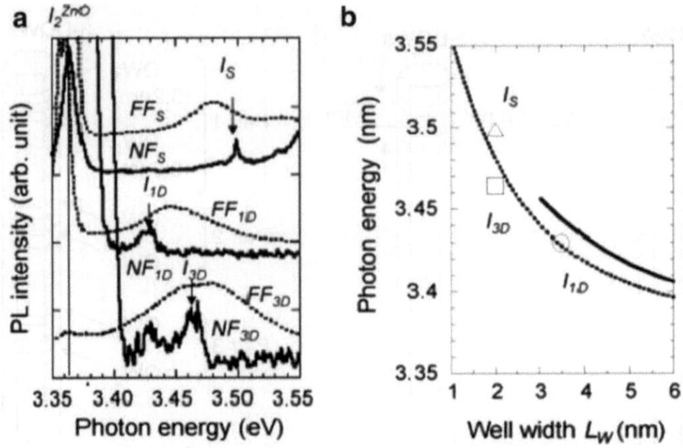

Fig. 10.7 Near-field time-resolved spectroscopy of ZnO nanorod DQWs at 15 K. (**a**) NF_S, NF_{1D}, NF_{3D} near-field PL spectra. FF_S, FF_{1D}, FF_{3D} far-field PL spectrum of ZnO SQWs (L_m = 2.0 nm), 1-DQWs (L_w = 3.5 nm, 6-nm separation), and 3-DQWs (L_w = 2.0 nm, 3-, 6-, and 10-nm separation), (**b**) Well-width dependence of the exciton ground state and the first excited state. SQWs: *open triangle*, 1-DQWs: *open circle*, 3-DQWs: open square

were fabricated on the ends of ZnO nanorods with a mean diameter of 80 nm using catalyst-free metal-organic vapor phase epitaxy [30]. The average concentration of Mg in the ZnMgO layers used in this study was determined to be 20 atm%.

The far-field PL spectra were obtained using a He–Cd laser (λ = 325 nm) before detection using near-field spectroscopy. The emission signal was collected with an achromatic lens (f = 50 mm). The incident beam spot size is about 10 mm in diameter. The near-field photoluminescence (NFPL) spectra were obtained using a He–Cd laser (λ = 325 nm), collected with a sharpened fiber probe with an aperture diameter of 30 nm, and detected using a cooled charge-coupled device through a monochromator. In addition to the emission from the bound exciton in ZnO nanorod (I_2^{ZnO}), blue-shifted PL peaks were observed at 3.499 (I_S), 3.429 ($I_{1\mathrm{D}}$), and 3.467 ($I_{3\mathrm{D}}$) eV in the far- and near-field PL spectra (Fig. 10.7a). These peaks originated from the respective ZnO QWs because their energies are comparable to the predicted ZnO well layer thicknesses of 1.7 (Is), 3.4 ($I_{1\mathrm{D}}$), and 2.2 ($I_{3\mathrm{D}}$) nm, respectively, calculated using the finite square-well potential of the quantum confinement effect in ZnO SQWs (see Fig. 10.7b) [30].

To confirm the near-field energy transfer between QWs, we compared the time-resolved near-field PL (TRNFPL) signals at the I_s, $I_{1\mathrm{D}}$, and $I_{3\mathrm{D}}$ peaks. For the time-resolved near-field spectroscopy, the signal was collected using a microchannel plate through a band-pass filter with 1-nm spectral width. Figure 10.8 shows the typical TRNFPL of SQWs (TR_S), 1-DQWs ($TR_{1\mathrm{D}}$), and 3-DQWs ($TR_{3\mathrm{D}}$).

We calculate the exciton dynamics using quantum mechanical density-matrix formalism [31, 32], where the Lindbrad-type dissipation is assumed for the relaxation due to exciton–photon and exciton–phonon couplings;

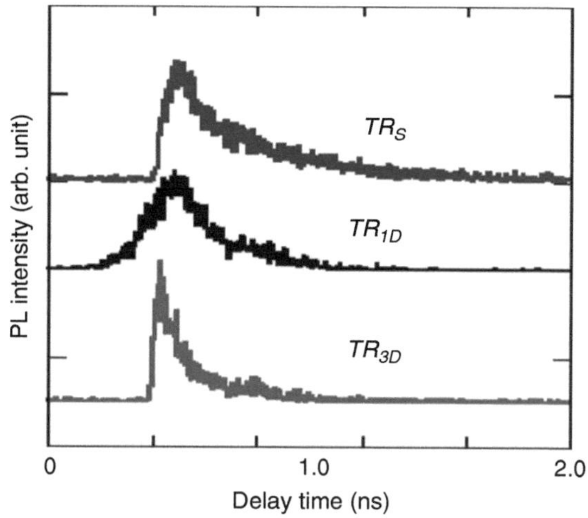

Fig. 10.8 TR_S, TR_{1D}, and TR_{3D} show TRNFPL signal obtained at I_s, I_{1D}, and I_{3D}, respectively, using the 4.025-eV ($\lambda = 308$ nm) light with a pulse of 10 ps duration to excite the barrier layers of ZnO QWs

$$\dot{\rho} = \frac{i}{\hbar}[H,\rho] + \sum_n \frac{\gamma_n}{2}\left(2A_n\rho A_n^\dagger - A_n^\dagger A_n\rho - \rho A_n^\dagger A_n\right) \tag{10.1}$$

where ρ is the density operator, H is the Hamiltonian in the considered system, $A_n^\dagger$ and A_n are the creation and annihilation operators for an exciton energy level labeled n, and γ_n is the photon or phonon relaxation constant for the energy level). The exciton population is calculated using matrix elements for all exciton states in the system considered. First, we apply the calculation to a three-level system of SQWs (Fig. 10.9), where the continuum state $\hbar\Omega_C$ is initially excited using a 10-ps laser pulse. Then, the initial exciton population in ZnO QWs is created in $\hbar\Omega_{C1S}$, where the energy transfer from $\hbar\Omega_C$ to $\hbar\Omega_{C1S}$ is expressed phenomenologically as a Gaussian input signal with a temporal width of $2\sigma_{1S}$ (an incoherent excitation term is added in (10.1)), because nonradiative relaxation paths via exciton–phonon coupling make a dephased input signal, statistically. Finally, an exciton carrier relaxes due to the electron–hole recombination with relaxation constant γ_{1S}. Figure 10.9b shows a numerical result and experimental data. Here, we used $2\sigma_{1S} = 100$ ps, and γ_{1S} was evaluated as 460 ps.

A similar calculation was applied for DQWs. We used two three-level systems, coupled via an optical near field with a coupling strength of U_{12} (Fig. 10.9a). Figure 10.9b shows the numerical results for the exciton population in QWA and the experimental data. Here, $2\sigma_{1D}$ and $2\sigma_{2D}$ were set at 200 ps, which is twice the value for SQWs, because the relaxation paths extend the barrier energy state in the two quantum wells (QWs). γ_{1D} and γ_{2D} are evaluated as 200 ps. We believe that the faster relaxation for DQWs compared with SQWs reflects the lifetime of the coupled states mediated by the optical near field (Fig. 10.10). Furthermore, the characteristic behavior that results from near-field coupling appears as the oscillatory decay in Fig. 10.9b, originated from the nutation of the exciton population between DQWs. This indicates that the timescale of the near-field coupling is shorter than

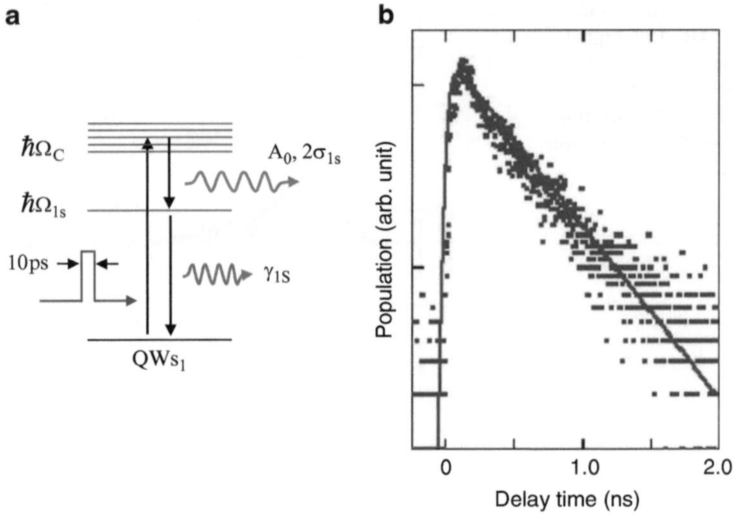

Fig. 10.9 (**a**) The schematic depict the SQWs system configurations. $\hbar\Omega_C$ barrier energy state with a central energy. (**b**) Theoretical results on the transient exciton population dynamics (*solid curves*) and experimental PL data (*filled squares*) of SQWs (same as curve TR_S in Fig. 10.8)

the decoherence time and that coherent coupled states, such as symmetric and antisymmetric states [33], determine the system dynamics. Furthermore, nutation never appears unless unbalanced initial exciton populations are prepared for $\hbar\Omega_{1D}$ and $\hbar\Omega_{2D}$ In the far-field excitation using the sharpened fiber probe, only the symmetric state is excited because the antisymmetric state is dipole-inactive. Then, the exciton populations of the two quantum wells are equal and they have the same decay rate. In contrast, in the near-field excitation, both the symmetric and antisymmetric states are excited due to the presence of a near-field probe. Since the symmetric and antisymmetric states have different eigen energies, the interference of these states generates a detectable beat signal. The unbalanced excitation rate is given by $A_1/A_2 = 10$ here. From the period of nutation, the strength of the near-field coupling is estimated to be $U_{12} = 7.7\,\text{ns}^{-1}$ ($= 4.9\,\mu$ eV).

To evaluate nutation frequencies from the time-resolved PL signal, we used Fourier analysis. In Fig. 10.11a, the power spectral density of SQWs (PSs) does not exhibit any peaks, indicating a monotonic decrease. In contrast, the power spectral density of 1-DQWs (PS_{1D}) had a strong peak at a frequency of $2.6\,\text{ns}^{-1}$. Furthermore, that of 3-DQWs (PS_{3D}) had three peaks at 1.9, 4.7, and $7.1\,\text{ns}^{-1}$. Since, the degree of the coupling strength, which is proportional to the frequency of the nutation, increases as the separation decreases, the three peaks correspond to the signals from DQWs with separations of 10, 6, and 3 nm, respectively. Since the coupling strength $\hbar U$ [eV] is given by $\hbar\pi f$ (f: nutation frequency), $\hbar U$ is estimated as 4.0, 9.9, and 14.2 μeV for DQWs with respective separations of 10, 6, and 3 nm. These values are comparable to that estimated above ($U_{12} = 4.9\,\mu$eV).

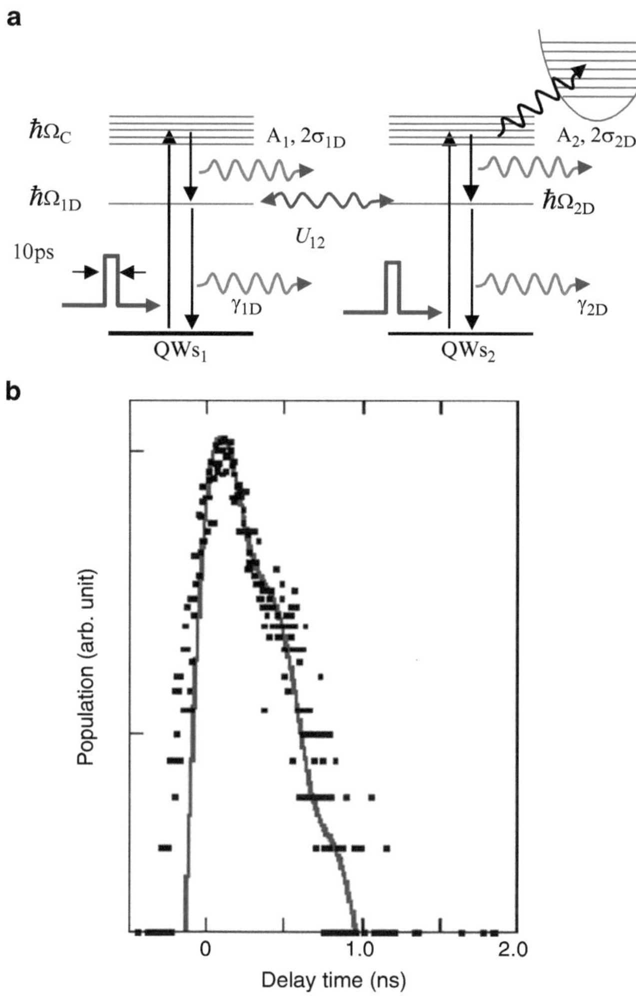

Fig. 10.10 (a) The schematic depict the 1-DQWs system configurations. $\hbar\Omega_C$ barrier energy state with a central energy. (**b**) Theoretical results on the transient exciton population dynamics (*solid curves*) and experimental PL data (*filled squares*) of 1-DQWs (same as curve TR_{1D} in Fig. 10.8)

Furthermore, the peak intensity for the DQWs with 3-nm separation is much lower than for those with 10-nm separation, which might be caused by decoherence of the exciton state due to penetration of the electronic carrier. Considering the carrier penetration depth, the strong peak of DQWs with 10-nm separation originates from the near-field coupling alone. The solid line in Fig. 10.11b shows the separation dependence of the peak frequency. The exponentially decaying dependence represented by this line supports the origin of the peaks in the power spectra from the localized near-field interaction between the QWs.

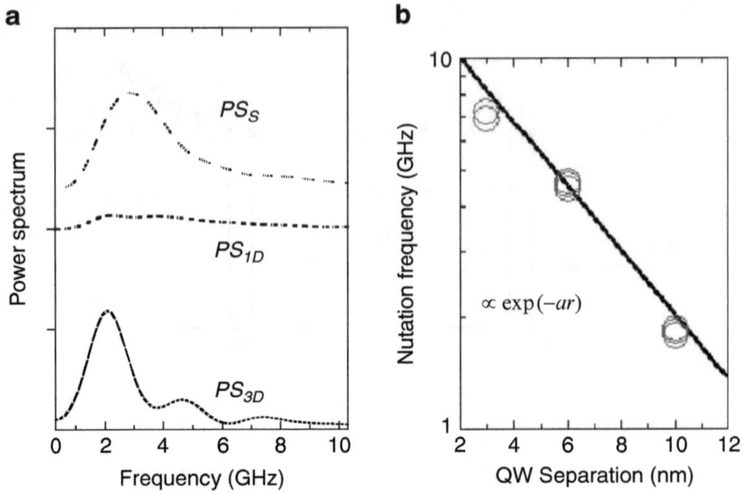

Fig. 10.11 Evaluation of the nutation frequencies between the QWs. (**a**) PSS, PS1D, and PS3D show the power spectrum of TR_S, TR_{1D}, and TR_{3D}, respectively. (**b**) Separation D dependence of frequency of the nutation

Next, we performed the switching operation. Figure 10.12a, b explains the "OFF" and "ON" states of the proposed nanophotonic switch, consisting of two coupled QWs. QWA and QWB are used as the input/output and control ports of the switch, respectively. Assuming $L_m = 3.2$ and 3.8 nm, the ground exciton state in QWA and the first excited state in QWB resonate. In the "OFF" operation (Fig. 10.12a), all the exciton energy in QWA is transferred to the excited state in the neighboring QWB and relaxes rapidly to the ground state. Consequently, no output signals are generated from QWA. In the "ON" operation (Fig. 10.12b), the escape route to QWB is blocked by the excitation of QWB owing to state filling in QWB on applying the control signal; therefore, an output signal is generated from QWA.

Figure 10.13 shows the NFPL for the three pairs of DQWs with $L_w = 3.2$ and 3.8 nm with different separations (3, 6, and 10 nm). Curve NF_{OFF} was obtained with continuous input light illumination from a He–Cd laser (3.814 eV). No emission was observed from the exciton ground state of QWA (E_{A1}) or the excited state of QWB (E_{B2}) at a photon energy of 3.435 eV, indicating that the excited energy in QWA was transferred to the excited state of QWB. Furthermore, the excited state of QWB is a dipole-forbidden level. Curve $NF_{Control}$ shows the NFPL signal obtained with control light excitation of 3.425 eV with a 10-ps pulse. Emission from the ground state of QWB at a photon energy of 3.425 eV was observed. Both input and control light excitation resulted in an output signal with an emission peak at 3.435 eV, in addition to the emission peak at 3.425 eV (curve NF_{ON}), which corresponds to the ground state of QWB. Since the excited state of QWB is a dipole-forbidden level, the observed 3.435-eV emission indicates that the energy transfer from the ground

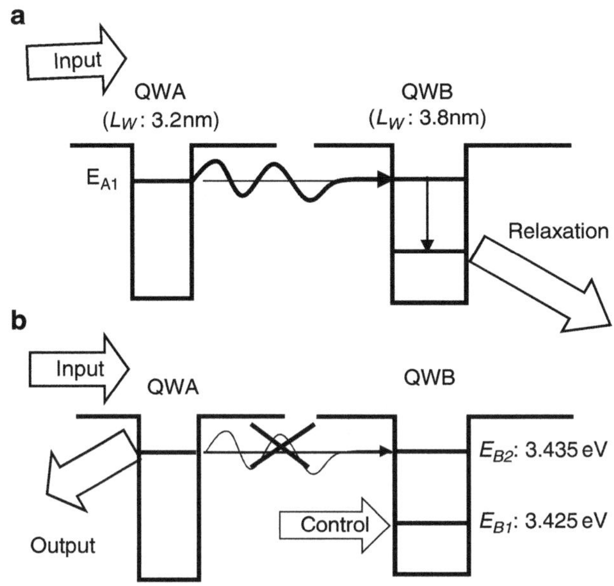

Fig. 10.12 The switching operation by controlling the exciton excitation. Schematic of the nanophotonic switch of (**a**) "OFF" state and (**b**) "ON" state

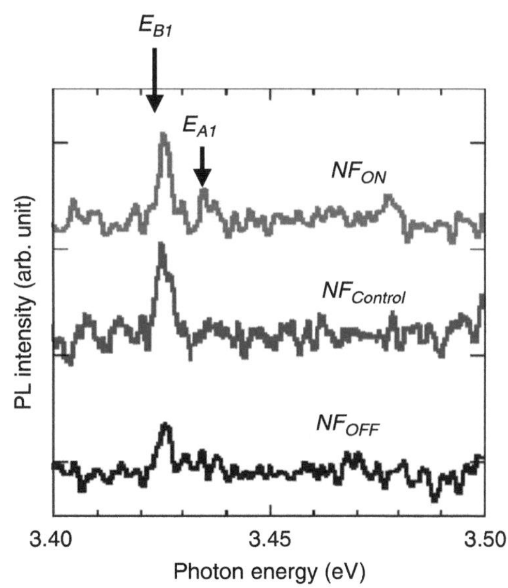

Fig. 10.13 NF_{ON}, $NF_{control}$, and NF_{OFF} show NF_{PL} signal obtained with the illumination of Input laser along, Control laser alone, and Input and Control laser, respectively

state of QWA to the excited state of QWB was blocked by the excitation of the ground state of QWB.

Finally, the dynamic properties of the nanophotonic switching were evaluated. We observed TRNFPL signals at 3.435 eV with both input and control laser

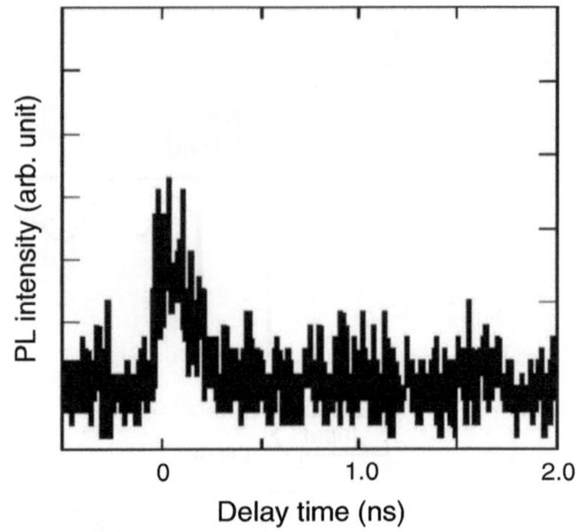

Fig. 10.14 Near-field time-resolved PL signal with "ON" state

excitation (see Fig. 10.14). The decay time constant was found to be 483 ps. The output signal increased synchronously, within 100 ps, with the control pulse. Since the rise time is considered equal to one quarter of the nutation period τ [34], the value agrees with those obtained for DQWs with the same well width in the range from $\tau/4 = 36$ ps (3-nm separation) to $\tau/4 = 125$ ps (10-nm separation).

10.5 Conclusions

This chapter reviewed nanophotonics, a novel optical nanotechnology utilizing local electromagnetic interactions between a small number of nanometric elements and an optical near field. Its potential for high integration beyond the diffraction limit of light can solve the technical problems of the future optical industry.

In order to confirm the possibility of using a nanometric ZnO nanorod heterostructures as a light emitter in a nanophotonic IC, we report on NOM of artificially fabricated ZnO/ZnMgO nanorod SQWs as a major breakthrough for realizing nanophotonic devices using a two-level system [35, 36].

We observed the nutation between DQWs and demonstrated the switching dynamics by controlling the exciton excitation in the QWs. Examination of the electronic coupling between QWs is now in progress to analyze the detailed switching dynamics. For room-temperature operation, since the spectral width reaches thermal energy (26 meV), a higher Mg concentration in the barrier layers and narrower L_w are required so that the spectral peaks of the first excited state (E_2) and ground state (E_1) do not overlap. This can be achieved by using two QWs with $L_m = 1.5$ nm (QWA) and 2 nm (QWB) with a Mg concentration of 50%, where the energy difference between E_2 and E_1 in QWB is 50 meV [37].

Acknowledgements We are grateful to Drs. Tadashi Kawazoe (The University of Tokyo), Suguru Sangu (Ricoh Company, Ltd.), and Prof. Kiyoshi Kobayashi (Yamanashi University) for many fruitful discussions. The authors thank Dr. Jinkyoung Yoo (Pohang University of Science and Technology) for sample preparation of ZnO nanorod and valuable discussions.

References

1. M. Ohtsu, K. Kobayashi (ed.) *Optical Near Fields* (Springer-Verlag, Berlin,2003)
2. M. Ohtsu (ed.) *Prefaces to Volume V* in Progress in Nano-Electro-Optics V (Springer-Verlag, Berlin, 2006)
3. M. Ohtsu, K. Kobayashi, T. Kawazoe, S. Sangu, T. Yatsui, IEEE J. Selected Topics in Quantum Electron. **8**, 839 (2002)
4. M.H. Huang, S. Mao, H. Feick, H. Yan, Y. Wu, H. Kind, E. Weber, R. Russo, P. Yang, Science, **292**, 1897 (2001)
5. A. Ohtomo, K. Tamura, M. Kawasaki, T. Makino, Y. Segawa, Z.K. Tang, G.K.L. Wong, Y. Matsumoto, H. Koinuma, Appl. Phys. Lett. **77**, 2204 (2000)
6. H.D. Sun, T. Makino, Y. Segawa, M. Kawasaki, A. Ohtomo, K. Tamura, H. Koinuma, J. Appl. Phys. **91**, 1993 (2002)
7. D.C. Reynolds, D.C. Look, B. Jogai, C.W. Litton, G. Cantwell, W.C. Harsch, Phys. Rev. B **60**, 2340 (1999)
8. Y. Wu, R. Fan, P. Yang, Nano Lett. **2**, 83 (2002)
9. M.T. Björk, B.J. Ohlsson, C. Thelander, A.I. Persson, K. Deppert, L.R. Wallenberg, L. Samuelson, Appl. Phys. Lett. **81**, 4458 (2003)
10. M.S. Gudiksen, L.J. Lauhon, J. Wang, D.C. Smith, C.M. Lieber, Nature **415**, 617 (2002)
11. W.I. Park, G.-C. Yi, M.Y. Kim, S.J. Pennycook, Adv. Mater. **15**, 526 (2002)
12. W.I. Park, S.J. An, J. Long, G.-C. Yi, S. Hong, T. Joo, M.Y. Kim, J. Phys. Chem. B **108**, 15457 (2004)
13. A. Tsukazaki, A. Ohtomo, T. Onuma, M. Ohtani, T. Makino, M. Sumiya, K. Ohtani, S.F. Chichibu, S. Fuke, Y. Segawa, H. Ohno, H. Koinuma, M. Kawasaki, Nature Materials, **4**, 42 (2005)
14. K. Matsuda, T. Saiki, S. Nomura, M. Mihara, Y. Aoyagi, Appl. Phys. Lett. **81**, 2291 (2002)
15. K. Matsuda, T. Saiki, S. Nomura, M. Mihara, Y. Aoyagi, S. Nair, T. Takagahara, Phys. Rev. Lett. **91**, 177401 (2003)
16. J.R. Guest, T.H. Stievater, G. Chen, E.A. Tabak, B.G. Orr, D.G. Steel, D. Gammon, D.S. Katzer, Science **293**, 2224 (2001)
17. T. Guenther, C. Lienau, T. Elsaesser, M. Clanemann, V.M. Axt, T. Kuhn, S. Eshlaghi, D. Wieck, Phys. Rev. Lett. **89**, 057401 (2002)
18. A. Ohtomo, M. Kawasaki, I. Ohkubo, H. Koinuma, T. Yasuda, Y. Segawa, Appl. Phys. Lett. **75**, 980 (1999)
19. T. Makino, A. Ohtomo, C.H. Chia, Y. Segawa, H. Koinuma, K. Kawasaki, Physica E **21**, 671 (2004)
20. M. Zerovos, L.-F. Feiner, J. Appl. Phys. **95**, 281 (2004)
21. M. Zamfirescu, A. Kavokin, B. Gil, G. Maplpuech, M. Kaliteevski, Phys. Rev. B **65**, 161205 (2001)
22. S.F. Chichibu, T. Sota, G. Cantwell, D.B. Eason, C.W. Litton, J. Appl. Phys. **93**, 756 (2003)
23. D.C. Reynolds, C.W. Litton, D.C. Look, J.E. Hoelscher, C. Claflin, T.C. Collins, J. Nause, B. Nemeth, J. Appl. Phys. **95**, 4802 (2004)
24. M. Tchounkeu, O. Briot, B. Gil, J.P. Alexis, R.L. Aulombard, J. Appl. Phys. **80**, 5352 (1996)
25. T. Wamura, Y. Masumoto, T. Kawamura, Appl. Phys. Lett. **59**, 1758 (1991)
26. T. Yatsui, J. Lim, M. Ohtsu, S.J. An, G.-C. Yi, Appl. Phys. Lett. **85**, 727 (2004)
27. T. Yatsui, M. Ohtsu, S.J. An, G.-C. Yi, Appl. Phys. Lett. **87**, 033101 (2005)

28. T. Yatsui, M. Ohtsu, S.J. An, J. Yoo, G.-C. Yi, Opt. Rev. **13**, 218 (2006)
29. T. Yatsui, S. Sangu, T. Kawazoe, M. Ohtsu, S.J. An, J. oo, G.-C. Yi, Appl. Phys. Lett. **90**, 223110 (2007)
30. W.I. Park, D.H. Kim, S.W. Jung, G.-C. Yi, Appl. Phys. Lett. **90**, 80 (2002)
31. B. Coffey, R. Friedberg, Phys. Rev. A **17**, 1033 (1978)
32. K. Kobayashi, S. Sangu, T. Kawazoe, M. Ohtsu, J. Lumin. **112**, 117 (2005)
33. S. Sangu, K. Kobayashi, A. Shojiguchi, M. Ohtsu, Phys. Rev. B **69**, 115334 (2005)
34. S. Sangu, K. Kobayashi, T. Kawazoe, A. Shojiguchi, M. Ohtsu, Trans. Mater. Res. Soc. Jap. **28**, 1035 (2003)
35. A. Zenner, E. Beham, S. Stufler, F. Findeis, M. Bichler, G. Abstreiter, Nature **418**, 612 (2002)
36. Z. Yuan, B.E. Kardynal, R.M. Stevenson, A.J. Shields, C.J. Lobo, K. Cooper, N.S. Beattie, D.A. Ritchie, M. Pepper, Science **295**, 102 (2002)
37. W.I. Park, G.-C. Yi, M. Jang, Appl. Phys. Lett. **79**, 2022 (2001)

Chapter 11
Semiconductor Nanowires for Solar Cells

S.T. Picraux, J. Yoo, I.H. Campbell, S.A. Dayeh, and D.E. Perea

Abstract This chapter discusses studies of semiconducting nanowire arrays for solar cells. The concept of 3D nanowire architectures for photovoltaic light harvesting to effectively decouple light absorption and carrier separation is presented. The available literature on semiconductor nanowire solar cell studies is summarized. Optical and electronic aspects specific to nanowires are discussed to illustrate the basic principles. Finally, issues related to integration for solar cell applications are highlighted.

11.1 Introduction

Nanoscale semiconductor materials have received great interest in recent years for their potential applications to solar energy harvesting. The primary motivation for this explosion of interest has been the ability to tailor materials properties at the nanoscale through size and structure in ways not possible at the macroscale. Key aspects being exploited are (1) nanostructuring to enhance the scattering, local field, absorption, and conversion of photons across the solar spectrum to generate charge carriers, and (2) nanoscale device designs to efficiently harvest the charge carriers for electrical energy generation or chemical fuel production. Areas of current research attention include arrays of radial junction nanowires to enhance photon absorption and provide efficient carrier collection [1], nanoparticle multi-exciton effects to create more than one electron–hole pair per absorbed photon [2,3], and mesoporous materials to enhance photochemical conversion in dye-sensitized solar cells [4]. In addition to the potential advantages of providing enhanced

S.T. Picraux (✉) · J. Yoo · I.H. Campbell · S.A. Dayeh · D.E. Perea
Center for Integrated Nanotechnologies, Los Alamos National Laboratory, Los Alamos, NM 87545, USA
e-mail: picraux@lanl.gov; trip297@gmail.com; campbell@lanl.gov; shadi@lanl.gov; d-perea@lanl.gov

G.-C. Yi (ed.), *Semiconductor Nanostructures for Optoelectronic Devices*, NanoScience and Technology, DOI 10.1007/978-3-642-22480-5_11,

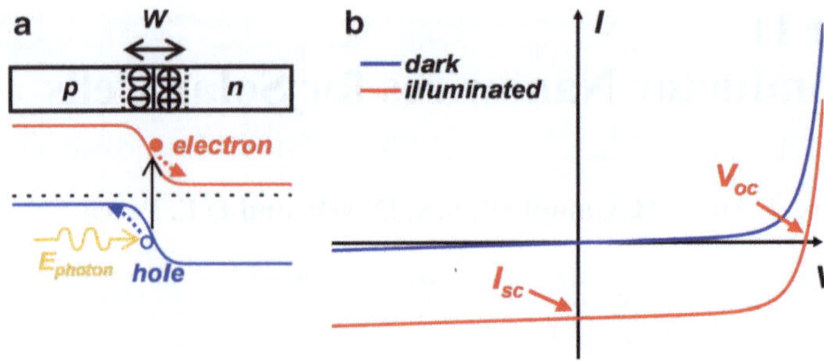

Fig. 11.1 (**a**) *Top* is schematic of a p–n junction showing depletion region (W) with ionized charges; *bottom* is the correspondent energy-band diagram. Incident light creates an electron–hole pair that is field swept to the charge-neutral regions. (**b**) Typical photodiode I–V characteristics under dark and illumination conditions

photon absorption, carrier generation, and carrier collection, nanoscale materials can be configured into more complex architectures, and synthesized directly on or transferred to low-cost substrates. In this chapter, we focus on nanowire solar cells. These structures comprise vertical arrays of semiconducting, coaxial p–n junction nanowires. The wires have typical diameters ranging from ~100 nm to ~2 μm; since they span the nano- to micrometer size range they are variously referred to as wires, pillars, nanowires, and microwires. Here, for simplicity, we exclusively use the term nanowires (NWs).

The solar cell principle, as illustrated in Fig. 11.1, is based on an unbiased p–n junction photodiode connected to a load impedance to generate power [5]. There is a region of high electric field, called the depletion region (labeled W), at the junction of the p- and n-type semiconductors where photogenerated electrons and holes are separated by the electric field, producing a current. In addition, carriers generated in either the p- or n-type material within a minority charge carrier diffusion length of the depletion region can be collected at the junction and contribute to the total current. Key criteria for effective solar cells are high absorption of the incident light in the active region of carrier collection and minimal loss of carriers due to recombination. Conventional crystalline silicon solar cells, as illustrated in Fig. 11.2a, rely primarily on thick diffusion regions (~200–250 μm), compared to the much thinner drift region, for carrier collection. This approach is due to the weak absorption of light by Si across the solar spectrum, necessitating a substantial thickness of Si to absorb the incident light. Thus, high-efficiency Si solar cells require very high-quality, defect-free material with large carrier mobilities and long carrier lifetimes. The current record efficiency under terrestrial illumination conditions for a Si cell is 25%, which is a significant fraction of the Shockley–Queisser limit of ~31% for single junction cells and was achieved over a decade ago [6]. For multijunction tandem cells composed of two or more different stacked semiconductors, the record is 42.8% (for Si/GaAs/GaInP) [6]. However, the manufacturing costs are high for such record efficiencies and commercial Si solar cells at the module level typically range from ~15 to 22% [7].

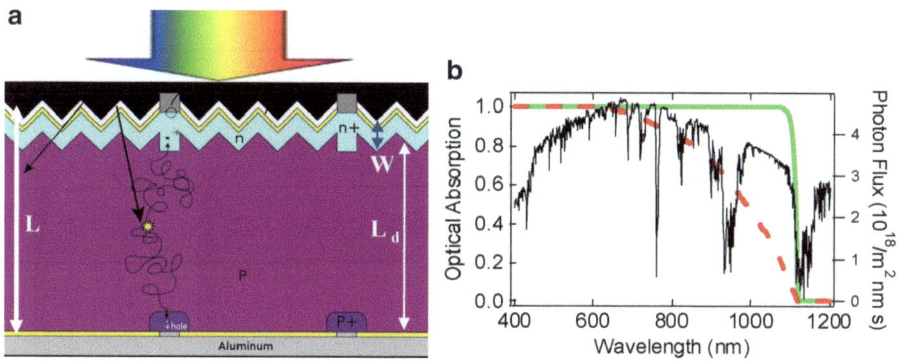

Fig. 11.2 (**a**) Schematic of a conventional Si solar cell with antireflection coating and surface texturing. Typical cell thickness, $L \sim 200\,\mu m$, is used to absorb the light; carrier collection is primarily from the diffusion region (L_d) where $L_d >> W$, the junction drift region. (**b**) Solar intensity spectrum (*black curve*) and the relative absorption for a 5 μm thick Si slab (*red dashed line*) and an array of 5 μm long Si NWs (*green line*)

The potential for improved performance and cost reductions of NW arrays over their bulk photovoltaic counterparts is primarily due to (1) increased absorption due to diffuse light scattering in NW arrays, (2) short collection lengths of minority carriers that are radially separated and collected normal to the light absorption direction, and (3) flexibility of cell integration on a variety of low-cost carrier substrates. A key point is that single-crystalline NWs can be grown by the vapor–liquid–solid (VLS) method, a process comparable to thin-film growth technology. In crystalline Si cells, the Si ingot growth and slicing account for up to 50% of the overall cost of wafer-Si cells. In contrast, thin-film solar cells are attractive due to their potential for reduced costs. However, after many years they continue to have limited cell efficiencies due to grain boundary and related carrier recombination loss issues. For example, for amorphous Si cells the highest reported efficiency is 10% [6]. Both CdTe and CIGS ($CuInGaSe_2$) thin-film cells have demonstrated higher efficiencies, but materials availability may limit their large-scale use.

With nanostructured 1D materials it is possible to achieve thin layers of single-crystal NWs grown on or transferred to low-cost substrates. Thus, a NW approach uses less high purity Si (~1/10) along with lower cost materials growth and device fabrication techniques. Although improvements in efficiency over conventional crystalline silicon solar cells are not anticipated, achieving reasonably good efficiencies (~15–20%) in combination with low-cost thin-film processing could dramatically impact the application of solar cells. Figure 11.3 illustrates this point by comparing the levelized cost of energy (LCOE) using the National Renewable Energy Laboratory "Solar Advisor Model" [8] vs. efficiency for several commercial crystalline Si and thin-film solar cells in comparison to estimated values that might be anticipated if a successful Si NW solar cell approach were achieved (P. Schuele, D. Evans, Sharp Laboratories of America, private communication).

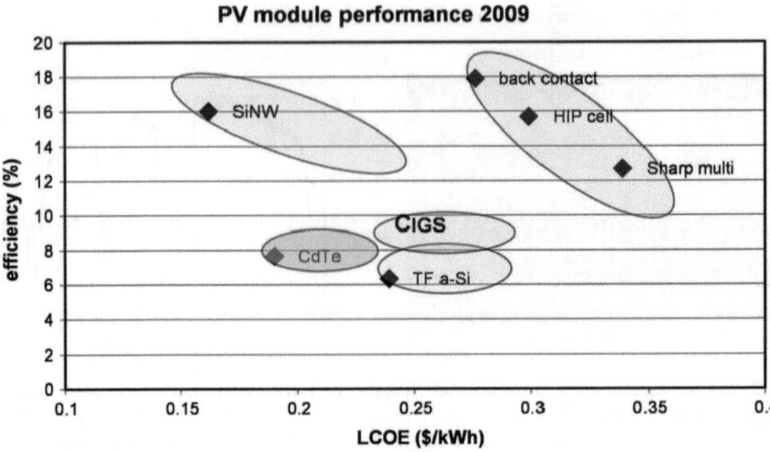

Fig. 11.3 Comparison of the levelized cost of energy model values of 2009 photovoltaic modules for several commercial Si crystalline cells (SunPower SPR-220 back contact and Sanyo HIP-190DA1 monocrystalline, and Sharp 165U1 multicrystalline) along with amorphous Si (Unisolar US64), CdTe, and CIGS thin-film cells to an estimated potential value for Si NW cells (P. Schuele, D. Evans, Sharp Laboratories of America, private communication)

11.2 Key Concepts

The orthogonalization of the light absorption and carrier collection directions is a key aspect of the radial p–n junction NW photovoltaic (PV) cell concept (Fig. 11.4). This approach is dependent on realizing three-dimensional (3D) structures. In 1994, a group at the University of New South Wales (UNSW) suggested the architecture of a parallel multijunction PV cell consisting of multilayered p–n junctions and metalized grooves [9]. The UNSW approach was to use the 3D structured cell to attain close to 100% carrier collection efficiency by dividing the PV cell into segments. Additionally, a 3D structure consisting of many etched Si stripes could provide photon recycling due to increased effective cell thickness [10]. In each Si stripe, light absorption occurs along the vertical direction, while the photogenerated carriers are extracted through the sidewalls of the Si stripe. In practice, the 3D structured PV cell based on etched Si stripes demonstrated up to 18.5% efficiency with production costs similar to surface textured Si solar cells. However, in bulk and thin-film PV cell research, the 3D structured PV cell has not been a major focus because of the difficulty of concurrent achievement of carrier loss minimization and light absorption maximization.

A radial p–n junction in an NW array architecture is an ideal structure for highly efficient p–n junction solar cells. The radial geometry of the p–n junction allows for the orthogonalization of the light absorption and carrier collection directions. In bulk and thin-film PV cells, the fabrication of radial or surrounding p–n junctions is not easily achieved, and requires micromachining processes, which increase production

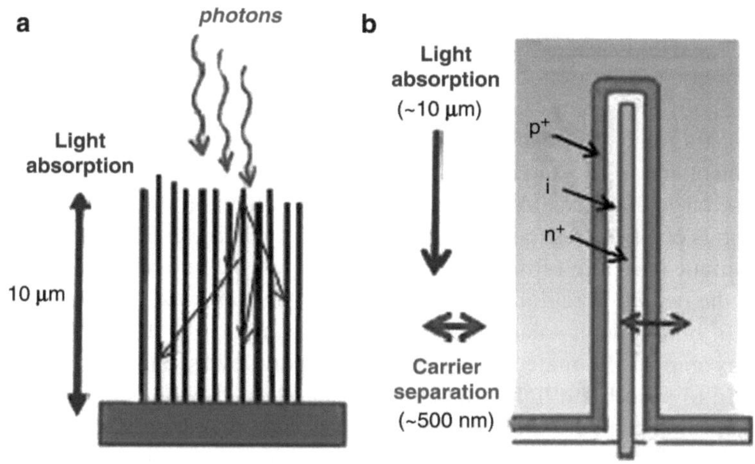

Fig. 11.4 (**a**) Diffuse scattering of photons by Si NWs greatly reduces vertical distance for light absorption. (**b**) Radial p–i–n NW junction for carrier separation and collection through substrate and top transparent contact

costs. Semiconductor NWs provide a suitable material system for radial p–n junction PV cells because as-grown semiconductor NWs are ready for radial p–n junctions. Furthermore, semiconductor NWs can overcome material compatibility issues critical to thin-film PV cells. For example, mismatches of thermal expansion coefficients and lattice constants between the substrate and thin films over large areas lead to stresses, whereas single-crystal semiconductor NWs can be prepared on various substrates by diverse methods with minimal accumulated stress. Minority carrier diffusion lengths (L_{diff}), an important material characteristic in determining solar cell efficiency, are in the range of ~100s of nanometers to several micrometers in thin films because they are limited by grain sizes, whereas values of L_{diff} in single-crystalline Si NWs have been reported in the range of 2 to > 20 μm.

The orthogonalization through 1D nanostructuring is a key factor in the NW approach to solar cells where the optical absorption and charge collection constraints are decoupled. In conventional crystalline silicon solar cells, a basic problem is that the optical absorption is weak. As a result thick Si regions are required to absorb the incident light and charge carrier generation occurs throughout the region (Fig. 11.2a). Thus, minority charge carriers must be transported by diffusion over large distances without recombination, necessitating the use of essentially perfect single-crystal materials. For example, for the highest silicon solar cell efficiencies ~260 μm thick region of high-quality single-crystal Si wafer is required. In contrast, crystalline Si NW arrays can absorb the light in ~10 μm and simultaneously allow independent control over the charge carrier separation (Fig. 11.4). The first key difference is that the light collection takes advantage of the diffuse scattering of photons by the NWs due to the dielectric constant difference with the surrounding medium to allow complete absorption and carrier excitation within a depth of

$\leq 10\,\mu m$. The second key difference is that rapid carrier separation ($\leq$ns) occurs by drift in the high electric field region within the radial p–n junction with distances of $\sim 1\,\mu m$ for the carriers to be collected by the single-crystal NW conducting core and shell. This contrasts to the process in conventional solar cells where slow carrier separation ($\sim\mu s$) occurs by diffusion with carriers collected over distances of $\geq 100\,\mu m$, requiring the absence of recombination centers (or charge traps) over these large distances. As a result the carrier collection times are dramatically reduced in NWs, from μs to ps, facilitating efficient charge collection.

The unique structure of the NW solar cell allows one to maximize the effective width of the depletion region and collect the carriers absorbed in the n- and p-type regions. In bulk silicon solar cells, the depletion region is $<10\,\mu m$ as limited by the energy gap of the material and residual doping levels. Since $\sim 200\,\mu m$ of Si is required to absorb the light efficiently, the depletion region is only $\sim 5\%$ of the absorbing volume. In contrast, in the NW structures the amount of Si required is much smaller and the depletion region can be made to be a large volume fraction, $\sim 50\%$, of the semiconductor carrier collection region. Similarly, because of the nanostructure design, the p, n, and any intrinsic regions are relatively thin in the radial direction, and carriers with effective minority carrier diffusion lengths of the order of $\sim 1\,\mu m$ will be effectively collected. This means the Si NWs could still function effectively with mobility-lifetime ($\mu\tau$) products four orders of magnitude smaller (lower quality material) than conventional Si solar cells. In addition to maximizing photon collection, the diode series resistance and defect-related surface states need to be minimized. Similar to bulk solar cells the series resistance can be made small by increasing the doping and size of the core and outer shell p+ and n+ regions. An important issue is that surface and interface states can act as recombination centers and reduce charge carrier collection. The adverse effects of such states must be minimized for these nanoscale architectures if acceptable efficiencies are to be achieved. Preliminary results, as summarized in Sect. 11.4, provide initial evidence for the benefits of NW array solar cells and highlight the need to develop methods to produce high performance, single-crystal p–i–n NW devices.

11.3 Nanowire Fabrication

The NW solar cell concept emerged from recent experience in the growth of semiconducting NWs by bottom-up techniques. The VLS technique, while first reported almost a half century ago [11], provides good control of bottom-up NW growth, and has been studied intensively over the last decade. VLS growth has become the preferred method for NW fabrication because it provides a versatile method to grow single-crystal wires of Si and other semiconductors based on chemical vapor deposition (CVD) methods [12]. Further, this growth method does not require crystalline substrates to achieve good single-crystal NWs. However, the NWs are randomly tilted when grown from a noncrystalline substrate, whereas

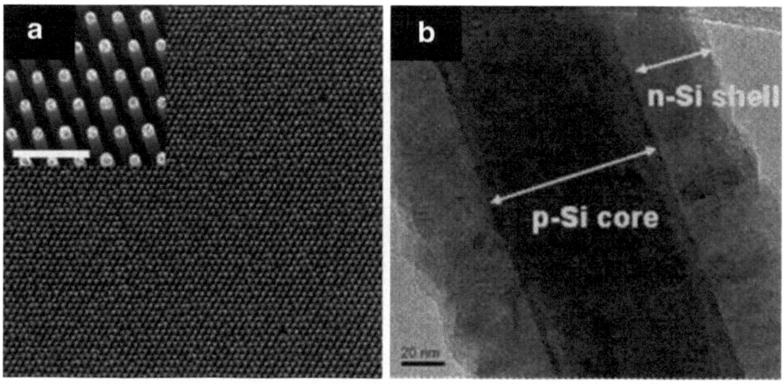

Fig. 11.5 (**a**) Lithographically patterned 400 × 400 μm VLS-grown Si NW array with diffused P shell on Si(111) substrate. *Inset* shows enlargement of 2.6 μm diameter NWs; *scale bar*: 20 μm [13]. (**b**) TEM image of Si NW with diffused shell; *scale bar*: 20 nm (image courtesy of Dr. Qi. Li and Prof. E.C. Dickey, www.personal.psu.edu/jmr31/index_files/Page1004.htm)

regular vertical arrays can be achieved generally for larger diameter NWs grown from crystalline Si(111) substrates. VLS growth proceeds from the substrate via a metal-catalyzed nanoparticle seed. Gold is most frequently used as the catalyst with either SiH_4 at temperatures ∼500°C or $SiCl_4$ at temperatures of ∼1,000°C used as the source gases. Figure 11.5a shows an array of Si NWs where the growth seeds were lithographically patterned to achieve a regular array [14]. In this case, the n-type shell was formed by P diffusion subsequent to $SiCl_4$ NW growth after removal of the Au growth seed. Cu has also been recently used as the growth catalyst, with improvements in the minority carrier diffusion lengths being reported for $SiCl_4$ growth [13]. To better understand the ultimate limits of efficiency for NW solar cells, the present authors have prepared Si NWs by a metal-free, top-down Bosch reactive ion etch process. Figure 11.6a shows a Si NW array on a thin Si substrate. Unlike other reports, these studies have achieved single-crystalline radial n^+ shell grown directly on the NWs by a vapor–solid method using SiH_4 and PH_3 precursors as demonstrated in Fig. 11.6b. Compound semiconductor NWs can also be grown by the bottom-up VLS process and radial vapor–solid shells have been grown. While improvements in the various processing steps for NW growth and radial junction formation are still needed for solar cell applications, the fabrication methods are encouraging.

11.4 Overview of Nanowire Solar Cell Studies

Despite the described advantages of using semiconducting NWs for solar cells, research on semiconductor NWs and semiconductor-based photovoltaics was essentially conducted separately until recently. Many researchers have addressed challenges related to NW materials science and device fabrication, including doping, axial, and radial heterostructure formation, electrical contact formation,

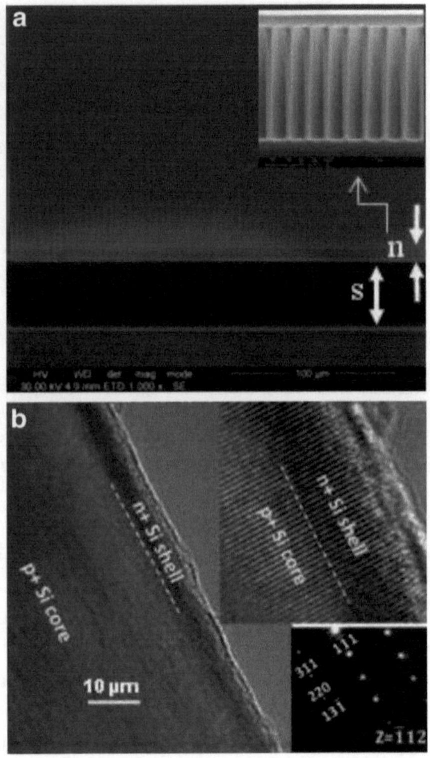

Fig. 11.6 (**a**) SEM tilted view of a large area top-down fabricated array of 10 μm tall Si NWs (n) on 40 μm thick Si substrate (s); *inset* shows close-up of Bosch-etched 500 nm diameter Si nanopillars. (**b**) TEM image of a single-crystalline epitaxial radial shell on a Si NW formed by vapor–solid growth of P-doped Si; *inset* shows HRTEM lattice image of the p+ core to n+ shell region and corresponding diffraction pattern

and integration of NWs on various substrates. These efforts have opened up opportunities for the realization of radial p–n junction PV cells. Research on semiconductor NW solar cells first received attention in 2005 when a group from Caltech published a paper that discussed important device physics aspects of a radial p–n junction nanorod geometry for efficient, low-cost solar cells [15]. In this paper, the authors formulated the advantages of radial p–n junctions compared to planar p–n junctions based on simultaneous maximization of light absorption and carrier collection through separation of the light absorption direction and carrier collection direction, and supported the proposed benefits with a model. The enhanced light absorption in Si NW arrays was formulated later by a group from MIT in 2007 [16], and has since been confirmed by many researchers [17–19]. A group from GE provided among the earliest demonstrations of Si NW solar cell operation using VLS growth of arrays of crystalline Si NW cores on metal foil followed by a radial shell of amorphous Si (a-Si) [20]. A Harvard group provided an early demonstration of the PV properties of single Si radial p–n and p–i–n junction NW solar cells [21].

Since 2007, the performance characteristics of radial NW p–n junction solar cells have been reported for various fabrication approaches as summarized in Table 11.1 for Si and Table 11.2 for compound semiconductors. In these tables, the experimental results are classified by material preparation process. The NW

Table 11.1 Measured properties of Si nanowire-based photovoltaic cell

Nanostructure formation	Radial p–n junction formation	Structure	Efficiency (%)	V_{oc}(V)	J_{sc}(mA/cm^2)	FF	References
VLS	Amorphous Si	ITO/n-type a-Si shell/p-Si NWs on metal foil	0.1	0.13	–	0.28	[20] – GE, USA (2007)
		n^+-Si/i-a-Si/i-Si NWs on glass		0.24	2.6	0.41	[22] – Ecole Polytechnique, France (2010)
	Polycrystalline Si	Single n-Si/i-Si/p-Si NW	3.4	0.26	$I_{sc} = 0.503$ nA	0.55	[21] – Harvard U., USA (2007)
		Al_2O_3/p-Si shell/n-Si NWs on n-Si, back contact	1.8	0.28	18	–	[23] – IBM, USA (2009)
		n-Si/p-Si NWs on p-Si	1.8	0.29	11	0.56	[24] – Penn. State U., USA (2010)
	Diffusion	n-Si shell/p-Si microwires on p-Si (p- and n-layers separated by oxide)	7.9	0.5	24	>0.65	[25] – Caltech, USA (2010)
		n-Si shell/p-Si microwires on p-Si	2.3	0.5	7.6	0.57	[13] – Penn. State U., USA (2010)
		Single silicon nitride/n-Si shell/p-Si microwires on p-Si (111)	9	0.54	23	0.75	[26] – Caltech, USA (2011)

(continued)

Table 11.1 (continued)

Nanostructure formation	Radial p–n junction formation	Structure	Efficiency (%)	V_{oc}(V)	J_{sc}(mA/cm^2)	FF	References
Dry etching	Amorphous Si	ITO/n–i–p *a*-Si/TiAuAg/etched and tapered Si nanopillars	8.2	0.89	13.9	0.66	[27] – Boston U./ Solasta, USA (2010)
	Ion implantation	p-Si shell/n-Si conical-frustrum array on n-Si	10.8	0.59	26.4	0.69	[28] – Cornell U., USA (2010)
	Diffusion	p-Si shell/n-Si nanopillars	5.3	0.52	16.8	0.61	[19] – U. C. Berkeley, USA (2010)
		n-Si shell/p-Si nanopillars	8.7	0.56	20	0.78	[29] – Penn. State U., USA (2010)
Wet etching	Polycrystalline Si	p-Si shell/n-Si NWs	0.46	0.29	4.3	0.33	[30] – U. C. Berkeley, USA (2008)
	Diffusion	n-Si shell/p-Si NWs on p-Si	7.19	0.5	20.6	0.70	[31] – Hanyang U., Republic of Korea (2010)
		n-Si shell/p-Si NWs on p-Si	1.47	0.45	6.3	0.53	[32] – NTU, Singapore (2010)
Conventional crystalline Si	Highest confirmed	Conventional cell	25.0	0.706	42.7	0.882	[6]
Conventional amorphous Si	Highest confirmed	Conventional cell	10.1	0.886	16.75	0.67	[6]

Table 11.2 Measured properties of compound semiconductor nanowire-based photovoltaic cell

Nanostructure formation (Material)	Radial p–n junction formation	Structure	Efficiency (%)	V_{oc} (V)	J_{sc}(mA/cm^2)	FF	References
VLS (GaAs)	Epitaxial grown (GaAs)	ITO/p-GaAs/n-GaAs NWs on n-GaAs(111)B	0.83	~0.2	$I_{sc} = 0.2$ mA	0.27	[33] – McMaster U., Canada (2009)
		Single n-GaAs/i-GaAs/p-GaAs NW	4.5				[34] – EPFL, Switzerland (2009)
		ITO/SiO_x/p-GaAs shell/n-GaAs NWs on n-GaAs substrate	1.09	0.14	$I_{sc} = 0.36$ mA	0.27	[35] – McMaster U., Canada (2011)
VLS (GaN)	Epitaxial grown (GaN)	Single p-GaN/InGaN/n-GaN NW	0.25	1.0	$I_{sc} = 58$ pA	0.56	[36] – Harvard U., USA (2009)
Electrodeposition (ZnO)	Electrodeposition (Cu_2O)	Au or Ag/p-Cu_2O/n-ZnO NW/ITO	0.47	0.28	4.4	0.39	[37] – Ludwig-Maximilians U., Germany/U. Cambridge, UK (2010)
Solution grown (ZnO)	Solution grown (Cu_2O)	Au or Al/p-Cu_2O NPs/n-ZnO NW/ITO	0.053	0.15	1.43	0.25	[38] – UC Berkeley, USA (2009)
VLS (CdS)	Thermal evaporation (CdTe)	Au/Cu/p-CdTe/n-CdS NW/Al	5.8	0.62	21	0.42	[39] – UC Berkeley/LBL, USA (2009)
ELCD (CdS)		Graphite/Cu/p-CdTe/n-CdS NW/ITO	6.5	0.71	25.3	0.36	[40] – U Kentucky, USA (2011)
Anodization (TiO_2)	Electrodeposition (Cu_2O)	Au/p-Cu_2O shell/n-TiO_2 nanotube/Ti foil	~0.01	0.25	0.33	0.27	[41] – UCLA, USA (2011)

formation process can be divided into bottom-up VLS growth and top-down dry or wet etching methods. For Si, the radial NW p–n junctions have been formed by dopant diffusion or vapor deposition, whereas cells with single-crystalline shells grown on NW cores have not been reported. To date, published results have reported NW solar cells with either diffused dopants to form single-crystalline radial shells on crystalline core or deposited vapor–solid amorphous or polycrystalline shells [20–23, 27, 30]. For compound semiconductor NWs, a variety of methods have been used to obtain doped shells including ion implantation, diffusion, and MBE [13, 19, 25, 26, 31, 32], with single-crystal core/shell structures demonstrated for GaN/InGaN [36].

In spite of the predicted advantages, NW radial p–n junction solar cell devices initially gave low efficiencies (<1%). The low efficiencies indicated that there are still many unresolved issues such as obtaining high-quality shell formation, reducing interfacial carrier recombination loss due to the large NW surface-to-volume ratio, and forming conformal transparent electrodes on the surface of radial p–n junction structures with low specific contact resistance. To resolve these issues, researchers have achieved increasingly improved crystalline shells by diffusion and ion implantation, surface passivation by dielectrics, and deposition of transparent conducting oxide as an electrode layer. These results are summarized in Table 11.1 for Si NWs, with several NW radial p–n junction Si solar cells having exhibited efficiencies in the range of 5–10%. These efficiency values are still far from those of conventional crystalline Si solar cells (see Table 11.1), but larger than that of other NW-based solar cell approaches, such as hybrid and dye-sensitized cells [42, 43].

Semiconductor NW radial p–n junction PV cells still have much room for improvement in their device performance. The process of improving device performance will help to clarify important issues in both the science and technology of nanoscale structures, both of which are closely related to fundamental understanding of how fabrication processes affect materials characteristics. On the science side, NW radial p–n junctions provide a novel platform to investigate carrier transport and photon management, as discussed in the following sections. One important material parameter for PV devices is the effective minority carrier diffusion length, and the values reported for Si NWs to date are summarized in Table 11.3 along with bulk values for Si at comparable doping levels. These effective diffusion lengths have been measured by electron beam induced current (EBIC) and scanning photocurrent microscopy (SPCM) techniques with the studies primarily conducted for axial NW junction configurations. Since the NW surface provides a favorable site for carrier recombination, it is useful to compare the measured effective diffusion length to the NW diameter. From Table 11.3, these ratios are seen to vary from $\sim$1 to 30. It is generally observed that these effective diffusion lengths are much lower than their bulk counterparts, but exact diameter dependences cannot be concluded at this time given the different doping levels and limited results. Maintaining relatively long diffusion lengths for the p–n junction NWs in solar cells will be important if reasonable efficiencies are to be achieved. In the following sections, insight into the various fundamental materials issues is discussed. Technologically, NW radial p–n

Table 11.3 Measured effective minority carrier diffusion lengths for nanowires

Material	Structure	Method	$N_A (cm^{-3})$	$N_D (cm^{-3})$	d (nm)	ℓ_p (nm)	ℓ_p/d	ℓ_n (nm)	ℓ_n/d	References
NW (Si)	Au/n-Si/Au	EBIC	–	$\sim 1 \times 10^{19}$	30–100	20–80	~1	–	–	[44] – Northwestern U., USA (2008)
	Al/n-Si/Al Schottky	SPCM	–	1 - 10×10^{17}	900	2,200	~2.3	–	–	[45] – Caltech, USA (2008)
	Al/p-Si/n-Si/Al Diffused radial p–n junctions	SPCM	–	$\sim 1 \times 10^{17}$	~1, 500	–	–	> 20,000	> 13	[26] – Caltech, USA (2011)
	Ni/n-Si/p-Si/Ni VLS axial p–n junctions	SPCM	2×10^{16}	2×10^{18}	80	660	~8	980	~12	A. Mohite et al. (in preparation) Los Alamos Natl. Lab., USA
NW (CdS)	Ti/i-CdS/Ti Schottky	SPCM	–	–	50	650	~10	1,470	~30	[46] – Northwestern U., USA (2006)
Bulk	Si	–	1×10^{17}	1×10^{17}	–	70,000	–	200,000	–	[47]
Bulk	Si	–	1×10^{18}	1×10^{18}	–	20,000	–	40,000	–	[47]
Bulk	Si	–	1×10^{19}	1×10^{19}	–	2,000	–	4,000	–	[47]

junction cell processing also presents many challenging issues, including large-scale processing, integration onto various substrates, small sizes, and highly sensitive surfaces. Integration of NW-based PV cells onto flexible substrates is a particularly important issue for the eventual achievement of highly efficient, low-cost solar cells and this area is discussed briefly at the end of the chapter.

11.5 Enhanced Optical Absorption in Nanowire Arrays

An efficient solar cell must absorb essentially all of the solar radiation above its energy gap and collect all the photogenerated charges. Increasing the effective optical absorption of the solar cell material allows one to use less and/or lower quality material. Less material can be used because the photons will be absorbed in a smaller volume. The resulting photogenerated charges are easier to collect because they are created closer to the device electrodes. Therefore, lower quality material with poorer charge collection properties, e.g., relatively small minority carrier diffusion lengths, can efficiently collect the charge. Using less and/or lower quality material decreases the absorbing layer cost which can be beneficial if these savings exceed the costs of the additional processing and materials used to increase the optical absorption.

The relative utility of using NWs to enhance light absorption depends significantly on the material being used. The most important material parameter is the optical absorption depth for the wavelengths of light in the solar spectrum (Fig. 11.2b). The smaller the absorption depth the less material is required to absorb the photons in a simple thin-film structure. Figure 11.7 shows the optical absorption depth as a function of wavelength for three representative solar cell materials: cadmium telluride (CdTe), gallium arsenide (GaAs), and crystalline

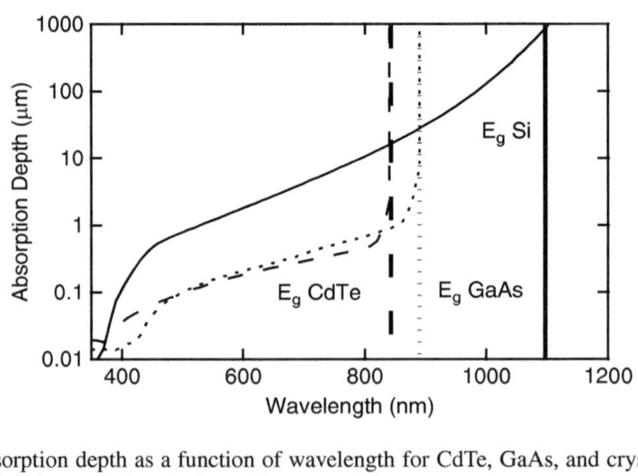

Fig. 11.7 Absorption depth as a function of wavelength for CdTe, GaAs, and crystalline Si. The energy gaps of each material are indicated by *vertical lines*

silicon (c-Si). Cadmium telluride is becoming the dominant thin-film solar material and GaAs is widely used in expensive, high-efficiency solar cells. Crystalline silicon is the dominant solar cell material accounting for ~90% of the world's solar cell production. Cadmium telluride and GaAs have direct band gaps of ~1.5 eV(830 nm) and ~1.42 eV(875 nm), respectively, and silicon has an indirect band gap of ~1.12 eV(1, 100 nm). For CdTe and GaAs, the absorption depths are <1 μm for all wavelengths from 350 nm to nearly their band gaps. In contrast, the absorption depth of Si is >1 μm for all wavelengths longer than ~550 nm and exceeds a depth of 100 μm for wavelengths up to ~100 nm above the energy gap. Because of this two orders of magnitude difference in absorption depth, most NW solar efforts are being focused on c-Si structures. Crystalline silicon NW structures have the potential to use both significantly less and lower quality material.

The effective absorption of solar cell material can be increased by three basic approaches: (1) decreasing the reflection of incident light; (2) increasing the optical path length in the material; and (3) increasing the optical intensity in the material. In many cases, the same physical structure produces combinations of these effects simultaneously. Conventional c-Si cells utilize antireflection (AR) coatings and surface texturing to decrease the reflection and increase the effective optical path length in the absorbing silicon layer [48]. A typical AR coating is a double layer MgF_2/ZnS thin-film structure [49]. The average reflectance from the AR-coated textured surface over the relevant 350–1,100 nm wavelength range is about 1%. The average path length enhancement of the cell, due to surface texturing, is a factor of 40 [50]. Due to these absorption enhancing designs, c-Si cells have been reduced to ~100 μm thickness, yet they efficiently collect charge where the fundamental material absorption depth is ~1, 000 μm [48]. The record c-Si solar cell is 25% efficient and commercial production cells are ~22.5% efficient [51]. The theoretical efficiency for these c-Si cells is ~28% so they are close to ideal [52]. The objective of Si NW array architectures is to achieve at least ~70% of the performance of these high-efficiency silicon cells, but at a much lower cost. From an optical perspective, the NW array can be designed to fulfill the functions of the antireflection coating and surface texturing while simultaneously using dramatically less silicon than a conventional cell.

11.5.1 Basic Principles of NW Array Optics

Inorganic solar cell materials have high indices of refraction, $n_s > 2.7$, and their bulk reflectivity is ≥30%. Figure 11.8 shows the reflectivity of bulk CdTe, GaAs, and c-Si as a function of wavelength. NW geometries can reduce this reflectivity substantially. For NW diameters less than the wavelength of light, the principal component of the decrease in reflectivity is due to the decrease in the effective material index. A NW array can be considered as a thin film consisting of a spatial distribution of columns (NWs) surrounded by void (refractive index = 1). The index of the NW film is determined by the index of the NW material, the volume fraction occupied by the NWs, and the alignment of the NWs with respect to the

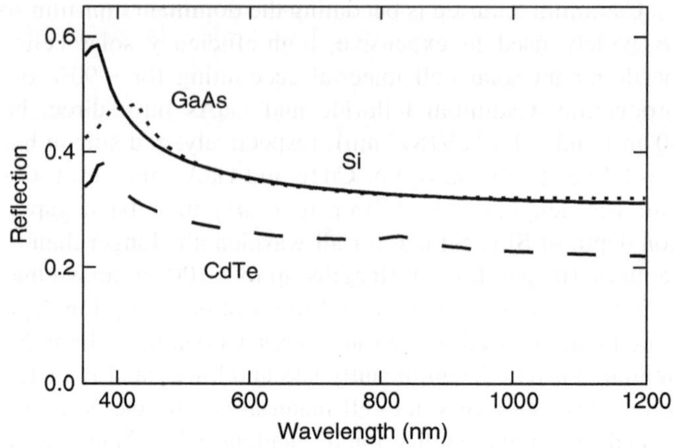

Fig. 11.8 Calculated reflection as a function of wavelength for bulk CdTe, GaAs, and crystalline Si

optical electric field. There are two limiting cases for the orientation of the NWs: parallel and perpendicular to the electric field. For perpendicular electric fields the film index, n_{f}, is approximately [53,54]

$$n_{\mathrm{f}}^2 = \frac{\left[(1-f)+(1+f)n_{\mathrm{s}}^2\right]}{\left[(1+f)+(1-f)n_{\mathrm{s}}^2\right]}, \tag{11.1}$$

where n_{s} is the refractive index of the semiconductor NW material and f is its volume fraction. For parallel electric fields, the index is [53,54]

$$n_{\mathrm{f}}^2 = fn_{\mathrm{s}}^2 + (1-f). \tag{11.2}$$

The resulting indices as a function of volume fraction are shown in Fig. 11.9 assuming c-Si, $n_{\mathrm{s}} = 3.5$. In the typical perpendicular case, the index at a volume fraction of 0.5 is only ~1.5, which is considerably lower than 2.25, the linear interpolation between the two indices. This decreased index alone reduces the reflection from ~31 to 4%. The calculated reflection for perpendicular electric fields is shown on the right vertical axis in Fig. 11.9. Lower NW volume fractions have even lower reflection. Further decrease in the surface reflection can be achieved using tapered, conical structures in which the effective volume fraction of the solar material is negligible at the first surface (point of the cone) and gradually increases into the NW array. Because the decrease in reflection comes from a decrease in the effective index of refraction, it is relatively insensitive to the angle of incidence, particularly for NW arrays with small volume fraction.

NW array geometries and particles which scatter the light can also lead to an increase in the effective optical path length. An NW array with features comparable or smaller to the wavelength of light is a discontinuous dielectric environment which leads to light scattering and diffraction. For wavelengths that require light trapping,

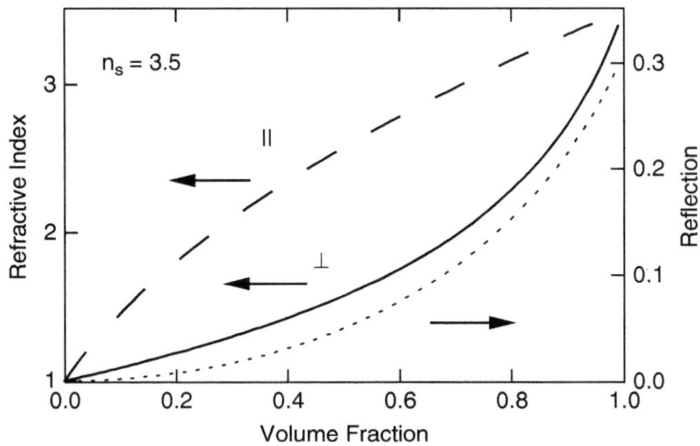

Fig. 11.9 Calculated refractive index for a thin film of columnar structures as a function of the volume fraction of columns for electric fields parallel and perpendicular to the column axis (*left vertical axis*). Calculated reflection for electric fields perpendicular to the columns (*right vertical axis*)

the NW array can be considered a weakly absorbing thin film. The basic approach to increasing the path length is to scatter light such that it propagates within the film which is then confined by total internal reflection. A widely considered ideal thin-film case is a perfectly transmissive, Lambertian (propagation direction randomized) top surface with a perfectly reflective bottom surface. In this case, the average path length is increased to

$$l = 4n_s^2 d, \tag{11.3}$$

where d is the film thickness [55]. For c-Si, $n_s = 3.5$, this corresponds to a path length of $\sim 50d$. This factor of 50 enhancement is only slightly more than the factor of 40 enhancement achieved in record efficiency c-Si cells. However, typical NW arrays have structures that vary with depth into the NW film either by design or as a result of fabrication variability. They thus scatter light throughout their volume as opposed to the model case of scattering by a single Lambertian surface. Thus, even larger path length enhancements should be achievable from NW films.

The scattering efficiency (scattering cross section/particle physical cross section) of particles less than or comparable to the optical wavelength can be obtained from Rayleigh and Mie theory. The most efficient scattering size (spherical particles) occurs for

$$\pi n_s^2 D \geq 5\lambda, \tag{11.4}$$

where D is the particle diameter and λ is the optical wavelength [56]. The scattering efficiency is essentially constant above this threshold. For c-Si at 1,100 nm, the scattering is most efficient for particles $> \sim 140$ nm. The total light scattering is a product of the scattering efficiency and the number of scatterers, so the best scattering is obtained for a dense array of ~ 100 nm particles.

So far we have considered ensemble NW effects and have only required that the individual NWs have diameters less than or comparable to the wavelength of light. For NWs of this size, there are also wave guiding and resonance effects determined by the individual NW geometry and optical properties. For wavelengths comparable to the NW diameter, the NWs act as efficient energy guides for propagation along their axis and the optical modes are largely confined to the NW. The NW guided modes are strongly confined for wavelengths

$$\lambda_g \sim n_s D, \tag{11.5}$$

where D is the NW diameter [57]. For Si NWs with $n_s = 3.5, 1,100$ nm light is strongly guided for NW diameters above ~300 nm. Thus for NW diameters ≥300 nm the NWs can act as waveguides for all wavelengths above the energy gap of Si relevant for solar cell performance. For light propagating perpendicular to the NW axis, the individual NWs also exhibit resonance effects that can alter the electric field strengths inside the wire. The resonance frequencies can be determined using the framework of leaky mode resonances [58, 59]. Silicon NWs of ~300 nm diameter can have resonances for optical wavelengths of ~1, 000 nm. Both of these wave guiding and resonance effects can increase the electric field in the NW above that in a thin-film structure and thus increase its relative absorption. These effects are usually small compared to the combination of reduced reflection and light scattering, but it may be possible to design structures that optimize these resonances particularly to enhance absorption for wavelengths around 1,000 nm where c-Si absorption is very weak.

These basic optical effects in NW arrays imply that optical absorption will be improved most in vertical NW arrays of small diameter (~100 nm). The vertical orientation is best for decreasing reflection and small diameters lead to the most efficient light scattering and trapping. All of these optical effects have been included in numerical optical modeling of the NW arrays [60–63] using finite difference time domain or transfer matrix methods. Theoretical calculations generally indicate that the optical performance of the NW array is relatively insensitive to the array periodicity for spacings in the 300–1,000 nm range for small wire volume fractions <0.3 [63].

11.5.2 Experimental Demonstrations of Increased Absorption

There has been a large amount of experimental work investigating the optical performance of NW arrays. It has been demonstrated that NW array structures, without additional antireflection coatings, can minimize reflection and increase the absorption (optical path length) [19,28]. The best structures can have <1% reflection and near 100% absorption throughout the 350–1,100 nm Si solar window. One example that highlights the optical efficiency of NW arrays is shown in Fig. 11.10 [28]. There the optical reflection, transmission, and absorption as a function of

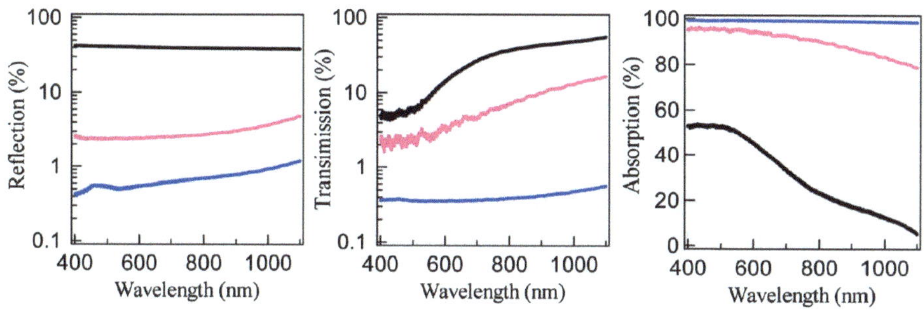

Fig. 11.10 Measured reflection (R), transmission (T), and absorption ($1-R-T$) as a function of wavelength for a nanocone array (*blue*), an NW array (*red*), and a reference thin c-Si film (*black*) [28]

wavelength for c-Si NW (red) and nanocone (blue) arrays and, for reference, for a 5 μm thin c-Si film (black) is shown. The nanocones had 340 nm top diameters and 800 nm bottom diameters, and the NWs had 340 nm diameters. The wires and cones were both 3.5 μm long arranged in a close packed array with 800 nm lattice constants and were supported on a 1.5 μm thin c-Si film (on 2 μm of SiO_2). The absorption was calculated from the measured reflection and transmission. The reflection from the c-Si thin film is similar to the bulk reflection shown in Fig. 11.8. This reflection is slightly larger because of the additional reflections from the c-Si/SiO_2 and SiO_2 /Air interfaces. The reflection from the NW array varies from about 2 to 4%, consistent with the simple estimates discussed above. The nanocone array has less than 1% reflection for all wavelengths. The transmission and calculated absorption for the thin film are consistent with the c-Si optical absorption depth shown in Fig. 11.7. The absorption for both arrays is dramatically superior to the reference thin film. The nanocone array has ~99% absorption throughout the spectrum and the NW array has absorption ranging from ~95 to 80%. These structures demonstrate the extremely efficient optical absorption that can occur in nanostructured Si arrays. However, they also serve to demonstrate that optical considerations alone do not determine solar cell performance. Solar cells were fabricated using these nanocone arrays and their efficiency was only 10.8% [28], much lower than the ~25% implied by the optical absorption alone.

In recognition of the electrical device problems of working with small diameter NWs, some groups have focused on using relatively large (several μm) diameter wires [25, 29]. The optical performance of arrays of such large diameter wires are quite poor; they have significant surface reflection and do not dramatically increase absorption. To improve the optical performance, one group works with long (25–50 μm) wires, coats the individual NWs with a thin film to serve as an antireflection coating, and adds ~80 nm diameter Al_2O_3 particles in between the wires to scatter the incoming light [25]. These techniques to improve the performance of large wire diameter arrays are consistent with the principles discussed above.

11.6 Optoelectronic Properties of Radial Nanowire Diodes

The electronic and photovoltaic properties of cylindrical NW diodes are significantly different from those of planar solar cells. These properties of p–n junction diodes can be analyzed by solving the coupled Poisson-continuity equations in order to determine depletion widths, carrier densities, and photo- and dark current densities. Analytic solutions of current densities are found by assuming that the depletion regions are free of mobile carriers, so no recombination occurs in the depletion region (depletion approximation) and that the recombination rates in the neutral regions are linear so that the photogeneration and bias factors in the continuity equations can be superimposed (superposition approximation) [5].

To gain insight into the differences in electronic properties and photovoltaic action in NW solar cells compared to planar cells, we briefly discuss solutions of Poisson's equation to determine depletion layer widths across the radial cell, given a certain doping density. While the effects here are not important from an optical absorption perspective, they must be considered electrically to prevent total depletion of either the NW core or shell, and to evaluate the series resistances of such regions.

A radial NW diode is shown schematically in Fig. 11.11a. Here, a is the radius of the n-type NW core, N_D^+,is the ionized donor doping density on the n-side of the junction with extent r_n into the NW core, and N_A^- is the ionized acceptor doping density on the p-side of the junction with extent r_p into the NW shell. The total depletion width is $W_{nanowire} = r_p - r_n$. Within the depletion approximation, the boundary conditions for solving Poisson's equation are (1) the electric field vanishes at both edges of the depletion region and (2) we assume $\varphi_n = V_{bi}$ and $\varphi_p = 0$, where φ is the electrostatic potential and V_{bi} is the built-in voltage. This leads to a characteristic equation for finding the shell depletion width, r_p:

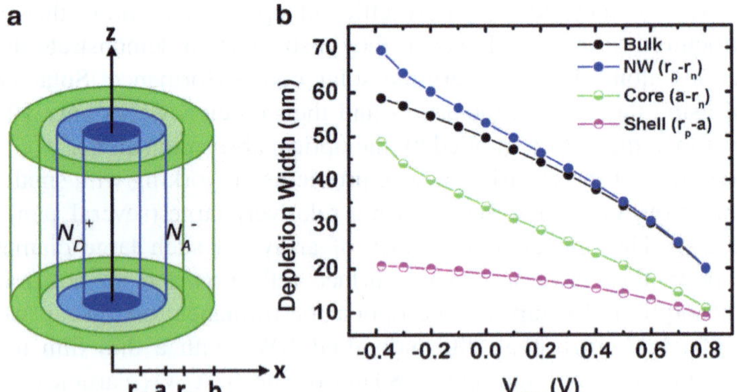

Fig. 11.11 (**a**) Schematic cross section of a radial NW p–n junction and (**b**) depletion width distribution in the core (*green*), shell (*magenta*), total depletion (*blue*) for a 50 nm diameter NW, compared to bulk (*black*), all calculated for a doping density of 10^{18} cm^{-3}

$$\frac{2\varepsilon_s V_{bi}}{qa^2} = \frac{N_A^-}{2}\left(\frac{r_p}{a}\right)^2 \ln\left(\frac{\left(\frac{r_p}{a}\right)^2}{\frac{N_A^-+N_D^+}{N_D^+} - \frac{N_A^-}{N_D^+}\left(\frac{r_p}{a}\right)^2}\right)$$

$$+\frac{1}{2}(N_A^- + N_D^+)\ln\left(\frac{N_A^- + N_D^+}{N_D^+} - \frac{N_A^-}{N_D^+}\left(\frac{r_p}{a}\right)^2\right), \qquad (11.6)$$

Where q is the electron charge and ε_s is the semiconductor dielectric constant. The charge neutrality condition on both sides of the junction gives $r_n = \sqrt{2a^2 - r_p^2}$. Figure 11.11b shows the calculated depletion widths in both the NW core and shell, and the total radial depletion width for a Si NW with a core diameter of 50 nm and doping densities of $N_D^+ = N_A^- = N = 10^{18}\,\text{cm}^{-3}$ as a function of applied bias (by replacing V_{bi} with $V_{bi} - V_{App}$ in (11.6)). Also shown in Fig. 11.11b is the depletion layer width of a bulk, planar p–n junction with $N = 10^{18}\,\text{cm}^{-3}$. The key difference from planar p–n junctions is that the depletion regions in the NW core and shell have different voltage dependences (the bulk junction depletion widths for planar structures would be equal and scale identically with bias). This difference is due to the increase with radius of the number of ionized charges contained in a cylindrical shell, so that for a larger radius a thinner depletion region is required to satisfy charge neutrality. As seen in Fig. 11.11b (1) the depletion region width in the NW core is larger than in the NW shell, and (2) going from forward (positive)- to reverse-bias (negative) voltages, the depletion region in the shell has a very weak dependence on applied voltage, whereas the depletion region in the core varies linearly followed by a superlinear increase just prior to total depletion, which for the example discussed here occurs at ~−0.4 V reverse bias. Because of the strong bias dependence of the core depletion width, the total NW depletion width strongly deviates from that of the bulk at reverse biases. It is critical to keep the nondepleted NW core diameter large enough to support the solar cell photocurrent with minimal voltage loss due to its series resistance. This is easily achieved for structures with doping densities in the 10^{18} to $10^{19}\,\text{cm}^{-3}$ range and wire lengths of ~5 μm. In addition, for proper operation of an NW solar cell with $N = 10^{18}\,\text{cm}^{-3}$ doping densities, the shell thickness has to exceed 20 nm in order to maintain p–n junction behavior and avoid Schottky contact effects on carrier extraction from the cell.

A typical Si solar cell consists of a diode formed by diffusion of n-type dopants into a thick p-type (~200 μm) substrate. The n-type layer is usually heavily doped ($10^{19}\,\text{cm}^{-3}$) in order to ensure low contact and series resistances but thin enough to be transparent to incident light. The p-type layer is thick to maximize light absorption and is lightly doped (~$10^{16}\,\text{cm}^{-3}$) to ensure long minority electron diffusion lengths to reach the top contact. The considerations for NW solar cell design are significantly different. As discussed in the optical absorption section, arrays of small diameter NWs only ~5–10 μm long are able to absorb most of the incoming solar radiation. Due to the short carrier transport length along the radial direction of the NW (<1 μm), the minority carrier diffusion lengths do not need to be very large to achieve high quantum efficiency. This significantly lowers the material quality required for an NW solar cell compared to conventional cells. However, surface recombination is more important in NW solar cells because the ratio of the surface area to the wire volume becomes larger as the NW diameter decreases.

To assess the influence of different device geometries and material parameters on NW solar cell performance, we need to calculate the electrical and photocurrent properties. Coupled solutions of the transport equations are generally obtained in semiconductor device simulators, and even simplified, decoupled solutions of these equations in cylindrical coordinates are generally complex and handled numerically [15]. For an intuitive perspective into NW solar cell performance, we utilize the fact that light absorption in NW arrays is essentially uniform throughout each individual NW since the absorption depth is larger than the NW diameter. We also use the cylindrical symmetry of the NW for an additional simplification to solve the 1D continuity equations that make use of depletion widths extracted from cylindrical coordinate solutions as discussed above. The boundary conditions for such solutions are well known and are applied here to the NW radial extent (Fig. 11.11a) with surface recombination boundary conditions at $r = b$ and a constant, photogenerated minority carrier density at the center of the NW. These solutions lead to the following current density equations:

$$J_{\mathrm{p_n}} = \frac{qD_{\mathrm{p}}}{L_{\mathrm{p}}}\left[p_{\mathrm{no}}\left(\mathrm{e}^{\frac{qV}{kT}}-1\right)-\frac{GL_{\mathrm{p}}^2}{D_{\mathrm{p}}}\right]\tan\mathrm{h}\left(\frac{r_{\mathrm{n}}}{L_{\mathrm{p}}}\right), \tag{11.7}$$

$$\begin{aligned} J_{\mathrm{n_p}} = {} & \frac{qn_{\mathrm{p0}}\left[D_{\mathrm{n}}^2L_{\mathrm{n}}^2\sin\mathrm{h}\left(\frac{(b-r_p)}{L_n}\right)-D_{\mathrm{n}}SL_{\mathrm{n}}^3\cos\mathrm{h}\left(\frac{(b-r_{\mathrm{p}})}{L_{\mathrm{n}}}\right)\right]^{\left(\mathrm{e}^{\frac{qV}{kT}}-1\right)}}{D_{\mathrm{n}}L_{\mathrm{n}}^3\cos\mathrm{h}\left(\frac{(b-r_{\mathrm{p}})}{L_{\mathrm{n}}}\right)-SL_{\mathrm{n}}^4\sin\mathrm{h}\left(\frac{(b-r_{\mathrm{p}})}{L_{\mathrm{n}}}\right)} \\ & +\frac{qGSL_{\mathrm{n}}^5\cos\mathrm{h}\left(\frac{b-r_{\mathrm{p}}}{L_{\mathrm{n}}}\right)-qGD_{\mathrm{n}}L_{\mathrm{n}}^4\sin\mathrm{h}\left(\frac{(b-r_{\mathrm{p}})}{L_{\mathrm{n}}}\right)-qSGL_{\mathrm{n}}^5}{D_{\mathrm{n}}L_{\mathrm{n}}^3\cos\mathrm{h}\left(\frac{b-r_{\mathrm{p}}}{L_{\mathrm{n}}}\right)-SL_{\mathrm{n}}^4\sin\mathrm{h}\left(\frac{(b-r_{\mathrm{p}})}{L_{\mathrm{n}}}\right)}, \end{aligned} \tag{11.8}$$

and

$$J_{\mathrm{dep}} = -qG(r_{\mathrm{p}}-r_{\mathrm{n}}), \tag{11.9}$$

where $J_{\mathrm{p_n}}$ is the current density due to minority holes in the n-type region, $J_{\mathrm{n_p}}$ is the current density of minority electrons in the p-type region, and J_{dep} is the current density due to photocarrier generation in the depletion region. The optical generation rate, G, is constant throughout the NW, S is the surface recombination velocity, and V is the applied voltage. The parameters n_{p0}, p_{n0}, L_{n}, L_{p}, D_{n}, and D_{p} are the equilibrium minority electron and hole densities, minority carrier diffusion lengths, and diffusion coefficients in the p- and n-type regions, respectively.

The first parts of (11.7) and (11.8) are the dark current components of the total current densities, and their latter parts are the G-dependent photocurrent components. The photocurrent components are largely determined by the minority

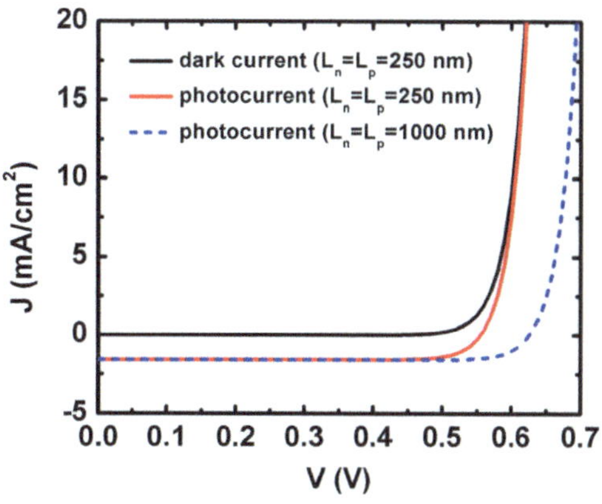

Fig. 11.12 Dark and photocurrent for a single NW solar cell with $N_A = N_D = 10^{18}\,\text{cm}^{-3}$ for $a = 50\,\text{nm}$, $b = 100\,\text{nm}$, and $G_L = 10^{21}\,\text{cm}^{-3}$

carrier diffusion lengths. For the same NW treated in Fig. 11.11, the total current density is plotted in Fig. 11.12 for dark and light conditions and for different minority carrier diffusion lengths. Here, the minority carrier diffusion lengths considered are both larger than the neutral portions of the n- and p-type regions of the NW, so the short circuit current density, J_{sc}, does not change when the diffusion length is increased. However, the larger diffusion length does produce a ~0.1 V increase in the open-circuit voltage, V_{oc}, of the cell. The increase in V_{oc} is caused by the decrease in the dark current of the diode with increasing minority carrier diffusion lengths. An increase in V_{oc} with longer minority carrier diffusion lengths in NWs has been observed experimentally [29]; however, since the diameters of the Si wires used in that work were either comparable to or larger than the minority carrier diffusion lengths, J_{sc} is also increased.

The quantum efficiency, Q.E., of the device is given by

$$\text{Q.E.} = \frac{J_{sc}}{qGb} = \frac{J_{p_n} + J_{n_p} + J_{dep}}{qGb}, \tag{11.10}$$

Figure 11.13 shows the quantum efficiency of a 100 nm diameter NW as a function of minority carrier diffusion length. Minority carrier diffusion lengths that are $\sim 2X$ the core diameter or shell thickness are sufficient to yield a Q.E. of ~1. Since optical absorption can be high with NW diameters of ~100 nm, these diffusion lengths can be significantly relaxed from that of bulk cells that require ~100 μm or more minority carrier diffusion lengths in equivalently thick Si absorption layers.

For larger diameter wires, the diffusion length must be comparable to the wire diameter to achieve high quantum efficiency. However, insertion of a thin intrinsic (undoped) absorption layer between the highly doped core/shell to create a p–i–n structure can improve the Q.E. of wires with much smaller minority carrier diffusion lengths. These effects are portrayed in Fig. 11.14 where the Q.E. is plotted against the thickness of the intrinsic layer, with fixed wire and core diameters of 500 nm and 50 nm, respectively, and for different diffusion lengths. Large minority carrier diffusion lengths produce high Q.E. at small intrinsic layer thicknesses. For small

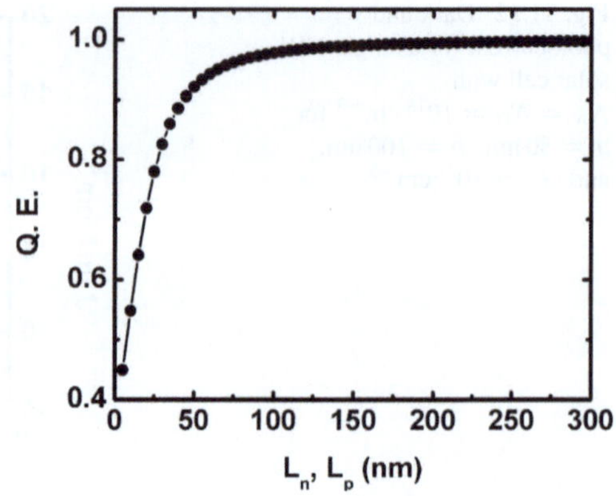

Fig. 11.13 Quantum efficiency as function of minority carrier diffusion lengths for $N_A = N_D = 10^{18}\,\text{cm}^{-3}$ for $a = 50\,\text{nm}$, $b = 100\,\text{nm}$, and $G_L = 10^{21}\,\text{cm}^{-3}$, and $S = 10^3\,\text{cm/s}$

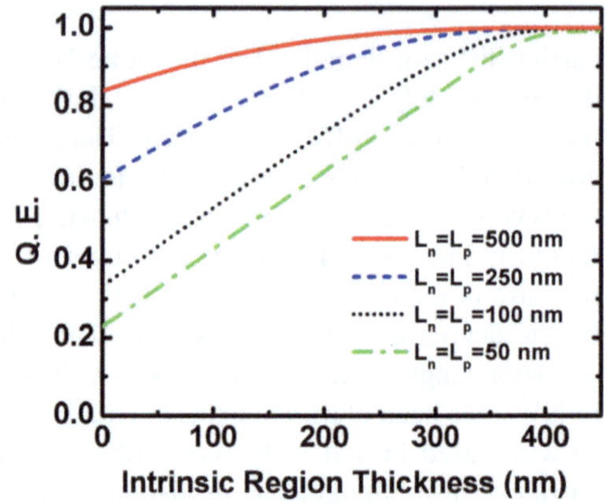

Fig. 11.14 Quantum efficiency as function of intrinsic layer thickness with different minority carrier diffusion lengths for a total NW diameter of 500 nm, $a = 50\,\text{nm}$, $N_A = N_D = 10^{18}\,\text{cm}^{-3}$, $G_L = 10^{21}\,\text{cm}^{-3}$, and $S = 10^3\,\text{cm/s}$

minority carrier diffusion lengths, the intrinsic layer thickness has to be increased such that the depletion current component becomes large enough to produce a high Q.E. This was observed experimentally in p–n and p–i–n core/shell NW solar cells with a few tens of nm core/intrinsic/shell thicknesses [21], indicating that minority carrier lengths are indeed short in the in situ grown polycrystalline intrinsic and n-shell layers.

11.7 Solar Cell Performance: Combined Optical and Electrical Properties

The two previous sections considered the optical properties of NW arrays and the photoelectric device properties of individual NWs. To predict the performance of an NW solar cell these properties must be combined. From an optical perspective, small diameter NWs have the best performance due to the efficiency of light scattering. Conversely, the electrical properties of large diameter NWs are superior due to the decrease in surface recombination loss. The best solar cell performance requires a compromise between these two diverging requirements or the addition of extra optical or device elements. Two limiting cases have been considered experimentally: small diameter wires with good optical absorption and large diameter wires with good electrical performance. The best NW solar cell results to date [28] have been achieved in the previously described conical structure with optical properties shown in Fig. 11.10. Because the nanocones are of small diameter, the uniform optical absorption approach of Sect. 11.5 is appropriate. The reported cell efficiency is 10.8% with a short circuit current of 26.4 mA/cm^2, a V_{oc} of 0.59 V and a fill factor of 0.69. From the short circuit current, we can estimate that the photocurrent quantum efficiency of this cell is about 60% [52], assuming it is constant as a function of wavelength throughout the c-Si optical absorption region. A 60% quantum efficiency would lead to about a 15% power efficiency device if all other device parameters were ideal. The open circuit voltage is about 0.05 V smaller than ideal and the fill factor is about 0.1 too small [60]. Thus, these devices suffer from a decrease in Q.E. from the absorption implied ~99 to ~60%, and from nonideal device *I–V* characteristics which reduce the open circuit voltage and fill factor. This structure employed 20 nm of wet oxide (SiO_2) as a surface passivation layer, which is one of the better passivation approaches. More detailed measurements of the individual nanocone performance are required to determine what properties need to be improved to increase the overall performance of these cells.

The best large diameter wire solar cell results have been achieved with 2–3 μm diameter, 45 μm long wires arranged on a square lattice with 7 μm spacing. For these large diameter wires, the optical absorption is not uniform throughout the wire and the absorption enhancement is relatively weak. The reported cell efficiency is 7.9% with a short circuit current of 24.3mA/cm^2, a V_{oc} of 0.5 V and a fill factor of 0.65 [25]. From the short circuit current, we can estimate that the integrated optical absorption efficiency of this cell is about 55% [52], assuming ~100% internal photocurrent quantum efficiency for each wire. Surprisingly, considering that the virtue of large diameter wires is their electrical device quality, the open circuit voltage is about 0.15 V smaller than ideal and the fill factor is about 0.15 too small. This structure employed an a-SiN_x:H film as a surface passivation and antireflection layer and had 80 nm diameter Al_2O_3 particles between the wires to increase the optical scattering. Additional measurements of the individual NW performance are required to determine what properties need to be improved to increase the overall performance of these cells.

The state of the art in NW array solar cells is still at a primitive state. The above record performance efficiencies of 10.8 and 7.9% have not been confirmed by a recognized testing laboratory such as the National Renewable Energy Laboratory (NREL).[1] The solar cells were both fabricated over extremely small areas, $\sim 2.5 \times 10^{-1}$ cm^2 and 2×10^{-3} cm^2, respectively. They are thus both smaller than the 1 cm^2

[1]The cell described in Yoon et al. [36] and listed as 8.7% efficient in Table 11.1 has a significant photovoltaic contribution from the substrate.

size considered the minimum required for testing of nonconcentrating solar cells [6]. Finally, there remain real questions about the areal scalability of these approaches. For NW arrays formed by VLS or etching techniques, the individual NW formation processes are largely independent and so defect occurrences must be considered. The NW arrays discussed above have pitches of $\sim$1–10 μm corresponding to NW densities of $10^8\,cm^{-2}$ to $10^6\,cm^{-2}$. Six sigma processes require less than $\sim$3.5 defects per million items produced. One thus might expect $\sim$3–300 defective wires/cm^2 even with six sigma uniformity. Since the individual NW diodes are connected in parallel, one low resistance/shorting diode could ruin an entire subsection of a solar cell. It remains unclear if it is possible to achieve better than six sigma process control with the low-cost substrates and inexpensive fabrication techniques required for solar cells.

11.8 Integration Strategies for Nanowire Solar Cells

A major barrier to the mainstream adoption of Si-based PV technology is the relatively high cost associated with the materials used to manufacture individual cells, which can account for >60% of the final cost (J. Lushetsky, presented at the $1/W Workshop: ARPA-E and Office of Energy Efficiency and Renewable Energy's (EERE) Solar Energy Technology Program (SETP), Washington, DC, 2010, unpublished; [64]). For Si-based solar cell technology, a direct correlation exists between the crystalline quality of Si, the solar energy conversion efficiency, and the overall manufacturing cost. Thus, PV cells made from high-quality single-crystalline Si tend to have the highest energy conversion efficiency, and also the highest cost. One strategy to mitigate manufacturing cost is the use of lower quality material such as polycrystalline Si substrates, but at the expense of lower solar energy conversion efficiency. Integrating nanostructured materials such as Si NW arrays directly onto low-cost alternative substrates such as metal foils, glass, or polymers provides an alternative approach. This approach potentially allows a relatively high solar energy conversion efficiency provided by the single-crystalline Si NWs along with a reduction in manufacturing cost [65], as discussed in the previous sections of this chapter. Additionally, NW array growth on metal foils and polymer transfer techniques provide opportunities to reduce cost while at the same time producing solar cells that are structurally flexible and able to mechanically conform to curved surfaces such as street lights or roofing tiles.

11.8.1 General Approaches

Currently there are two basic approaches to integrating semiconductor NWs onto low-cost alternative substrates for solar cell applications. One approach is direct NW synthesis using CVD. However, the choice of an alternative substrate material

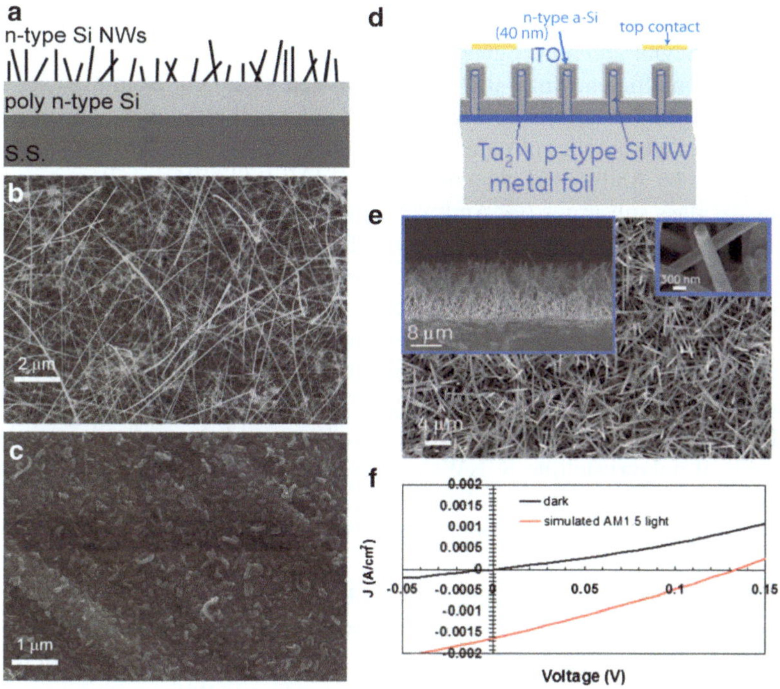

Fig. 11.15 Si NW growth on S.S. foil substrates. (**a**) Schematic of substrate structure. (**b**) SEM image of Si NWs grown on poly-Si thin film on top of S.S. (**c**) SEM image of silicide nanoworm formation by direct synthesis on S.S. (**d**) Schematic of Si NW-based solar cell on S.S. foil. (**e**) SEM image of Si NWs grown on Ta2N layer on S.S. foil. (**f**) Dark and light current–voltage characteristics of a Si NW solar cell shown in (**d**). Panels (**d**)–(**f**) after [20]

is limited by the synthesis and processing requirements. Considering the typical CVD growth temperatures of $\sim$450–550°C for Au-catalyzed VLS Si NW growth with SiH_4 and $\sim$800–1,000°C with $SiCl_4$, suitable alternative substrate materials are primarily restricted to growth on metal foils such as stainless steel (S.S.) or glass substrates. Shown in Fig. 11.15a is a schematic illustrating the growth of Si nanowires on an S.S. substrate. Silicon NWs are being grown by the VLS method directly on a plasma-enhanced CVD polycrystalline Si thin-film buffer layer on S.S. (Fig. 11.15b) by the present authors. A 2 nm Au film is used as the catalyst for the NW growth. The presence of the polycrystalline Si layer served as a diffusion barrier, and also prevents unwanted silicide "nanoworm" formation (Fig. 11.15c). The thicknesses of $\sim$100 μm(4 mil) S.S. and $\sim$400 nm polycrystalline Si were optimized in order to eliminate delamination of the polycrystalline Si layer as a result of the thermal mismatch induced stresses during the NW growth.

Researchers at G.E. have demonstrated a complete Si NW-based solar cell on S.S [20]. In their design (Fig. 11.15d), Si NWs were fabricated via VLS growth from a Au seed layer on a Ta_2N thin diffusion barrier on S.S. (Fig. 11.15e). Figure 11.15f shows the resultant dark and light current–voltage curves. The device shows a rectifying behavior and power generation is observed; however, the authors report a low conversion efficiency of ~0.1% for a Si-based solar cell. Both high series and low shunt resistances appear to limit the NW solar cell efficiency and suggest the presence of a high number of parasitic short circuit paths. Similar results have also been reported by Andra et al. who demonstrated VLS Si NW growth on crystalline Si films on glass substrates (G. Andra, M. Pietsch, T. Stelzner, A. Gawlik, E. Ose, S. Christiansen, F. Falk, presented at the IEEE Photo. Spec Conf, San Diego CA, 2008, unpublished; [66]). The difficulty in obtaining good conformal coverage of thin-film overlayers on the NWs is exacerbated by the fact that NWs grown from catalyst films on polycrystalline surfaces tend to grow in a more-or-less random orientation with respect to the substrate surface. One strategy to alleviate this is by synthesis of well-defined vertical NW arrays by top-down etching techniques. Using wet electroless chemical etching, Sivakov et al. were able to synthesize a solar cell from vertical polycrystalline Si NWs arrays on glass substrates which exhibited a power conversion efficiency of 4.4% [67].

Another approach to integrating NWs onto low-cost alternative substrates is by polymer transfer of NW arrays by means of a flexible polymer film with the NWs imbedded. This technique was first reported by Plass et al. where an array of VLS-grown Si wires shown in Fig. 11.16b was suspended in a PDMS film [68]. This process is depicted in Fig. 11.16a where after NW synthesis, a PDMS film was cast at the base of the array. Subsequently, a free-standing PDMS film with suspended NWs was created by mechanically peeling the film from the Si substrate. The SEM

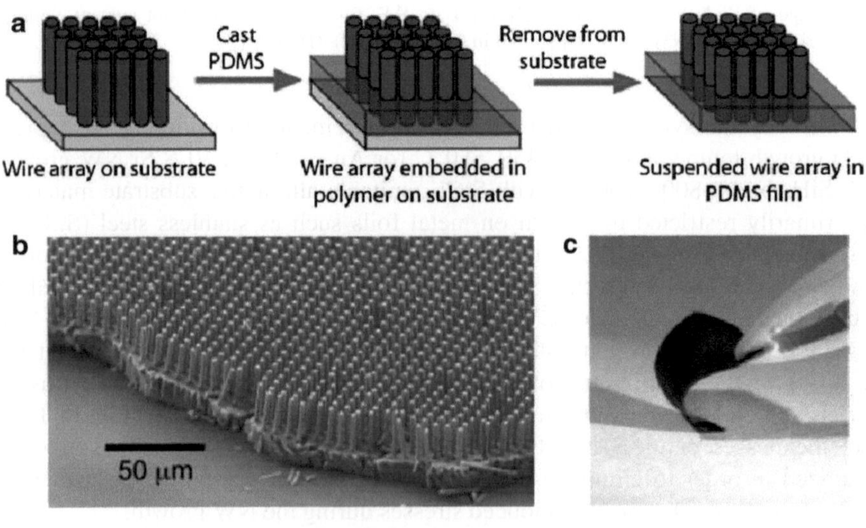

Fig. 11.16 Transfer of vertical array of Si microwires into free-standing polymer films. (**a**) Schematic representation of process to imbed and remove microwire array. (**b**) SEM image of vertical Si microwire array suspended in a PDMS film. (**c**) A flexible free-standing PDMS film with suspended vertical array of Si microwires. Panels (**a**)–(**c**) after [68]

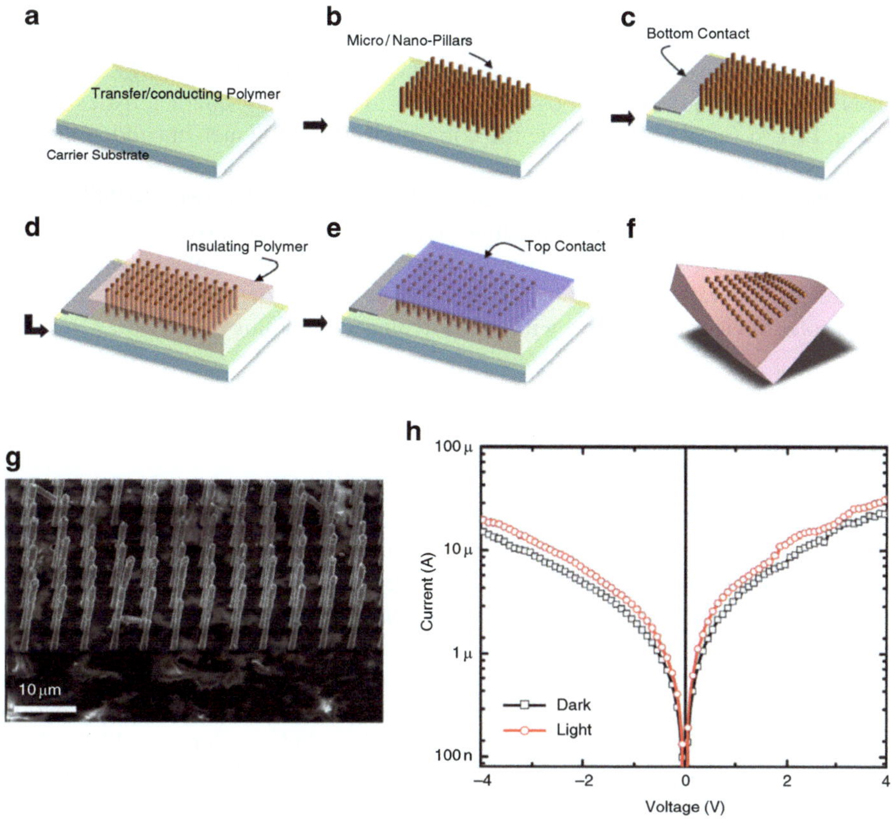

Fig. 11.17 (**a**) Vertical Si wire arrays are mechanically imbedded into a conducing polymer film supported on a solid substrate (e.g., glass). A lateral force is then applied to the wire substrate to shear fracture the wires, (**b**) leaving behind the vertical NW array imbedded in the conducting polymer. (**c**) A bottom contact metal is evaporated on the conducing polymer outside of the device region. (**d**) An insulating layer (e.g., PMMA) is infiltrated to fill the space between the wires. (**e**) After an etch-back step, a transparent contact (e.g., ITO) is then evaporated on top. (**f**) flexible Si wire array after release from carrier substrate (**g**) SEM image of wires imbedded in conducting polymer. (**h**) Electrical I–V measurement of the transferred Si nanowire photoconductor with and without optical illumination. Panels (**g**, **h**) after [71]

image in Fig. 11.16b clearly shows the vertical array of Si wires suspended in the released PDMS film. The advantage of having the NWs suspended in a polymer film is that the polymer allows a wide range of structural flexibility as demonstrated in Fig. 11.16c. Currently, there are very few reports on the performance of free-standing polymer-supported Si wire films for solar cell applications. In one report Spurgeon et al. created a photoelectrochemical cell from a polymer-supported Si wire film and demonstrated that the free-standing wire film could potentially be made into a solar cell without sacrificing solar energy conversion efficiency compared to substrate-supported NW solar cells [69].

Alternatively, a free-standing polymer-supported NW film could be fabricated by direct transfer of wire arrays into a preexisting polymer films and subsequently made into a working device [68]. This approach [70,71] is outlined in Fig. 11.17. Although the NWs were of uniform composition, the light and dark I–V characteristics shown in Fig. 11.17h further demonstrate the concept of using polymer-imbedded Si wire arrays as solar cell devices.

Despite the demonstration of a range of strategies to integrate Si NWs onto low-cost alternative substrates, much work is yet to be done. For example, finding solutions to reduce the contact resistance between conducting contact layers and the Si wires is imperative to improve energy conversion efficiency and make Si NW-based solar devices completive with existing planar technology [68].

11.9 Conclusions

Semiconducting NWs present a fascinating potential opportunity for developing high-efficiency, low-cost solar cells by decoupling light absorption and carrier separation. Significant recent advances have been made and these studies are leading to new understanding of both NW fabrication challenges and nanomaterial property control. However despite the many interesting features of NW radial p–n junction PV cells, the costs of electricity production and of PV module production remain to be determined. According to life cycle assessments of NW radial p–n junction solar cells, the electricity generation cost of NW radial p–n junction PV cells is likely to be comparable to second-generation PV cells, such as thin-film cells in Green's classification [72]. While this assessment sheds favorable light on the future of NW radial p–n junction PV cells, many challenging nanomaterials issues remain to be solved, including fundamental understanding of carrier transport, photon management, 3D architecture fabrication, and long-term device reliability.

Acknowledgements This research was funded in part by the Laboratory Directed Research and Development Program at Los Alamos National Laboratory and by the Department of Energy EERE (EB2102010) Work was performed in part at the Center for Integrated Nanotechnologies, a U.S. Department of Energy, Office of Basic Energy Sciences, user facility at Los Alamos National Laboratory (Contract DE-AC52–06NA25396).

References

1. E.C. Garnett, M.L. Brongersma, Y. Cui, M. McGehee, Annu. Rev. Mater. Res. DOI: 10.1146/annurev-matsci-062910--100434
2. V.I. Klimov, J. Phys. Chem. B **110**, 16827 (2006)
3. A.J. Nozik, Nano Lett. **10**, 2735 (2010)
4. A. Hagfeldt, G. Boschloo, L. Sun, L. Kloo, H. Pettersson, Chem. Rev. **110**, 6595 (2010)
5. J. Nelson, *Physics of Solar Cells* (Imperial College Press, London, 2003)

6. M.A. Green, K. Emery, Y. Hishikawa, W. Warta, Prog. Photovolt. **19**, 84 (2011)
7. C.A. Wolden, J. Kurtin, J.B. Baxter, I. Repins, S.E. Shaheen, J.T. Torvik, A.A. Rockett, J.M. Fthenakis, E.S. Aydil, J. Vac. Sci. Technol. A **29**, 030801 (2011)
8. http://www.nrel.gov/analysis/sam/
9. M.A. Green, S.R. Wenham, Appl. Phys. Lett. **65**, 2907 (1994)
10. K.J. Weber, A.W. Blakers, M.J. Stocks, J.H. Babaei, V.A. Everett, A.J. Neuendorf, P.J. Verlinden, IEEE Electron Dev. Lett. **25**, 37 (2004)
11. R.S. Wagner, W.C. Ellis, Appl. Phys. Lett. **4**, 89 (1964)
12. S.T. Picraux, S.A. Dayeh, P. Manandhar, D.E. Perea, S.G. Choi, JOM **62**, 35 (2010)
13. C.E. Kendrick, H.P. Yoon, Y.A. Yuwen, G.D. Barber, H.T. Shen, T.E. Mallouk, E.C. Dickey, T.S. Mayer, J.M. Redwing, Appl. Phys. Lett. **97**, 143108 (2010)
14. B.M. Kayes, M.A. Filler, M.C. Putnam, M.D. Kelzenberg, N.S. Lewis, H.A. Atwater, Appl. Phys. Lett. **91**, 103110 (2007)
15. B.M. Kayes, H.A. Atwater, N.S. Lewis, J. Appl. Phys. **97**, 114302 (2005)
16. L. Hu, G. Chen, Nano Lett. **7**, 3249 (2007)
17. J. Zhu, Z.F. Yu, G.F. Burkhard, C.M. Hsu, S.T. Connor, Y.Q. Xu, Q. Wang, M. McGehee, S.H. Fan, Y. Cui, Nano Lett. **9**, 279 (2009)
18. Z.Y. Fan, R. Kapadia, P.W. Leu, X.B. Zhang, Y.L. Chueh, K. Takei, K. Yu, A. Jamshidi, A.A. Rathore, D.J. Ruebusch, M. Wu, A. Javey, Nano Lett. **10**, 3823 (2010)
19. E. Garnett, P.D. Yang, Nano Lett. **10**, 1082 (2010)
20. L. Tsakalakos, J. Balch, J. Fronheiser, B.A. Korevaar, O. Sulima, J. Rand, Appl. Phys. Lett. **91**, 23317 (2007)
21. B. Tian, X.L. Zheng, T.J. Kempa, Y.J. Fang, N. Yu, G. Yu, J. Huang, C.M. Lieber, Nature **449**, 885–890 (2007)
22. L.W. Yu, B. O'Donnell, P.J. Alet, P.R.I. Cabarrocas, Solar Energy Mater. Solar Cells **94**, 1855 (2010)
23. O. Gunawan, S. Guha, Solar Energy Mater. Solar Cells **93**, 1388 (2009)
24. C.E. Kendrick, S.M. Eichfeld, Y. Ke, X.J. Weng, X. Wang, T.S. Mayer, J.M. Redwing, Proc. SPIE **7768**, 77680I (2010)
25. M.C. Putnam, S.W. Boettcher, M.D. Kelzenberg, D.B. Turner-Evans, J.M. Spurgeon, E.L. Warren, R.M. Briggs, N.S. Lewis, H.A. Atwater, Energy Environ. Sci., **3**, 1037 (2010)
26. M.D. Kelzenberg, D.B. Turner-Evans, M.C. Putnam, S.W. Boettcher, R.M. Briggs, J.Y. Baek, N.S. Lewis, H.A. Atwater, Energy Environ. Sci. **4**, 866 (2011)
27. M.J. Naughton, K. Kempa, Z.F. Ren, Y. Gao, J. Rybczynski, N. Argenti, W. Gao, Y. Wang, Y. Peng, J.R. Naughton, G. McMahon, T. Paudel, Y.C. Lan, M.J. Burns, A. Shepard, M. Clary, C. Ballif, F.J. Haug, T. Soderstrom, O. Cubero, C. Eminian, Phys. Status Solidi – RRL **4**, 181 (2010)
28. Y.R. Lu, A. Lal, Nano Lett. **10**, 4651 (2010)
29. H.P. Yoon, Y.A. Yuwen, C.E. Kendrick, G.D. Barber, N. Podraza, J.M. Redwing, T.E. Mallouk, C.R. Wronski, T.S. Mayer, Appl. Phys. Lett. **96**, 213503 (2010)
30. E.C. Garnett, P.D. Yang, J. Am. Chem. Soc. **130**, 9224 (2008)
31. J.Y. Jung, Z. Guo, S.W. Jee, H.D. Um, K.T. Park, M.S. Hyun, J.M. Yang, J.H. Lee, Nanotechnology **21**, 445303 (2010)
32. X. Wang, K.L. Pey, C.H. Yip, E.A. Fitzgerald, D.A. Antoniadis, J. Appl. Phys. **108**, 124303 (2010)
33. J.A. Czaban, D.A. Thompson, R.R. LaPierre, Nano Lett. **9**, 148 (2009)
34. C. Colombo, M. Heiss, M. Gratzel, A.F.I. Morral, Appl. Phys. Lett. **94**, 173108 (2009)
35. N. Tajik, Z. Peng, P. Kuyanov, R.R. LaPierre, Nanotechnology **22**, 225402 (2011)
36. Y. Dong, B. Tian, T.J. Kempa, C.M. Lieber, Nano Lett. **9**, 2183 (2009)
37. K.P. Musselman, A. Wisnet, D.C. Iza, H.C. Hesse, C. Scheu, J.L. MacManus-Driscoll, L. Schmidt-Mende, Adv. Mater. **22**, E254 (2011)
38. B.D. Yuhas, P.D. Yang, J. Am. Chem. Soc. **131**, 3756 (2009)

39. Z. Fan, H. Razavi, J.-w. Do, A. Moriwaki, O. Ergen, Y.-L. Chueh, P.W. Leu, J.C. Ho, T. Takahashi, L.A. Reichertz, S. Neale, K. Yu, M. Wu, J.W. Ager, A. Javey, Nat. Mater. **8**, 648 (2009)
40. P. Liu, V.P. Singh, C.A. Jarro, S. Rajaputra, Nanotechnology **22**, 145304 (2011)
41. D. Li, C.-J. Chien, S. Deora, P.-C. Chang, E. Moulin, J.G. Lu, Chem. Phys. Lett. **501**, 446 (2011)
42. A.A.D.T. Damitha, D.M.N.M. Dissanayake, S.R.P. Silva, IEEE J. Sel. Topics Quantum Electron. **16**, 1595 (2010)
43. Q. Zhang, G. Cao, Nano Today **6**, 91 (2011)
44. J.E. Allen, E.R. Hemesath, D.E. Perea, J.L. Lensch-Falk, J.Y. Li, F. Yin, M.H. Gass, P. Wang, A.L. Bleloch, R.E. Palmer, L.J. Lauhon, Nat. Nanotechnol. **3**, 168 (2008)
45. M.D. Kelzenberg, D.B. Turner-Evans, B.M. Kayes, M.A. Filler, M.C. Putnam, N.S. Lewis, H.A. Atwater, Nano Lett. **8**, 710 (2008)
46. Y. Gu, J.P. Romankiewicz, J.K. David, J.L. Lensch, L.J. Lauhon, Nano Lett. **6**, 948 (2006)
47. www.ioffe.rssi.ru/SVA/NSM
48. M.A. Green, Semicond. Sci. Technol. **8**, 1 (1993)
49. J. Zhao, M.A. Green, IEEE Trans. Electron Dev. **38**, 1925 (1991)
50. J. Zhao, A. Wang, P. Altermatt, M.A. Green, Appl. Phys. Lett. **66**, 3636 (1995)
51. (2011–12–5)
52. S.M. Sze, *Physics of Semiconductor Devices*, 2nd edn. (John Wiley & Sons, New York, NY, 1981)
53. M. Harris, H.A. Macleod, S. Ogura, Thin Solid Films **57**, 173 (1979)
54. H.A. Macleod, J. Vac. Sci. Technol. A **4**, 418 (1986)
55. P. Campbell, M.A. Green, J. Appl. Phys. **62**, 243 (1987)
56. A.J. Cox, A.J. DeWeerd, J. Linden, Am. J. Phys. **70**, 620 (2002)
57. L. Tong, J. Lou, E. Mazur, Opt. Express **12**, 1025 (2004)
58. W.F. Liu, J.I. Oh, W.Z. Shen, IEEE Electron Dev. Lett.**32**, 45 (2011)
59. L. Cao, J.S. White, J.-S. Park, J.A. Schuller, B.M. Clemens, M.L. Brongersma, Nat. Mater. **8**, 643 (2009)
60. M.D. Kelzenberg, M.C. Putnam, D.B. Turner-Evans, N.S. Lewis, H.A. Atwater, in Proceedings of the 34th IEEE PVSC, USC 1948 (2009)
61. H. Bao, X. Ruan, Opt. Lett. **35**, 3378 (2010)
62. C. Lin, M.L. Povinelli, Opt. Express **17**, 19371 (2009)
63. J. Li, H.Y. Yu, S.M. Wong, X. Li, G. Zhang, P.G.-Q. Lo, D.-L. Kwong, Appl. Phys. Lett. 95, 243113 (2009)
64. H.M. Lee, Phys. High Technol. (Korean Phys. Soc.) **17**, 15 (2008)
65. N.S. Lewis, Science **315**, 798 (2007)
66. T. Stelzner, M. Pietsch, G. Andra, F. Falk, E. Ose, S. Christiansen, Nanotechnology **19**, 295203 (2008)
67. V. Sivakov, G. Andra, A. Gawlik, A. Berger, J. Plentz, F. Falk, S.H. Christiansen, Nano Lett. **9**, 1549 (2009)
68. K.E. Plass, M.A. Filler, J.M. Spurgeon, B.M. Kayes, S. Maldonado, B.S. Brunschwig, H.A. Atwater, N.S. Lewis, Adv. Mater. **21**, 325 (2009)
69. J.M. Spurgeon, S.W. Boettcher, M.D. Kelzenberg, B.S. Brunschwig, H.A. Atwater, N.S. Lewis, Adv. Mater. **22**, 3277 (2010)
70. V.J. Logeeswaran, A.M. Katzenmeyer, M.S. Islam, IEEE Trans Electron Dev. **57**, 1856 (2010)
71. V.J. Logeeswaran, M. Ombaba, A.M. Katzenmeyer, M. S. Islam, Energy harvesting and storage: Materials, devices, and applications. Proc. SPIE **7683**, 768313 (2010)
72. M.A. Green, Prog. Photovolt. **9**, 123–135 (2001)

Index

G.-C. Yi (ed.), *Semiconductor Nanostructures for Optoelectronic Devices*, NanoScience and Technology, DOI 10.1007/978-3-642-22480-5,

MIX
Papier aus verantwortungsvollen Quellen
Paper from responsible sources
FSC® C105338

If you have any concerns about our products,
you can contact us on
ProductSafety@springernature.com

In case Publisher is established outside the EU,
the EU authorized representative is:
Springer Nature Customer Service Center GmbH
Europaplatz 3, 69115 Heidelberg, Germany

Printed by Libri Plureos GmbH
in Hamburg, Germany